STUDENT WORKBOOK

7th Grade / Jr. High   1 Credit

Includes: Answer Keys

BOOK 1
PRINCIPLES OF MATHEMATICS
BIBLICAL WORLDVIEW CURRICULUM

[Katherine A. Loop]

First printing: May 2015

Copyright © 2015 by Katherine A. Loop. All rights reserved. No part of this book may be used or reproduced in any manner whatsoever without written permission of the publisher, except in the case of brief quotations in articles and reviews. For information write:

Master Books®, P.O. Box 726, Green Forest, AR 72638

Master Books® is a division of the New Leaf Publishing Group, Inc.

ISBN: 978-0-89051-876-2

Cover by Diana Bogardus

Unless otherwise noted, Scripture quotations are from the King James Version of the Bible.

Some problems are used/adapted from the following resources:

Katherine Loop, *Revealing Arithmetic: Math Concepts from a Biblical Worldview* (Fairfax, VA: Christian Perspective, 2010).

Eugene Henry Barker, *Applied Mathematics for Junior High Schools and High Schools* (Boston: Allyn and Bacon, 1920). Available on Google Books, http://books.google.com/books?id=-t5EAAAAIAAJ&vq=3427&pg=PR2#v=onepage&q&f=false

John C. Stone and James F. Millis, *A Secondary Arithmetic: Commercial and Industrial for High, Industrial, Commercial, Normal Schools, and Academies* (Boston: Benj. H. Sanborn & Co., 1908). Available on Google Books, http://books.google.com/books?id=RtYGAAAAYAAJ&pg=PP1#v=onepage&q&f=false

Joseph Victor Collins, *Practical Algebra: First Year Course* (New York: American Book Co., 1910). Available on Google Books, http://google.com/books?id=hNdHAAAAIAAJ&pg=PP1#v=onepage&q&f=false

We have attempted to mark all the problems adapted from early 1900s textbooks for easy reference; however, they inspired other problems as well.

For the most part, units are based on the official standards given in Tina Butcher, Linda Crown, Rick Harshman, and Juana Williams, eds. *NIST Handbook 44: 97th National Conference on Weights and Measures 2012*, 2013 ed. (Washington: U. S. Department of Commerce, 2012), Appendix C. Found on http://www.nist.gov/pml/wmd/pubs/h44-13.cfm (Accessed 10/6/2014)

Please consider requesting that a copy of this volume be purchased by your local library system.

Printed in the United States of America

Please visit our website for other great titles:
www.masterbooks.com

For information regarding author interviews,
please contact the publicity department at (870) 438-5288.

How to Use This Course

Get ready to discover math from a biblical worldview! Designed for use alongside the *Principles of Mathematics Student Textbook*, this *Student Workbook*, contains worksheets, quizzes, and tests to help build both a biblical worldview of math and mathematical skills. The problems incorporate a lot of history, science, and real-life examples, helping students master the skills, build problem-solving skills, and learn to use math as a real-life tool to both explore God's creation and serve Him.

- **Suggested Schedule** — Check out the Suggested Daily Schedule (page 6–13) or the Accelerated Daily Schedule (page 14–18) for an easy-to-follow daily plan. Most days, students will be instructed to read a lesson in the Student Textbook and then work a worksheet in this Student Workbook. Quizzes, tests, study days, and days off are also built into the schedule. Feel free to adapt as needed.

- **Tear-Out Worksheets, Quizzes, and Tests** — The worksheets, quizzes, and tests are all perforated, so you can easily tear them out for use. Students may solve problems directly on the sheets, using additional notebook paper if needed.

- **Answer Key** — Answers to all the problems given are included in the back of this book. Grading suggestions are given on pages 391-392.

Note to Parent/Teacher: God has created each person individually, so please modify and adapt this curriculum as needed.

Supplies Needed

- *Principles of Mathematics Student Textbook Book 1*
- Binder with Notebook Paper — Students will need to tear out the reference section from this book and put it in the binder, as well as add notes to it during the course.
- Abacus — You can either make your own (instructions are given on Worksheet 1.3), use a premade one, or use an online abacus (see www.christianperspective.net/math/pom1).
- Blank Index Cards to use in making flashcards.
- Calculator
- Graph Paper
- Compass
- Measuring Tape with both Metric and U.S. Customary markings
- Ruler with Metric and U.S. Customary markings
- Protractor

Additional Ideas and Support

For additional math ideas and resources, please check out www.ChristianPerspective.net. You'll find links to helpful supplemental resources there (including links to online fact sheets for students needing more drill), as well as ways to stay connected and ask questions.

Problems from the Early 1900s

History...in math? Why not! Throughout the text, we've sprinkled in some math problems from history, often with significant adaptation. The sources are listed here for your reference. Feel free to look up the books and have fun with additional problems!

The following problems were adapted from Eugene Henry Barker, *Applied Mathematics for Junior High Schools and High Schools* (Boston: Allyn and Bacon, 1920). Available on Google Books, http://books.google.com/books?id=-t5EAAAAIAAJ&vq=3427&pg=PR2#v=onepage&q&f=false

Worksheet 2.6, problem 5; Worksheet 4.1, problems 3, 4, and 5; Worksheet 4.5, problem 1; Worksheet 5.2, problem 6; Worksheet 5.7, problem 8; Worksheet 6.5, problem 3; Worksheet 8.2, problem 2b and 3; Worksheet 8.3, problems 5 and 6; Worksheet 8.6, problems 3a and 3b; Worksheet 9.1, problem 8; Worksheet 14.3, problem 5; Worksheet 18.1B, problem 4a; Worksheet 18.4, problems 1e, 3a, and 3b; Worksheet 18.6, problem 3; Worksheet 21.1, problem 2a and 7; Worksheet 21.4B, problem 14; Quiz 3, problem 1; Quiz 6, problem 3a; Test 5; extra credit problems

The following problems were adapted from John C. Stone and James F. Millis, *A Secondary Arithmetic: Commercial and Industrial for High, Industrial, Commercial, Normal Schools, and Academies* (Boston: Benj. H. Sanborn & Co., 1908). Available on Google Books, http://books.google.com/books?id=RtYGAAAAYAAJ&pg=PP1#v=onepage&q&f=false

Worksheet 12.3, problem 3; Worksheet 12.6, problem 1; Worksheet 12.7, problem 4; Worksheet 12.8, problem 7; Worksheet 15.4, problem 4; Worksheet 16.4, problem 8; Worksheet 18.1B, problem 4b; Worksheet 18.2B, problems 2a and 2b; Worksheet 18.3, problem 2a; Worksheet 18.5, problem 3; Worksheet 18.6, problem 4; Worksheet 21.2, problem 6

This problem was adapted from Joseph Victor Collins, *Practical Algebra: First Year Course* (New York: American Book Co., 1910). Available on Google Books, http://google.com/books?id=hNdHAAAAIAAJ&pg=PP1#v=onepage&q&f=false

Worksheet 10.7, problem 2b

General Instructions for Worksheets, Quizzes, and Tests

- **Review** — If at any point you hit a concept that does not make sense, back up and review the preceding concepts. Be sure to take advantage of the reference notebook you'll be instructed to start in Worksheet 1.2.

- **Calculator** — Anytime you see a 🖩, you are permitted to use a calculator to solve the problem (instructions on using a calculator can be found in Lesson 4.5). Unless instructed otherwise by your parent/teacher, all other problems should be solved without the use of a calculator, as you won't always have a calculator with you when you need to solve a problem in real life.

- **Word Problems** — Except for when told to solve mentally, you should show your work on all word problems — meaning, write down enough steps of what you did that someone can see how you solved the problem (what you added, subtracted, etc.). Unless otherwise specified, it does not matter how you show your work (it doesn't have to be as in-depth as the answer key)—the important thing is that you can see how you obtained your answer.

  While showing your work may seem like busy work on simple problems, forming the habit of organizing your steps on paper from the beginning will greatly help you when you come to in-depth problems involving numerous steps.

- **Units** — If a unit is given in the problem (dollars, feet, etc.), you should always include it in your answer. From Worksheet 16.1 on, you should always list square units using exponents.

- **Decimals** — From Worksheet 7.4 on, decimal answers should be rounded to the hundredth digit unless otherwise specified.

- **Fractions** — From Worksheet 5.3 on, fractional answers should be in simplest terms unless otherwise specified. This includes rewriting mixed numbers as improper fractions. If a question is asked using only fractions, your answer should be listed as a fraction. If a question includes both decimals and fractions, you can pick which notation to use, unless otherwise specified.

General Instructions for Study Days

The study days built into the schedule are designed to give you a chance to study on your own! While you can study different ways, here are a few suggestions:

- Look at the "Chapter Synopsis" for each chapter you need to study. Read the text, and review any aspects you may have forgotten. Look at the problems too, especially at concepts with which you frequently struggled. Do you know how to solve them now?

- Go back over the quizzes for the chapters you're studying. Again, look at what you got wrong—do you know how to solve it now? How about the ones you got right?

- Review any concepts you know were hard for you.

- Go through your math notebook and look at the notes you've taken and review any flashcards.

Suggested Daily Schedule
(to complete Year 1 in a school year)

Date	Day	Assignment	Due Date	✓	Grade
\multicolumn{6}{First Semester—First Quarter}					
Week 1	Day 1	Lesson 1.1 (*Student Textbook*, pages 13–14) Worksheet 1.1 (*Student Workbook*, page 21)			
	Day 2	Lesson 1.2 (*Student Textbook*, pages 15–17) Worksheet 1.2 (*Student Workbook*, page 23)			
	Day 3	Lesson 1.3 (*Student Textbook*, pages 18–22) Worksheet 1.3 (*Student Workbook*, pages 25–26)*			
	Day 4	Lesson 1.4 (*Student Textbook*, pages 22–27) Worksheet 1.4 (*Student Workbook*, page 27)			
	Day 5	Lesson 1.5 (*Student Textbook*, pages 27–30; Worksheet 1.5 (*Student Workbook*, pages 29–31)			
Week 2	Day 6	Lesson 1.6 (*Student Textbook*, pages 31–35) Worksheet 1.6 (*Student Workbook*, pages 33–36)			
	Day 7	Lesson 2.1 (*Student Textbook*, pages 37–42) Worksheet 2.1 (*Student Workbook*, page 37)			
	Day 8	Lesson 2.2 (*Student Textbook*, pages 42–45) Worksheet 2.2 (*Student Workbook*, pages 39–40)*			
	Day 9	Lesson 2.3 (*Student Textbook*, pages 46–51) Worksheet 2.3 (*Student Workbook*, pages 41–42)			
	Day 10	Lesson 2.4 (*Student Textbook*, pages 52–56) Worksheet 2.4 (*Student Workbook*, pages 43–44)			
Week 3	Day 11	Lesson 2.5 (*Student Textbook*, pages 56–58) Worksheet 2.5 (*Student Workbook*, pages 45–46)			
	Day 12	Lesson 2.6 (*Student Textbook*, pages 58–63) Worksheet 2.6 (*Student Workbook*, pages 47–50)			
	Day 13	Lesson 2.7 (*Student Textbook*, pages 63–64) Quiz 1 (*Student Workbook*, pages 335)			
	Day 14	Lesson 3.1 (*Student Textbook*, pages 65–66) Worksheet 3.1 (*Student Workbook*, page 51)			
	Day 15	Lesson 3.2 (*Student Textbook*, pages 67–68) Worksheet 3.2 (*Student Workbook*, pages 53–54)			
Week 4	Day 16	Lesson 3.3 (*Student Textbook*, pages 68–73) Worksheet 3.3 (*Student Workbook*, pages 55–56)			
	Day 17	Lesson 3.4 (*Student Textbook*, pages 73–76) Worksheet 3.4 (*Student Workbook*, page 57–58)			
	Day 18	Lesson 3.5 (*Student Textbook*, pages 76–81) Worksheet 3.5 (*Student Workbook*, pages 59–60)			
	Day 19	Lesson 3.6 (*Student Textbook*, pages 81–83) Worksheet 3.6 (*Student Workbook*, pages 61–62)			
	Day 20	Lesson 3.7 (*Student Textbook*, pages 83–84) Study Day			

* Worksheet 1.3 includes instructions on building an abacus. To build an abacus, students will need an 8 x 10 or larger picture frame, multi-color pony beads, wire, needle-nose pliers, and carpet tacks/small nails. Alternately, students can use an online or premade abacus.
* Worksheet 2.2 includes extra-credit assignment to research the history of time zones.

Suggested Daily Schedule
(to complete Year 1 in a school year)

Date	Day	Assignment	Due Date	✓	Grade
Week 5	Day 21	Quiz 2 (*Student Workbook*, pages 337–338)			
	Day 22	Lesson 4.1 (*Student Textbook*, pages 85–91) Worksheet 4.1 (*Student Workbook*, pages 63–64)			
	Day 23	Lesson 4.2 (*Student Textbook*, pages 92–94) Worksheet 4.2 (*Student Workbook*, pages 65–66)			
	Day 24	Lesson 4.3 (*Student Textbook*, pages 95–99) Worksheet 4.3 (*Student Workbook*, pages 67–68)			
	Day 25	Lesson 4.4 (*Student Textbook*, pages 99–102) Worksheet 4.4 (*Student Workbook*, pages 69–70)			
Week 6	Day 26	Lesson 4.5 (*Student Textbook*, pages 102–105) Worksheet 4.5 (*Student Workbook*, pages 71–72)			
	Day 27	Lesson 4.6 (*Student Textbook*, pages 105–107) Worksheet 4.6 (*Student Workbook*, pages 73–75)*			
	Day 28	Quiz 3 (*Student Workbook*, pages 339–340)			
	Day 29	Lesson 5.1 (*Student Textbook*, pages 109–114) Worksheet 5.1 (*Student Workbook*, pages 77–78)			
	Day 30	Lesson 5.2 (*Student Textbook*, pages 114–117) Worksheet 5.2 (*Student Workbook*, pages 79–80)			
Week 7	Day 31	Lesson 5.3 (*Student Textbook*, pages 117–120) Worksheet 5.3 (*Student Workbook*, pages 81–82)			
	Day 32	Lesson 5.4 (*Student Textbook*, pages 121–123) Worksheet 5.4 (*Student Workbook*, pages 83–84)			
	Day 33	Lesson 5.5 (*Student Textbook*, page 124–125) Worksheet 5.5 (*Student Workbook*, pages 85–86)			
	Day 34	Lesson 5.6 (*Student Textbook*, pages 125–129) Worksheet 5.6 (*Student Workbook*, pages 87–88)			
	Day 35	Lesson 5.7 (*Student Textbook*, pages 130–131) Worksheet 5.7 (*Student Workbook*, pages 89–92)			
Week 8	Day 36	Lesson 5.8 (*Student Textbook*, pages 131–133) Quiz 4 (*Student Workbook*, pages 341–343)			
	Day 37	Lesson 6.1 (*Student Textbook*, pages 135–138) Worksheet 6.1 (*Student Workbook*, pages 93–94)			
	Day 38	Lesson 6.2 (*Student Textbook*, pages 138–141) Worksheet 6.2 (*Student Workbook*, pages 95–96)			
	Day 39	Lesson 6.3 (*Student Textbook*, pages 142–145) Worksheet 6.3 (*Student Workbook*, pages 97–98)			
	Day 40	Lesson 6.4 (*Student Textbook*, pages 146–147) Worksheet 6.4 (*Student Workbook*, page 99)			
Week 9	Day 41	Lesson 6.5 (*Student Textbook*, pages 148–152) Worksheet 6.5 (*Student Workbook*, pages 101–102)			
	Day 42	Lesson 6.6 (*Student Textbook*, page 152) Worksheet 6.6 (*Student Workbook*, pages 103–104)*			
	Day 43	Worksheet 6.7 (*Student Workbook*, pages 105–108)			
	Day 44	Study Day			
	Day 45	Test 1 (*Student Workbook*, pages 373–374)			

* Worksheet 4.6 includes hands-on activity with gas prices and extra-credit assignment to make Napier's rods.
* Worksheet 6.6 includes an assignment to half or double a recipe.

Suggested Daily Schedule
(to complete Year 1 in a school year)

Date	Day	Assignment	Due Date	✓	Grade
		First Semester—Second Quarter			
Week 1	Day 46	Lesson 7.1 (*Student Textbook*, pages 153–158) Worksheet 7.1 (*Student Workbook*, pages 109–110)			
	Day 47	Lesson 7.2 (*Student Textbook*, pages 158–161) Worksheet 7.2 (*Student Workbook*, pages 111–112)			
	Day 48	Lesson 7.3 (*Student Textbook*, pages 162–163) Worksheet 7.3 (*Student Workbook*, pages 113–114)			
	Day 49	Lesson 7.4 (*Student Textbook*, pages 164–166) Worksheet 7.4 (*Student Workbook*, pages 115–116)*			
	Day 50				
Week 2	Day 51	Lesson 7.5 (*Student Textbook*, pages 166–168) Worksheet 7.5 (*Student Workbook*, pages 117–119)			
	Day 52	Lesson 7.6 (*Student Textbook*, page 168) Worksheet 7.6 (*Student Workbook*, pages 121–122)*			
	Day 53	Quiz 5 (*Student Workbook*, pages 345–346)			
	Day 54	Lesson 8.1 (*Student Textbook*, pages 169–171) Worksheet 8.1 (*Student Workbook*, pages 123–125)			
	Day 55	Lesson 8.2 (*Student Textbook*, pages 171–173) Worksheet 8.2 (*Student Workbook*, pages 127–128)			
Week 3	Day 56	Lesson 8.3 (*Student Textbook*, pages 174–175) Worksheet 8.3 (*Student Workbook*, pages 129–130)			
	Day 57	Lesson 8.4 (*Student Textbook*, pages 176–179) Worksheet 8.4 (*Student Workbook*, pages 131–132)*			
	Day 58	Lesson 8.5 (*Student Textbook*, pages 179–181) Worksheet 8.5 (*Student Workbook*, pages 133–134)			
	Day 59	Lesson 8.6 (*Student Textbook*, pages 181–182) Worksheet 8.6 (*Student Workbook*, pages 135–136)			
	Day 60	Quiz 6 (*Student Workbook*, pages 347–348)			
Week 4	Day 61	Lesson 9.1 (*Student Textbook*, pages 183–186) Worksheet 9.1 (*Student Workbook*, pages 137–138)			
	Day 62	Lesson 9.2 (*Student Textbook*, page 186–187) Worksheet 9.2 (*Student Workbook*, pages 139–140)			
	Day 63	Lesson 9.3 (*Student Textbook*, pages 187–189) Worksheet 9.3 (*Student Workbook*, pages 141–142)			
	Day 64	Lesson 9.4 (*Student Textbook*, pages 189–192) Worksheet 9.4 (*Student Workbook*, pages 143–144)			
	Day 65	Study Day			

* Worksheet 7.4 includes assignment to round purchases at a store.
* Worksheet 7.6 includes computer assignment.
* Worksheet 8.4 includes assignment to make a scale drawing of bookcase and look at a home blueprint.

Suggested Daily Schedule
(to complete Year 1 in a school year)

Date	Day	Assignment	Due Date	✓	Grade
Week 5	Day 66	Lesson 9.5 (*Student Textbook*, page 192–193) Worksheet 9.5 (*Student Workbook*, pages 145–146)*			
	Day 67	Lesson 9.6 (*Student Textbook*, pages 193–194) Study Day			
	Day 68	Quiz 7 (*Student Workbook*, pages 349–350)			
	Day 69	Lesson 10.1 (*Student Textbook*, pages 195–197) Worksheet 10.1 (*Student Workbook*, pages 147–148)			
	Day 70	Lesson 10.2 (*Student Textbook*, pages 197–200) Worksheet 10.2A (*Student Workbook*, pages 149–150)			
Week 6	Day 71	Worksheet 10.2B (*Student Workbook*, pages 151–152)			
	Day 72	Lesson 10.3 (*Student Textbook*, pages 201–203) Worksheet 10.3 (*Student Workbook*, pages 153–154)			
	Day 73	Lesson 10.4 (*Student Textbook*, pages 203–206) Worksheet 10.4 (*Student Workbook*, pages 155–156)			
	Day 74	Lesson 10.5 (*Student Textbook*, pages 206–208) Worksheet 10.5 (*Student Workbook*, pages 157–158)			
	Day 75				
Week 7	Day 76	Lesson 10.6 (*Student Textbook*, pages 208–210) Worksheet 10.6 (*Student Workbook*, pages 159–160)			
	Day 77	Lesson 10.7 (*Student Textbook*, pages 210–212) Worksheet 10.7 (*Student Workbook*, pages 161–162)			
	Day 78	Lesson 10.8 (*Student Textbook*, page 212–213) Worksheet 10.8 (*Student Workbook*, pages 163–164)			
	Day 79	Lesson 10.9 (*Student Textbook*, pages 213–214)			
	Day 80	Quiz 8 (*Student Workbook*, pages 351–352)			
Week 8	Day 81	Lesson 11.1 (*Student Textbook*, pages 215–219) Worksheet 11.1 (*Student Workbook*, pages 165–166)*			
	Day 82	Lesson 11.2 (*Student Textbook*, pages 219–222) Worksheet 11.2 (*Student Workbook*, page 167)			
	Day 83	Lesson 11.3 (*Student Textbook*, pages 222–224) Worksheet 11.3 (*Student Workbook*, pages 169–170)			
	Day 84	Lesson 11.4 (*Student Textbook*, pages 224–226) Worksheet 11.4 (*Student Workbook*, pages 171–172)*			
	Day 85				
Week 9	Day 86	Lesson 11.5 (*Student Textbook*, page 226) Worksheet 11.5 (*Student Workbook*, pages 173–174)			
	Day 87	Quiz 9 (*Student Workbook*, pages 353–354)			
	Day 88	Study Day			
	Day 89	Study Day			
	Day 90	Test 2 (*Student Workbook*, pages 375–376)			
		Midterm Grade			

* Worksheet 9.5 includes assignment to locate a percent in a newspaper.
* Worksheet 11.4 includes assignment to find and count change.

Suggested Daily Schedule
(to complete Year 1 in a school year)

Date	Day	Assignment	Due Date	✓	Grade
		Second Semester—Third Quarter			
Week 1	Day 91	Lesson 12.1 (*Student Textbook*, pages 227–229) Worksheet 12.1 (*Student Workbook*, pages 175)			
	Day 92	Lesson 12.2 (*Student Textbook*, pages 229–231) Worksheet 12.2 (*Student Workbook*, pages 177–178)*			
	Day 93	Lesson 12.3 (*Student Textbook*, pages 232–236) Worksheet 12.3 (*Student Workbook*, pages 179–180)			
	Day 94	Lesson 12.4 (*Student Textbook*, pages 236–237) Worksheet 12.4 (*Student Workbook*, pages 181–182)*			
	Day 95	Lesson 12.5 (*Student Textbook*, pages 238–241) Worksheet 12.5 (*Student Workbook*, pages 183–184)			
Week 2	Day 96	Lesson 12.6 (*Student Textbook*, pages 241–244) Worksheet 12.6 (*Student Workbook*, pages 185–187)			
	Day 97	Lesson 12.7 (*Student Textbook*, pages 244–248) Worksheet 12.7 (*Student Workbook*, pages 189–191)			
	Day 98	Lesson 12.8 (*Student Textbook*, pages 249–250) Worksheet 12.8 (*Student Workbook*, pages 193–194)			
	Day 99	Lesson 12.9 (*Student Textbook*, pages 250–251)			
	Day 100	Quiz 10 (*Student Workbook*, pages 355)*			
Week 3	Day 101	Lesson 13.1 (*Student Textbook*, pages 253–256) Worksheet 13.1 (*Student Workbook*, pages 195–196)			
	Day 102	Lesson 13.2 (*Student Textbook*, pages 256–260) Worksheet 13.2 (*Student Workbook*, pages 197–198)			
	Day 103	Lesson 13.3 (*Student Textbook*, pages 260–263) Worksheet 13.3 (*Student Workbook*, pages 199–200)			
	Day 104	Lesson 13.4 (*Student Textbook*, pages 263–266) Worksheet 13.4 (*Student Workbook*, pages 201–203)			
	Day 105	Lesson 13.5 (*Student Textbook*, pages 267–272) Worksheet 13.5 (*Student Workbook*, pages 205–206)*			
Week 4	Day 106	Lesson 13.6 (*Student Textbook*, pages 272–273) Worksheet 13.6 (*Student Workbook*, pages 207–208)			
	Day 107	Quiz 11 (*Student Workbook*, pages 357–358)			
	Day 108	Lesson 14.1 (*Student Textbook*, pages 275–281) Worksheet 14.1 (*Student Workbook*, pages 209–210)			
	Day 109	Lesson 14.2 (*Student Textbook*, pages 281–283) Worksheet 14.2 (*Student Workbook*, pages 211–212)			
	Day 110	Lesson 14.3 (*Student Textbook*, pages 283–286) Worksheet 14.3 (*Student Workbook*, pages 213–215)			

* Worksheet 12.2 includes extra-credit assignment to read *How to Lie with Statistics*.
* Worksheet 12.4 includes extra-credit assignment to make a graph on the computer.
* Quiz 10 includes assignment to write a three-paragraph analysis of a real-life graph.
* Worksheet 13.5 tells students to pick one of these assignments: make a design on graph paper, find a wallpaper pattern online, or look at quilt patterns.

Suggested Daily Schedule
(to complete Year 1 in a school year)

Date	Day	Assignment	Due Date	✓	Grade
Week 5	Day 111	Lesson 14.4 (*Student Textbook*, pages 286–287) Worksheet 14.4 (*Student Workbook*, pages 215–216)			
	Day 112	Lesson 14.5 (*Student Textbook*, pages 288–290) Worksheet 14.5 (*Student Workbook*, pages 217–219)*			
	Day 113	Lesson 14.6 (*Student Textbook*, pages 291–292) Worksheet 14.6 (*Student Workbook*, pages 221–223)*			
	Day 114	Lesson 14.7 (*Student Textbook*, pages 292–294) Worksheet 14.7 (*Student Workbook*, pages 225–226)			
	Day 115				
Week 6	Day 116	Lesson 14.8 (*Student Textbook*, pages 294–297) Worksheet 14.8 (*Student Workbook*, pages 227–228)			
	Day 117	Lesson 14.9 (*Student Textbook*, pages 297–298) Study Day			
	Day 118	Quiz 12 (*Student Workbook*, pages 359–360)			
	Day 119	Lesson 15.1 (*Student Textbook*, pages 299–302) Worksheet 15.1 (*Student Workbook*, pages 229–230)			
	Day 120	Lesson 15.2 (*Student Textbook*, pages 303–305) Worksheet 15.2 (*Student Workbook*, pages 231–232)			
Week 7	Day 121	Lesson 15.3 (*Student Textbook*, pages 305–309) Worksheet 15.3 (*Student Workbook*, pages 233–234)			
	Day 122	Lesson 15.4 (*Student Textbook*, pages 309–311) Worksheet 15.4 (*Student Workbook*, pages 235–236)			
	Day 123	Lesson 15.5 (*Student Textbook*, pages 311–313) Study Day			
	Day 124	Quiz 13 (*Student Workbook*, pages 361–362)			
	Day 125	Lesson 16.1 (*Student Textbook*, pages 315–318) Worksheet 16.1 (*Student Workbook*, pages 237–238)			
Week 8	Day 126	Lesson 16.2 (*Student Textbook*, pages 319–321) Worksheet 16.2 (*Student Workbook*, pages 239–240)			
	Day 127	Lesson 16.3 (*Student Textbook*, pages 322–324) Worksheet 16.3 (*Student Workbook*, pages 241–242)			
	Day 128	Study Day			
	Day 129	Lesson 16.4 (*Student Textbook*, pages 324–328) Worksheet 16.4 (*Student Workbook*, pages 243–244)			
	Day 130	Lesson 16.5 (*Student Textbook*, pages 328–330) Worksheet 16.5 (*Student Workbook*, pages 245–246)			
Week 9	Day 131	Lesson 16.6 (*Student Textbook*, page 331)			
	Day 132	Quiz 14 (*Student Workbook*, pages 363–364)			
	Day 133	Worksheet 16.6 (*Student Workbook*, pages 247–250)			
	Day 134	Study Day			
	Day 135	Test 3 (*Student Workbook*, pages 377–380)			

* Worksheet 14.5 includes assignment to measure height and someone else's height.
* Worksheet 14.6 includes extra-credit assignment to watch the suggested online video.

Suggested Daily Schedule
(to complete Year 1 in a school year)

Date	Day	Assignment	Due Date	✓	Grade
		Second Semester—Fourth Quarter			
Week 1	Day 136	Lesson 17.1 (*Student Textbook*, pages 333–335) Worksheet 17.1 (*Student Workbook*, pages 251–252)			
	Day 137	Lesson 17.2 (*Student Textbook*, pages 336–338) Worksheet 17.2A (*Student Workbook*, pages 253–254)			
	Day 138	Worksheet 17.2B (*Student Workbook*, pages 255–257)			
	Day 139	Lesson 17.3 (*Student Textbook*, pages 338–342) Worksheet 17.3 (*Student Workbook*, pages 259–260)			
	Day 140				
Week 2	Day 141	Lesson 17.4 (*Student Textbook*, page 343–344) Worksheet 17.4 (*Student Workbook*, pages 261–262)			
	Day 142	Lesson 17.5 (*Student Textbook*, pages 344–346) Study Day			
	Day 143	Quiz 15 (*Student Workbook*, pages 365–366)			
	Day 144	Lesson 18.1 (*Student Textbook*, pages 347–349) Worksheet 18.1A (*Student Workbook*, pages 263–264)			
	Day 145	Worksheet 18.1B (*Student Workbook*, pages 265–266)			
Week 3	Day 146	Lesson 18.2 (*Student Textbook*, pages 350–356) Worksheet 18.2A (*Student Workbook*, pages 267–268)			
	Day 147	Worksheet 18.2B (*Student Workbook*, pages 269–270)			
	Day 148	Lesson 18.3 (*Student Textbook*, pages 356–358) Worksheet 18.3 (*Student Workbook*, pages 271–272)			
	Day 149	Lesson 18.4 (*Student Textbook*, pages 358–360) Worksheet 18.4 (*Student Workbook*, pages 273–276)			
	Day 150	Lesson 18.5 (*Student Textbook*, pages 361–362) Worksheet 18.5 (*Student Workbook*, pages 277–278)			
Week 4	Day 151	Lesson 18.6 (*Student Textbook*, pages 362–363) Worksheet 18.6 (*Student Workbook*, pages 279–280)			
	Day 152	Lesson 18.7 (*Student Textbook*, pages 364–365) Worksheet 18.7 (*Student Workbook*, pages 281–284)*			
	Day 153	Lesson 18.8 (*Student Textbook*, pages 365–366) Study Day			
	Day 154	Quiz 16 (*Student Workbook*, pages 367–368)			
	Day 155	Lesson 19.1 (*Student Textbook*, pages 367–370) Worksheet 19.1 (*Student Workbook*, pages 285–286)			

* Worksheet 18.7 includes assignment to locate food items with specific units of measure.

Suggested Daily Schedule
(to complete Year 1 in a school year)

Date	Day	Assignment	Due Date	✓	Grade
Week 5	Day 156	Lesson 19.2 (*Student Textbook*, pages 370–374) Worksheet 19.2 (*Student Workbook*, pages 287–289)			
	Day 157	Lesson 19.3 (*Student Textbook*, pages 374–376) Worksheet 19.3 (*Student Workbook*, pages 291–292)			
	Day 158	Lesson 19.4 (*Student Textbook*, pages 376–379) Worksheet 19.4 (*Student Workbook*, pages 293–295)*			
	Day 159	Lesson 19.5 (*Student Textbook*, pages 379–380) Worksheet 19.5 (*Student Workbook*, pages 297–298)			
	Day 160	Quiz 17 (*Student Workbook*, pages 369–370)			
Week 6	Day 161	Lesson 20.1 (*Student Textbook*, pages 381–383) Worksheet 20.1 (*Student Workbook*, pages 299–300)			
	Day 162	Lesson 20.2 (*Student Textbook*, pages 384–390) Worksheet 20.2 (*Student Workbook*, pages 301–302)			
	Day 163	Lesson 20.3 (*Student Textbook*, pages 390–392) Worksheet 20.3 (*Student Workbook*, pages 303–304)			
	Day 164	Lesson 20.4 (*Student Textbook*, pages 392–394) Worksheet 20.4 (*Student Workbook*, pages 305–306)			
	Day 165	Lesson 20.5 (*Student Textbook*, pages 395–396) Worksheet 20.5 (*Student Workbook*, pages 307–308)			
Week 7	Day 166	Lesson 20.6 (*Student Textbook*, pages 396–397) Worksheet 20.6 (*Student Workbook*, pages 309–310)			
	Day 167	Quiz 18 (*Student Workbook*, pages 371–372)			
	Day 168	Worksheet 20.7 (*Student Workbook*, pages 311–314)			
	Day 169	Study Day			
	Day 170	Test 4 (*Student Workbook*, pages 381–384)			
Week 8	Day 171	Lesson 21.1 (*Student Textbook*, page 399) Worksheet 21.1 (*Student Workbook*, pages 315–318)			
	Day 172	Lesson 21.2 (*Student Textbook*, pages 400–402) Worksheet 21.2 (*Student Workbook*, pages 319–321)			
	Day 173	Lesson 21.3 (*Student Textbook*, page 400–402) Worksheet 21.3 (*Student Workbook*, page 323); Report Assigned			
	Day 174	Research and Write Report			
	Day 175	Lesson 21.4 (*Student Textbook*, pages 403–405) Worksheet 21.4A (*Student Workbook*, pages 325–327)			
Week 9	Day 176	Research and Write Report			
	Day 177	Worksheet 21.4B (*Student Workbook*, pages 329–332)			
	Day 178	Study Day			
	Day 179	Study Day			
	Day 180	Test 5 (Final) (*Student Workbook*, pages 385–390)*			
		Final Grade			

* Worksheet 19.4 includes experiment with flashlight and mirror.
* Test 5 includes extra-credit problems.

Accelerated Daily Schedule
(to complete Year 1 in a semester)

Date	Day	Assignment	Due Date	✓	Grade
		First Semester—First Quarter			
Week 1	Day 1	Lessons 1.1–1.3 (*Student Textbook*, pages 13–22) Worksheets 1.1–1.3 (*Student Workbook*, pages 21–26)*			
	Day 2	Lessons 1.4–1.5 (*Student Textbook*, pages 22–31) Worksheets 1.4–1.5 (*Student Workbook*, pages 27–31)			
	Day 3	Lesson 1.6 (*Student Textbook*, pages 31–35) Worksheet 1.6 (*Student Workbook*, pages 33–36)			
	Day 4	Lessons 2.1–2.2 (*Student Textbook*, pages 37–45) Worksheets 2.1–2.2 (*Student Workbook*, pages 37–40)*			
	Day 5	Lessons 2.3–2.5 (*Student Textbook*, pages 46–58) Worksheets 2.3–2.5 (*Student Workbook*, pages 41–46)			
Week 2	Day 6	Lessons 2.6–2.7 (*Student Textbook*, pages 58–64) Worksheets 2.6 (*Student Workbook*, pages 47–50) Quiz 1 (*Student Workbook*, pages 335)			
	Day 7	Lessons 3.1–3.2 (*Student Textbook*, pages 65–68) Worksheets 3.1–3.2 (*Student Workbook*, pages 51–54)			
	Day 8	Lessons 3.3–3.4 (*Student Textbook*, pages 68–76) Worksheets 3.3–3.4 (*Student Workbook*, pages 55–58)			
	Day 9	Lessons 3.5–3.6 (*Student Textbook*, pages 76–83) Worksheets 3.5–3.6 (*Student Workbook*, pages 59–62)			
	Day 10	Lesson 3.7 (*Student Textbook*, pages 83–84) Quiz 2 (*Student Workbook*, pages 337–338)			
Week 3	Day 11	Lessons 4.1–4.2 (*Student Textbook*, pages 85–94) Worksheets 4.1–4.2 (*Student Workbook*, pages 63–66)			
	Day 12	Lessons 4.3–4.4 (*Student Textbook*, pages 95–102) Worksheets 4.3–4.4 (*Student Workbook*, pages 67–70)			
	Day 13	Lessons 4.5–4.6 (*Student Textbook*, pages 102–107) Worksheets 4.5–4.6 (*Student Workbook*, pages 71–75)*			
	Day 14	Quiz 3 (*Student Workbook*, pages 339–340); Lesson 5.1 (*Student Textbook*, pages 109–114) Worksheet 5.1 (*Student Workbook*, pages 77–78)			
	Day 15	Lessons 5.2–5.3 (*Student Textbook*, pages 114–120) Worksheets 5.2–5.3 (*Student Workbook*, pages 79–82)			
Week 4	Day 16	Lesson 5.4 (*Student Textbook*, pages 121–123) Worksheet 5.4 (*Student Workbook*, pages 83–84)			
	Day 17	Lesson 5.5 (*Student Textbook*, page 124–125) Worksheet 5.5 (*Student Workbook*, pages 85–86)			
	Day 18	Lesson 5.6 (*Student Textbook*, pages 125–130) Worksheet 5.6 (*Student Workbook*, pages 87–88)			
	Day 19	Lesson 5.7 (*Student Textbook*, page 130–131) Worksheet 5.7 (*Student Workbook*, pages 89–92)			
	Day 20	Lesson 5.8 (*Student Textbook*, pages 131–133) Quiz 4 (*Student Workbook*, pages 341–343)			

* Worksheet 1.3 includes instructions on building an abacus. To build an abacus, students will need an 8 x 10 or larger picture frame, multi-color pony beads, wire, needle-nose pliers, and carpet tacks/small nails. Alternately, students can use an online or premade abacus.

* Worksheet 2.2 includes extra-credit assignment to research the history of time zones.

* Worksheet 4.6 includes hands-on activity with gas prices and extra-credit assignment to make Napier's rods.

Accelerated Daily Schedule
(to complete Year 1 in a semester)

Date	Day	Assignment	Due Date	✓	Grade
Week 5	Day 21	Lessons 6.1–6.2 (*Student Textbook*, pages 135–141) Worksheets 6.1–6.2 (*Student Workbook*, pages 93–96)			
	Day 22	Lessons 6.3–6.4 (*Student Textbook*, pages 142–147) Worksheets 6.3–6.4 (*Student Workbook*, pages 97–99)			
	Day 23	Lessons 6.5–6.6 (*Student Textbook*, pages 148–152) Worksheets 6.5–6.6 (*Student Workbook*, pages 101–104)*			
	Day 24	Worksheet 6.7 (*Student Workbook*, pages 105–108); Study Day			
	Day 25	Test 1 (*Student Workbook*, pages 373–374)			
Week 6	Day 26	Lesson 7.1 (*Student Textbook*, pages 153–158) Worksheet 7.1 (*Student Workbook*, pages 109–110)			
	Day 27	Lessons 7.2–7.3 (*Student Textbook*, pages 158–163) Worksheets 7.2–7.3 (*Student Workbook*, pages 111–114)			
	Day 28	Lessons 7.4–7.5 (*Student Textbook*, pages 164–168) Worksheets 7.4–7.5 (*Student Workbook*, pages 115–119)*			
	Day 29	Lesson 7.6 (*Student Textbook*, page 168) Worksheet 7.6 (*Student Workbook*, pages 121–122)*			
	Day 30	Quiz 5 (*Student Workbook*, pages 345–346); Lessons 8.1–8.2 (*Student Textbook*, pages 169–173) Worksheets 8.1–8.2 (*Student Workbook*, pages 123–128)			
Week 7	Day 31	Lessons 8.3–8.4 (*Student Textbook*, pages 174–179) Worksheets 8.3–8.4 (*Student Workbook*, pages 129–132)*			
	Day 32	Lessons 8.5–8.6 (*Student Textbook*, pages 179–181) Worksheets 8.5–8.6 (*Student Workbook*, pages 133–136)			
	Day 33	Quiz 6 (*Student Workbook*, pages 347–348); Lessons 9.1–9.2 (*Student Textbook*, pages 183–187) Worksheets 9.1–9.2 (*Student Workbook*, pages 137–140)			
	Day 34	Lessons 9.3–9.4 (*Student Textbook*, pages 187–192) Worksheets 9.3–9.4 (*Student Workbook*, pages 141–144)			
	Day 35	Lessons 9.5–9.6 (*Student Textbook*, pages 192–194) Worksheets 9.5 (*Student Workbook*, page 145–146); Quiz 7 (*Student Workbook*, pages 349–350)*			
Week 8	Day 36	Lessons 10.1–10.2 (*Student Textbook*, pages 195–200) Worksheets 10.1–10.2A (*Student Workbook*, pages 147–150)			
	Day 37	Lessons 10.3–10.4 (*Student Textbook*, pages 201–206) Worksheets 10.2B–10.4 (*Student Workbook*, pages 151–156)			
	Day 38	Lessons 10.5–10.6 (*Student Textbook*, pages 206–210) Worksheets 10.5–10.6 (*Student Workbook*, pages 157–160)			
	Day 39	Lessons 10.7–10.8 (*Student Textbook*, pages 210–213) Worksheets 10.7–10.8 (*Student Workbook*, pages 161–164)			
	Day 40	Lessons 10.9–11.1 (*Student Textbook*, pages 213–219) Quiz 8 (*Student Workbook*, pages 351–352); Worksheets 11.1 (*Student Workbook*, pages 165–166)			

* Worksheet 6.6 includes an assignment to half or double a recipe.
* Worksheet 7.4 includes assignment to round purchases at a store.
* Worksheet 7.6 includes computer assignment.
* Worksheet 8.4 includes assignment to make a scale drawing of bookcase and look at a home blueprint.
* Worksheet 9.5 includes assignment to locate a percent in a newspaper.

Accelerated Daily Schedule
(to complete Year 1 in a semester)

Date	Day	Assignment	Due Date	✓	Grade
Week 9	Day 41	Lessons 11.2–11.3 (*Student Textbook*, pages 219–224) Worksheets 11.2–11.3 (*Student Workbook*, pages 167–170)			
	Day 42	Lessons 11.4–11.5 (*Student Textbook*, pages 224–226) Worksheets 11.4–11.5 (*Student Workbook*, pages 171–174);* Quiz 9 (*Student Workbook*, pages 353–354)*			
	Day 43	Study Day			
	Day 44	Study Day			
	Day 45	Test 2 (*Student Workbook*, pages 375–376)			
First Semester—Second Quarter					
Week 1	Day 46	Lessons 12.1–12.2 (*Student Textbook*, pages 227–231) Worksheets 12.1–12.2 (*Student Workbook*, pages 175–178)*			
	Day 47	Lessons 12.3–12.4 (*Student Textbook*, pages 232–237) Worksheets 12.3–12.4 (*Student Workbook*, pages 179–182)*			
	Day 48	Lessons 12.5–12.6 (*Student Textbook*, pages 238–244) Worksheets 12.5–12.6 (*Student Workbook*, pages 183–187)			
	Day 49	Lessons 12.7–12.8 (*Student Textbook*, pages 244–250) Worksheets 12.7–12.8 (*Student Workbook*, pages 189–194)			
	Day 50	Lesson 12.9 (*Student Textbook*, pages 250–251; Quiz 10 (*Student Workbook*, pages 355)*			
Week 2	Day 51	Lessons 13.1–13.2 (*Student Textbook*, pages 253–260) Worksheets 13.1–13.2 (*Student Workbook*, pages 195–198)			
	Day 52	Lessons 13.3–13.4 (*Student Textbook*, pages 260–266) Worksheets 13.3–13.4 (*Student Workbook*, pages 199–203)			
	Day 53	Lessons 13.5–13.6 (*Student Textbook*, pages 267–273) Worksheets 13.5–13.6 (*Student Workbook*, pages 205–208)*			
	Day 54	Lesson 14.1 (*Student Textbook*, pages 275–281) Quiz 11 (*Student Workbook*, pages 357–358); Worksheet 14.1 (*Student Workbook*, pages 209–210)			
	Day 55	Lessons 14.2–14.3 (*Student Textbook*, pages 281–286) Worksheets 14.2–14.3 (*Student Workbook*, pages 211–214)			

* Worksheet 11.4 includes assignment to find and count change.
* Worksheet 12.2 includes extra-credit assignment to read *How to Lie with Statistics*.
* Worksheet 12.4 includes extra-credit assignment to make a graph on the computer.
* Quiz 10 includes assignment to write a three-paragraph analysis of a real-life graph.
* Worksheet 13.5 tells students to pick one of these assignments: make a design on graph paper, find a wallpaper pattern online, or look at quilt patterns.

Accelerated Daily Schedule
(to complete Year 1 in a semester)

Date	Day	Assignment	Due Date	✓	Grade
Week 3	Day 56	Lessons 14.4–14.5 (*Student Textbook*, pages 286–290) Worksheets 14.4–14.5 (*Student Workbook*, pages 215–219)*			
	Day 57	Lessons 14.6–14.7 (*Student Textbook*, pages 291–294) Worksheets 14.6–14.7 (*Student Workbook*, pages 221–226)*			
	Day 58	Lessons 14.8–14.9 (*Student Textbook*, pages 294–298) Worksheets 14.8 (*Student Workbook*, pages 227–228) Quiz 12 (*Student Workbook*, pages 359–360)			
	Day 59	Lessons 15.1–15.2 (*Student Textbook*, pages 299–305) Worksheets 15.1–15.2 (*Student Workbook*, pages 229–232)			
	Day 60	Lessons 15.3–15.5 (*Student Textbook*, pages 305–313) Worksheets 15.3–15.4 (*Student Workbook*, pages 233–236)			
Week 4	Day 61	Lesson 16.1 (*Student Textbook*, pages 315–318) Quiz 13 (*Student Workbook*, pages 361–362); Worksheet 16.1 (*Student Workbook*, pages 237–238)			
	Day 62	Lesson 16.2 (*Student Textbook*, pages 319–321) Worksheet 16.2 (*Student Workbook*, pages 239–240)			
	Day 63	Lesson 16.3 (*Student Textbook*, pages 322–324) Worksheet 16.3 (*Student Workbook*, pages 241–242)			
	Day 64	Lessons 16.4–16.6 (*Student Textbook*, pages 324–331) Worksheets 16.4–16.6 (*Student Workbook*, pages 243–250)			
	Day 65	Quiz 14 (*Student Workbook*, pages 363–364); Study Day			
Week 5	Day 66	Test 3 (*Student Workbook*, pages 377–380)			
	Day 67	Lesson 17.1 (*Student Textbook*, pages 333–335) Worksheet 17.1 (*Student Workbook*, pages 251–252)			
	Day 68	Lesson 17.2 (*Student Textbook*, pages 336–338) Worksheet 17.2A and B (*Student Workbook*, pages 253–257)			
	Day 69	Lessons 17.3–17.5 (*Student Textbook*, pages 338–346) Worksheets 17.3–17.4 (*Student Workbook*, pages 259–262)			
	Day 70	Lesson 18.1 (*Student Textbook*, pages 347–349) Worksheet 18.1A and B (*Student Workbook*, pages 263–266); Quiz 15 (*Student Workbook*, pages 365–366)			
Week 6	Day 71	Lesson 18.2 (*Student Textbook*, pages 350–356) Worksheet 18.2A and B (*Student Workbook*, pages 267–270)			
	Day 72	Lessons 18.3–18.4 (*Student Textbook*, pages 356–360) Worksheets 18.3–18.4 (*Student Workbook*, pages 271–276)			
	Day 73	Lesson 18.5 (*Student Textbook*, page 361–362) Worksheet 18.5 (*Student Workbook*, pages 277–278)			
	Day 74	Lessons 18.6–18.7 (*Student Textbook*, pages 362–365) Worksheets 18.6–18.7 (*Student Workbook*, pages 279–284)*			
	Day 75	Lesson 18.8 (*Student Textbook*, pages 365–366) Quiz 16 (*Student Workbook*, pages 367–368)			

* Worksheet 14.5 includes assignment to measure height and someone else's height.
* Worksheet 14.6 includes extra-credit assignment to watch the suggested online video.
* Worksheet 18.7 includes assignment to locate food items with specific units of measure.

Accelerated Daily Schedule
(to complete Year 1 in a semester)

Date	Day	Assignment	Due Date	✓	Grade
Week 7	Day 76	Lessons 19.1–19.2 (*Student Textbook*, pages 367–374) Worksheets 19.1–19.2 (*Student Workbook*, pages 285–289)			
	Day 77	Lessons 19.3–19.4 (*Student Textbook*, pages 374–379) Worksheets 19.3–19.4 (*Student Workbook*, pages 291–295)*			
	Day 78	Lesson 19.5 (*Student Textbook*, pages 379–380) Worksheets 19.5 (*Student Workbook*, pages 297–298) Quiz 17 (*Student Workbook*, pages 369–370)			
	Day 79	Lessons 20.1–20.2 (*Student Textbook*, pages 381–390) Worksheets 20.1–20.2 (*Student Workbook*, pages 299–302)			
	Day 80	Lessons 20.3–20.5 (*Student Textbook*, pages 390–396) Worksheets 20.3–20.5 (*Student Workbook*, pages 303–308)			
Week 8	Day 81	Lesson 20.6 (*Student Textbook*, pages 396–397) Worksheet 20.6 (*Student Workbook*, pages 309–310); Quiz 18 (*Student Workbook*, pages 371–372)			
	Day 82	Worksheet 20.7 (*Student Workbook*, pages 311–314); Study Day			
	Day 83	Test 4 (*Student Workbook*, pages 381–384)			
	Day 84	Lesson 21.1 (*Student Textbook*, page 399) Worksheet 21.1 (*Student Workbook*, pages 315–318)			
	Day 85	Lesson 21.2 (*Student Textbook*, page 400) Worksheet 21.2 (*Student Workbook*, pages 319–321)			
Week 9	Day 86	Lesson 21.3 (*Student Textbook*, pages 400–402) Worksheet 21.3 (*Student Workbook*, page 323); Report Assigned			
	Day 87	Lesson 21.4 (*Student Textbook*, pages 403–405) Worksheet 21.4A (*Student Workbook*, pages 325–327); Work on Report			
	Day 88	Worksheet 21.4B (*Student Workbook*, pages 329–332); Work on Report			
	Day 89	Study Day; Work on Report			
	Day 90	Test 5 (Final) (*Student Workbook*, pages 385–390);* Report Due			
		Final Grade			

* Worksheet 19.4 includes experiment with flashlight and mirror.
* Test 5 includes extra-credit problems.

Worksheets

PRINCIPLES OF MATHEMATICS

CHAPTER 1. Introduction and Place Value
LESSON 1. Math Misconceptions

Worksheet 1.1

1. **Numbers Everywhere** — Take a piece of paper and write down every different use for math you encounter today. Try to find at least 10 different ways math (including numbers) is used outside a textbook. Ask your parents how they use math if you get stumped. (*Hint*: Look for numbers on phones, exit signs, at stores, etc.)

    a.

    b.

    c.

    d.

    e.

    f.

    g.

    h.

    i.

    j.

2. **Definition** — Look up the word "worldview" in a dictionary and write out the definition you find.

3. **Misconceptions** — List the three common misconceptions about math covered in Lesson 1.1.

    a.

    b.

    c.

4. **Question** — When are you allowed to use a calculator in this course? (See the "How to Use This Curriculum" section at the beginning of this *Student Workbook*—be sure to read it if you haven't yet!)

CHAPTER 1. Introduction and Place Value
LESSON 2. What Is Math?

Worksheet 1.2

1. **Math in Action** — Give 5 examples of how math is used outside a textbook that are different than the uses you listed in Worsheet 1.1.

    a.

    b.

    c.

    d.

    e.

2. **Notebook Preparation** — Tear out the Reference Sheet Section from this *Student Workbook* and place it inside a binder, along with some lined paper you can use to add additional notes as you study. Taking notes of key information as you go will help you both remember the information and find it easily when you forget.

3. **Math Defined** — What is math and why does math work outside of a textbook?

PRINCIPLES OF MATHEMATICS

CHAPTER 1. Introduction and Place Value
LESSON 3. The Spiritual Battle in Math

Worksheet 1.3

1. **Question** — How would you define the spiritual battle in math?

2. **Definition** — Look up the words "naturalism" and "humanism" in a dictionary and write out the definitions you find.

3. **Preparing Your Abacus** — Some of the problems in the upcoming lessons will require the use of an abacus. You can either make your own, use a premade one if you have one, or use an online abacus (see www.christianperspective.net/math/pom1). Today's the day to decide and either find or make one! The instructions for making one are below if you choose to assemble your own.

   WARNING: These abacuses contain small parts (beads) that can be a choking hazard as well as wires/nails that could hurt if handled inappropriately; please be careful if using around young children.

   Supplies:
   - **Wooden frame** — You will need an 8 x 10 or larger picture frame with the glass removed, or make your own frame out of 1 x 2s.
   - **Multicolor beads** — Basic pony beads will work—look in the craft section of your local department or craft store. The number of beads you need depends on the size of your frame. You need 50 beads for an 8 x 10 frame.
   - **Wire** — You can use plant wire, stripped electrical wire, or any sort of thin, flexible wire you can wrap around a carpet tack/small nail. Alternately, if you have a thick enough picture frame to drill holes into, you can use any sort of thick wire that is sturdy enough to insert into drilled holes.
   - **Needle-nose pliers and carpet tacks/small nails**, or, if using thicker wire, a **drill**

   Instructions:
   1. Cut the wire into strips a few inches longer than the width of your frame. Five is a good number of rows for most medium frames and the minimum required for the problems in this text; really large frames can handle more.
   2. Mark the frame at evenly spaced intervals along both sides where you want your rows to be.

PAGE 25

3. Prepare the frame for the wire by either inserting carpet tacks or tiny nails at each of the marks, or else drilling holes in the frame. A lot will depend on what type of frame and wire you have. You must have a sturdy frame and wire to drill holes; otherwise, you will need to use the carpet tacks or tiny nails.
4. Secure one end of the wire by wrapping it around the carpet tacks/tiny nails, or by pushing a thicker wire into the drilled holes.
5. Add the beads to the first row of the abacus. Alternate between 5 beads of one color and 5 beads of another color (grouping makes it easier to see the quantity represented). You should have at least 10 beads on each row.
6. Secure the second end of the wire to the frame the same way you did in step 4.
7. Repeat steps 4–6 until you have completed all the rows.

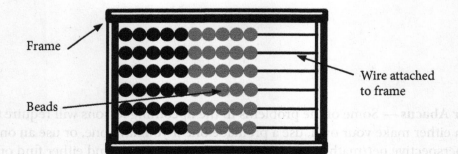

PRINCIPLES OF MATHEMATICS

CHAPTER 1. Introduction and Place Value
LESSON 4. Numbers, Place Value, and Comparisons

Worksheet 1.4

1. **Writing Numbers**[1] — Write out the following quantities using our place-value system.

    a. 2011 Population of the U.S.: three hundred eleven million, fifty thousand, nine hundred seventy-seven

    b. 2010 U.S. National Debt: thirteen trillion, five hundred sixty-one billion, six hundred million

    c. 2011 Population of China: one billion, three hundred thirty-six million, seven hundred eighteen thousand, fifteen

2. **Reading Numbers**[2] — Write the words you would use to read these numbers.

    a. 2010 Population of California: 27,253,956

    b. 2010 Population of Texas: 25,145,561

    c. 2010 Population of New York: 19,378,102

3. **Greater Than, Less Than, or Equal To** — Put the appropriate symbol (>, <, or =) in between each pair to show how they relate.

    a.   1,589   1,590
    b. 445,020,008   445,008,500
    c.   3,427   3,359

4. **History Check** — Use one of the historic equal signs shown in today's text to show 5 = 5.

---
[1] Facts from Sarah Janssen, sr. ed., M. L. Liu, Shmuel Ross, and Nan Badgett, eds., *The World Almanac and Book of Facts, 2012* (Infobase Learning, NY: 2012), pp. 63, 734.
[2] Facts from Ibid., p. 607.

PRINCIPLES OF MATHEMATICS | CHAPTER 1. Introduction and Place Value
LESSON 5. Different Number Systems | Worksheet 1.5

1. **Reading an Abacus** — Identify the following quantities and record the quantity using the decimal system.

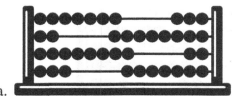
a.

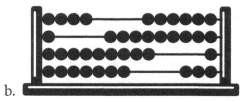
b.

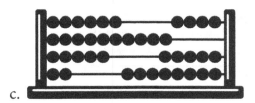
c.

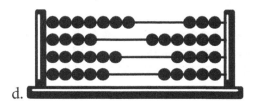
d.

2. **Abacus/Place Value**[1] — Use the abacus you made or located (see Worksheet 1.3) to form the following 2010 populations. (If you do not have an abacus or access to one online, draw one on paper for each problem.)

   a. Population of Bismarck, ND: 61,272

   b. Population of Dickinson, ND: 17,727

   c. Population of Amherst, OH: twelve thousand, twenty-one

   d. Population of Mansfield, OH: forty-seven thousand, eight hundred twenty- one

3. **Reading and Writing Numbers** — Express the first two quantities in the last problem (2a and 2b) with words, and the last two (2c and 2d) in the decimal system.

   a.

   b.

   c.

   d.

---
1 Facts from Sarah Janssen, sr. ed., M. L. Liu, Shmuel Ross, and Nan Badgett, eds., *The World Almanac and Book of Facts, 2012* (Infobase Learning, NY: 2012), p. 639.

4. **Comparing on an Abacus** — Put the appropriate symbol in between each pair of abacuses to show how the quantities they represent relate.

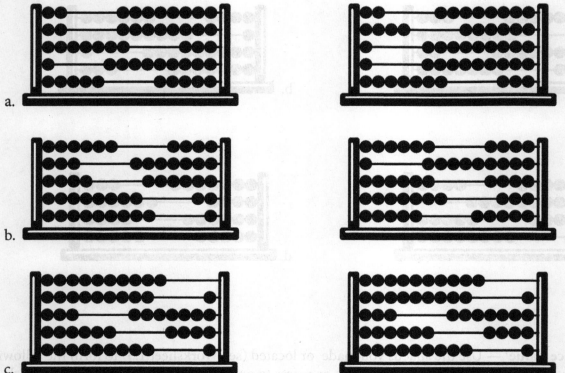

a.

b.

c.

5. **Question** — What do we call the number system we use today?

6. **Thinking It Through** — If one city has a population of 102,300 people, and another has a population of 123,000, which city has the greater population?

7. **Question** — Describe in your own words how place value works.

8. **Egyptian Hieroglyphics** — Looking at the figures presented in this lesson, do your best to represent the following quantities using Egyptian hieroglyphics (don't worry if you're not sure of a detail—just try to use the necessary symbols to convey the correct quantity and don't forget to put the smaller quantities on the left, opposite the way we do in our place value system).

   a. 26 b. 75 c. 89

9. **Numerals**

   a. Finish labeling this clock using Roman numerals to mark each hour.

   b. Books will sometimes list their publication date in Roman numerals. Suppose one says it was published in MCMXCVIII. What year is that in decimal notation? *Hint*: Work from left to right.

   c. In music, Roman numerals are used to number chords. The V chord (read "fifth chord") is the chord based off the fifth note of a scale. Knowing this, take a guess at what the IV chord means.

   d. Sundials keep track of time using the sun's shadow as the "hour" hand. Notice that the shadow on this sundial is falling near the spot labeled II. What hour is the sundial indicating?

10. **Question** — How do different numbering systems help us see our place-value system from a biblical worldview?

PRINCIPLES OF MATHEMATICS

CHAPTER 1. Introduction and Place Value
LESSON 6. Binary and Hexadecimal Place-value Systems

Worksheet 1.6

1. **Binary** — The following numbers are written in binary. Translate them into the decimal system by filling in the blanks.

    a. 1100

    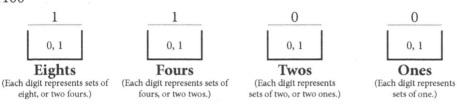

    Meaning:

    ___ set(s) of 8   = ___ x 8 = _____
    ___ set(s) of 4   = ___ x 4 = _____
    ___ set(s) of 2   = ___ x 2 = _____
    ___ set(s) of 1   = ___ x 1 = _____

    1100 in binary is the same as _____ in the decimal system.

    b. 10000

    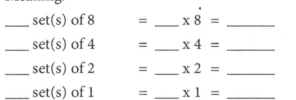

    Meaning:

    ___ set(s) of 16   = ___ x 16 = _____
    ___ set(s) of 8   = ___ x 8 = _____
    ___ set(s) of 4   = ___ x 4 = _____
    ___ set(s) of 2   = ___ x 2 = _____
    ___ set(s) of 1   = ___ x 1 = _____

    10000 in binary is the same as _____ in the decimal system.

    c. 10100

    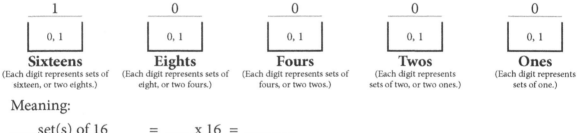

    Meaning:

    ___ set(s) of 16   = ___ x 16 = _____
    ___ set(s) of 8   = ___ x 8 = _____
    ___ set(s) of 4   = ___ x 4 = _____
    ___ set(s) of 2   = ___ x 2 = _____
    ___ set(s) of 1   = ___ x 1 = _____

    10100 in binary is the same as _____ in the decimal system.

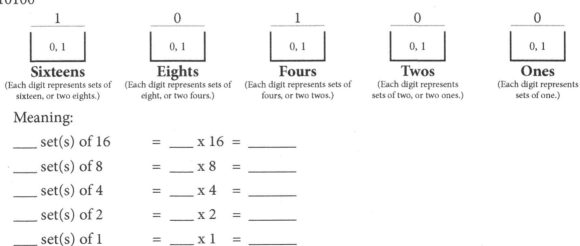

Hexadecimal System (Base 16)
16 Symbols: 0, 1, 2, 3, 4, 5, 6, 7, 8, 9, A, B, C, D, E, F

A represents the decimal value of 10.
B represents the decimal value of 11.
C represents the decimal value of 12.
D represents the decimal value of 13.
E represents the decimal value of 14.
F represents the decimal value of 15.

2. **Hexadecimal Number and Color** — Website programmers often specify colors using hexadecimal numbers in the RGB color system. *RGB* stands for *Red*, *Green*, and *Blue*. We can represent the intensity of each color using a scale, with 0 being none of the color and values increasing from there. A color with 0 red would have no red in it.

   We use two hexadecimal digits for each color. For example, in 8EC5E9 the 8E tells us the amount of red in the color, the C5 the amount of green, and the E9 the amount of blue. When all these colors mix together, we get a specific shade of blue.

8E	C5	E9
Red	Green	Blue

   Use what you know about the hexadecimal system to answer the question.

   Example: Write the amount of red—hexadecimal number 8E—using the decimal system.

8	E
0, 1, 2, 3, 4, 5, 6, 7, 8, 9, A, B, C, D, E, F	0, 1, 2, 3, 4, 5, 6, 7, 8, 9, A, B, C, D, E, F
Sixteens (Each digit represents sets of sixteen, or sixteen ones.)	**Ones** (Each digit represents sets of one.)

   8 set(s) of 16 = 8 x 16 = *128*
   14 set(s) of 1 = 14 x 1 = *14*
   128 + 14 = 142

   a. Write the amount of green—hexadecimal number C5—using the decimal system.

C	5
0, 1, 2, 3, 4, 5, 6, 7, 8, 9, A, B, C, D, E, F	0, 1, 2, 3, 4, 5, 6, 7, 8, 9, A, B, C, D, E, F
Sixteens (Each digit represents sets of sixteen, or sixteen ones.)	**Ones** (Each digit represents sets of one.)

   ___ set(s) of 16 = ___ x 16 = _____
   ___ set(s) of 1   = ___ x 1   = _____

   C5 in hexadecimal is the same as _____ in the decimal system.

b. Write the amount of blue—hexadecimal E9—using the decimal system.

E
| 0, 1, 2, 3, 4, 5, 6, 7, 8, 9, A, B, C, D, E, F |

Sixteens
(Each digit represents sets of sixteen, or sixteen ones.)

9
| 0, 1, 2, 3, 4, 5, 6, 7, 8, 9, A, B, C, D, E, F |

Ones
(Each digit represents sets of one.)

___ set(s) of 16 = ___ x 16 = _____

___ set(s) of 1   = ___ x 1 = _____

E9 in hexadecimal is the same as _____ in the decimal system.

c. Find the value of hexadecimal FF. *Note*: FF is the highest hexadecimal value we could form using just two digits; it is thus the max amount of red, green, or blue we could represent in the RGB color system.

F
| 0, 1, 2, 3, 4, 5, 6, 7, 8, 9, A, B, C, D, E, F |

Sixteens
(Each digit represents sets of sixteen, or sixteen ones.)

F
| 0, 1, 2, 3, 4, 5, 6, 7, 8, 9, A, B, C, D, E, F |

Ones
(Each digit represents sets of one.)

___ set(s) of 16 = ___ x 16 = _____

___ set(s) of 1   = ___ x 1 = _____

FF in hexadecimal is the same as _____ in the decimal system.

3. **Comparing Numbers** — Put a comparison symbol (<, >, =) to show how the quantities compare. *Hint*: You don't actually need to convert these binary numbers to our decimal system in order to tell how they compare. Instead, just look at where the 1s and 0s are and use place value to tell you which one must represent the greater quantity.

    a. 1100   1000

    b. 10000   1000

    c. 1010   1011

    d. 1111   1111

4. **Writing Numbers** — Write these numbers using our decimal system.[1]

    a. Land and water area for 50 states and Washington, D.C., in square miles: three million, seven hundred ninety-six thousand, seven hundred forty-two

    b. Distance to the sun in miles: ninety-two million, nine hundred sixty thousand

    c. Mean radius of the sun in miles: four hundred thirty-two thousand, two hundred

---

[1] Facts from Sarah Janssen, sr. ed., M. L. Liu, Shmuel Ross, and Nan Badgett, eds, *The World Almanac and Book of Facts, 2012* (Infobase Learning, NY: 2012), pp. 428, 344.

5. **Reading an Abacus** — Identify the following quantities and record the quantity using the decimal system.

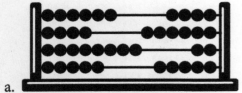
a.

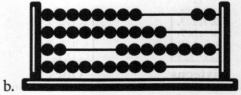
b.

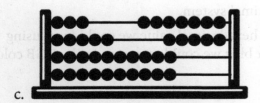
c.

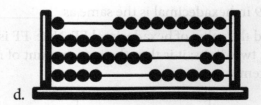
d.

6. **Roman numerals** — Express these quantities using Roman numerals.
    a. 2014

    b. 1,076

    c. 592

7. **Questions**
    a. What would it mean if you were told a number was written in a base-5 place-value system?

    b. How many digits would you need to write a number in a base-5 place-value system? *Hint*: Think through what you learned about the base-10 (decimal), base-2 (binary), and base-16 (hexadecimal) systems.

PRINCIPLES OF MATHEMATICS

CHAPTER 2. Operations, Algorithms, and Problem Solving
LESSON 1. Addition and Subtraction

Worksheet 2.1

1. **Terms** — Identify the addends, sum, minuend, subtrahend, and difference in the problems below.

   Example: 2 + 3 = 5
   2 and 3 are the addends; 5 is the sum.

   a. 4 + 9 = 13
   b. 15 − 9 = 6
   c. 8 + 5 = 13
   d. 17 − 6 = 11

2. **Practicing the Skill** — Use addition or subtraction as needed to find the missing numbers in these equations.

   a. 4 + 7 = _____
   b. 9 + ___ = 16
   c. 12 − 8 = _____
   d. _____ − 5 = 5
   e. IV + VI = _____
   f. X + _____ = XVIII
   g. XVI − VII = _____

3. **Comparing with Addition and Subtraction** — Perform the addition and subtraction first, compare the results, and then use the appropriate sign (<, >, or =) to show how the results compare.

   a. 17 − 9   8
   b. 15 − 6   8
   c. 2 + 3   IV
   d. 5 + 3   7 + 2
   e. V + VI   IV + IX

4. **Reflection** — Copy Hebrews 1:3 and Jeremiah 33:25-26 into your math notebook. As you do, take a moment today to thank the Lord for placing within math the reminder that He is a covenant-keeping God. Thank Him for holding all things together by the power of His Word.

1. **Terms** — Identify the addends, sum, minuend, subtrahend, and difference in the problems below.

   Example: 2 + 3 = 5
   2 and 3 are the addends; 5 is the sum.

   a. 4 + 9 = 13
   b. 15 − 9 = 6
   c. 8 + 5 = 13
   d. 17 − 6 = 11

2. **Practicing the Skill** — Use addition or subtraction as needed to find the missing numbers in these equations.

   a. 4 + 7 = _____
   b. 9 + _____ = 16
   c. 12 − 8 = _____
   d. _____ − 5 = 15
   e. IV + VI = _____
   f. X + _____ = XVIII
   g. XVI − VII = _____

3. **Comparing with Addition and Subtraction** — Perform the addition and subtraction first, compare the results, and then use the appropriate sign (<, >, or =) to show how the results compare.

   a. 17 − 9   8
   b. 15 − 6   8
   c. 2 + 3   IV
   d. 5 + 3   7 + 2
   e. V + VI   IV + IX

4. **Reflection** — Copy Hebrews 1:3 and Jeremiah 33:25-26 into your math notebook. As you do, take a moment today to thank the Lord for placing within math the reminder that He is a covenant-keeping God. Thank Him for holding all things together by the power of His Word.

CHAPTER 2. Operations, Algorithms, and Problem Solving
LESSON 2. Applying Basic Addition and Subtraction to Time
Worksheet 2.2

1. **Applying Addition and Subtraction: Keeping Track of Time**

    a. If a 3-hr wedding celebration starts at 5 p.m., at what time will it finish?

    b. If a 2-hr class starts at 10 a.m., at what time will it finish?

    c. If a luncheon starts at 11:15 a.m. and lasts 2 hours, when will it end?

    d. If the clock on an old building reads IX, and you have to get back to your hotel by 1 p.m., how much time do you have to get back?

2. **Understanding Time Biblically** — Look up the Scripture verses given and use them to answer the questions.

    a. Who was in the beginning and created day and night? Genesis 1:1–5

    b. Will time as we know it (with day and night, the sun and the moon) have an end? Isaiah 60:19; Revelation 21:1, 23; Revelation 22:5

    c. Will Heaven and Hell have an end? Revelation 22:5; 2 Thessalonians 1:8–9; Matthew 25:41

    d. How should we live in light of what God tells us about time and eternity? 2 Peter 3:10-14

3. **Applying Addition and Subtraction: Time Zones** — Use the chart to answer the questions. Notice how you're using basic addition and subtraction.

a. If you're on Eastern Standard Time and it's 9 a.m., what time is it on the west coast (Pacific Standard Time)?

b. If a show you want to watch is airing live at 8 p.m. EST (Eastern Standard Time), and you're in Mountain Standard Time (MST), what time will it be showing for you?

c. Say you're in EST. You have a business meeting at 2 p.m. PST. What time in your time zone is the meeting?

d. If you're in Central Standard Time (CST) and it's 4 p.m., what time is it in EST?

e. If a help desk closes at 5 p.m. EST and it's 1 p.m. MST where you are, for how many hours will the help desk still be open?

f. Add the time zones and their abbreviations to your notebook and make a flashcard to help you learn them. You will need to know the time differences for the continental United States on future worksheets, quizzes, and/or tests.

Extra Credit — Research the history of time zones (a quick Internet search should yield plenty of information) and write down one interesting tidbit.

PRINCIPLES OF MATHEMATICS

CHAPTER 2. Operations, Algorithms, and Problem Solving
LESSON 3. Multi-digit Addition and Subtraction

Worksheet 2.3

You may find it easier to rewrite problems vertically so you can line up the places better.

Example: 2,589 + 89

Rewrite vertically:

```
  2,5 8 9
+     8 9
```

1. **Abacus Addition** — Solve the following problems on an abacus.

   a. 18 + 9

   b. 205 + 807

   c. 378 + 879

2. **Abacus Subtraction** — Solve the following problem on an abacus.

   a. 807 − 99

   b. 1,008 − 560

   c. 2,000 − 899

3. **Traditional Addition and Subtraction** — Solve the following problems using the traditional written methods.

   a. 1,489 + 2,008

   b. 89,871 + 689

   c. 450,123 + 589,256

   d. 40,259 + 25,879

e. 206,845 + 1,234

f. 750 − 67

g. 456 − 88

h. 230 − 150

i. 15,567 + ____ = 16,000

j. 985 − ____ = 900

4. **Adding and Subtracting Roman Numerals** — Add LXVI + CXXX using an abacus. Notice how Roman numerals don't lend themselves well to a written method.

5. **Time for Time**
   a. Convert 1530 (military time) to the 12-hour clock.

   b. Convert 21:30 (24-hour clock) to the 12-hour clock.

   c. Convert 17:20 to the 12-hour clock.

   d. Convert 1 p.m. to military time.

   e. Convert 9 p.m. to military time.

6. **Comparing Numbers** — Use the symbols <, >, or = to show how these quantities compare.
   a. 756 + 98   800 − 50

   b. 265 + 874   789 + 450

7. **Question** — Why are addition and subtraction so consistent?

PRINCIPLES OF MATHEMATICS
CHAPTER 2. Operations, Algorithms, and Problem Solving
LESSON 4. Keeping and Balancing a Checkbook Register
Worksheet 2.4

1. **Keeping a Checkbook** — Below is a list of transactions and a blank checkbook register. Input the transactions into the checkbook register, updating the balance column after each new transaction. Remember, it's not that important exactly what you put in the memo field — it just needs to be enough to remind someone of where the money was spent or received.

    9/1 – Started with an opening balance of $3,456.
    9/10 – Donated $150 to My Favorite Charity using check 250.
    9/10 – Deposited a $59 check for babysitting.
    9/17 – Deposited a $9 check for raking the neighbor's yard.
    9/25 – Deposited a $20 check and a $25 check for mowing the neighbors' yards. Both checks were deposited as one deposit.
    9/26 – Bought a new bike at Dillon's Department Store for $179 using check 251.
    9/30 – Bought a new laptop computer at C's Computers for $1,549 with check 252.

Check Number	Date	Memo	Payment Amount	Deposit Amount	$ Balance

2. **Balancing Your Checkbook** — Reconcile your checkbook from the previous problem against the statement shown by following the given steps. See Lesson 2.4 in the *Student Textbook* for more details and an example if needed.

    a. Make sure that everything the bank statement shows is in your checkbook. Use a highlighter to highlight each transaction in your register that's on the statement, or put a checkmark next to it.
    b. Find the current balance of your checkbook.
    c. Add deposits and subtract payments that haven't been processed yet to the statement's balance.
    d. Work until reconciled; mark that you've reconciled the statement.

    Your Hometown Bank
    Anytown, USA

    Statement
    Dates: 9/1–9/30
    Beginning Balance: $3,456
    Ending Balance: $3,241

    Payments
9/10	Check 250	My Favorite Charity	$150
9/26	Check 251	Dillon's Department Store	$179

    Deposits
9/10	Counter Deposit	$59
9/17	Counter Deposit	$9
9/25	Counter Deposit	$45
9/30	Interest	$1

3. **Time for Time**

   a. If it's 4 p.m. PST, what time is it on the East Coast?

   b. If a meeting you're attending in a foreign country is taking place at 17:00, what time is that in the 12-hour clock?

   c. Review time zone flashcard.

PRINCIPLES OF MATHEMATICS

CHAPTER 2. Operations, Algorithms, and Problem Solving
LESSON 5. Different Addition and Subtraction Methods

Worksheet 2.5

1. **Bhāskara Addition** — Solve the following problem using the Bhāskara method.
    a. 20 + 35
    b. 16 + 15

2. **Standard Addition and Subtraction** — Solve the following problems on paper using the traditional addition and subtraction methods.

    a. 456 + 289

    b. 1,205 + 477

    c. 250 + 556

    d. 1,987 − 191

    e. 251,890 − 240,962

    f. 400,000 − 25,680

3. **Time for Time**
    a. If a meeting is occurring at 3 p.m. MST, and it's currently 2 p.m. EST, how long do you have before the meeting?

    b. Suppose a military friend asks to meet you somewhere at 1830. What time is that in the 12-hour clock?

    c. Review time zone flashcard.

4. **Question** — In order for an addition or subtraction method to work, what must it accurately do?

PAGE 45

5. **Keeping a Checkbook** — Below is a list of transactions and a blank checkbook register. Input them into the checkbook register appropriately, updating the balance column after each new transaction.

    11/1 – Started with a beginning balance of $3,025.
    11/2 – Spent $45 at Sporting Goods with check 150.
    11/6 – Spent $27 at grocery store with check 151.
    11/8 – An automatic deposit of $124 was made to your account.
    11/19 – Deposited a $15 rebate.
    11/28 – Withdrew $25 cash.
    12/3 – Paid cell phone bill for $54 with check 152.
    12/7 – An automatic deposit of $124 was made to your account.

Check Number	Date	Memo	Payment Amount	Deposit Amount	$ Balance

6. **Balancing Your Checkbook** — Reconcile your checkbook from the previous problem against the statement shown by following the given steps.

    a. Make sure that everything the bank statement shows is in your checkbook. Use a highlighter to highlight each transaction in your checkbook that's on the statement, or put a checkmark next to it.
    b. Find the balance of your checkbook.
    c. Add deposits and subtract payments that haven't been processed yet to the statement's balance.
    d. Work until reconciled; mark that you've reconciled the statement.

Your Hometown Bank Anytown, USA				**Statement** Dates: 11/1–11/31 Beginning Balance: $3,025 Ending Balance: $3,068
Payments				
	11/2	Check 150	Sporting Goods Store	$45
	11/6	Check 151	Grocery Store	$27
	11/28		Withdrawal	$25
Deposits				
	11/8	Auto Deposit		$124
	11/19	Counter Deposit		$15
	11/30	Interest		$1

    * Note: You have received your bank statement on 12/8, so there are some December entries in your check register that are not on your statement.

PRINCIPLES OF MATHEMATICS

CHAPTER 2. Operations, Algorithms, and Problem Solving
LESSON 6. Problem Solving: An Introduction

Worksheet 2.6

> **Remember to follow these steps for word problems:**
> - **Define** the information given in the problem (feel free to use abbreviations or letters).
> - **Plan** how the problem should be solved by expressing the relationship between the information you've been given using words or letters.
> - **Execute** your plan, showing your math.
> - **Check** your answer. Does it make sense? If you reached it via addition, check it via subtraction, and vice versa.

Practicing the Process — *While these word problems are ones you could probably solve in your head, use them to practice the process of defining, planning, executing, and checking. You'll be very glad you understand the process once you get into more complicated problems!*

1. **Dinner Chairs** — Suppose you had 5 people in your family, and you invited a family of 8 over for dinner. You want to find how many chairs you need for everyone to get a chair.

    a. Define:

    b. Plan:

    c. Execute:

    d. Check:

2. **Party Bags** — If you have made 12 party favor bags, and only 9 people show up, how many bags will you have left over?

    a. Define:

    b. Plan:

    c. Execute:

    d. Check:

3. **Gifts** — One gift you're considering for a family member costs $45, while another costs $37. How much more expensive is the $45 one?

    a. Define:

    b. Plan:

    c. Execute:

    d. Check:

More Problems — On these next problems, you only need to show on paper the numbers you added and subtracted.

4. **Shopping** — Suppose you went to the store with $80. You saw a shirt you wanted to buy for $13.

    a. How much money would you have left if you bought the shirt?

    b. What if you bought both the shirt and a candy bar that cost $3?

5. **Distance Planning** — Use the following information to answer the questions. The abbreviation *mi* stands for *miles*.

    From San Francisco to New York via Straits of Magellan — 13,083 mi

    From San Francisco to Liverpool via Straits of Magellan — 13,630 mi

    From San Francisco to New York via Panama — 5,300 mi

    From San Francisco to Liverpool via Panama — 7,954 mi

    a. How much distance is saved between San Francisco and New York if the Panama route is taken instead of the Straits of Magellan route?

    b. How much distance is saved between San Francisco and Liverpool by taking the Panama route instead of the Straits of Magellan route?

6. **You Pick It!** — It's your turn to write a word problem! Write one problem that involves addition and another that involves subtraction. *Hint*: If you aren't sure what to write about, think about something you enjoy and write a problem about it, or look at the word problems you solved on this worksheet for inspiration.

    a. Addition Word Problem:

    b. Subtraction Word Problem:

PAGE 49

7. **You Pick It! Part 2** — Solve both the word problems you wrote in the previous problem.
   a. Addition Word Problem:

   b. Subtraction Word Problem:

8. **Time for Time**
   a. If it's 1330 (military time) and you have a meeting at 1500, how long do you have before your meeting?

   b. What time in the 12-hour clock is your meeting at?

   c. Review time zone flashcard.

9. **Notebook Time** — Add the problem-solving steps you learned about today to your math notebook so you can easily refer to them.

10. **Question** — To what did Walter W. Sawyer compare mathematics?

PAGE 50

PRINCIPLES OF MATHEMATICS

CHAPTER 3. Mental Math and More Operations
LESSON 1. Mental Math

Worksheet 3.1

1. **Adding 10** — Notice how easy it is to add by 10!
    a. 25 + 10
    b. 205 + 10
    c. 7 + 10

2. **Complements** — Mentally find the complement (the number you have to add to reach 10) for each number listed below.
    a. 1 + ____ = 10
    b. 4 + ____ = 10
    c. 7 + ____ = 10
    d. 8 + ____ = 10
    e. 3 + ____ = 10
    f. 4 + ____ = 10
    g. 5 + ____ = 10
    h. 7 + ____ = 10
    i. 2 + ____ = 10

3. **Mental Subtraction** — Mentally subtract the problems below. Be sure to try adding from the subtrahend.
    a. 27 – 19 =
    b. 35 – 27 =
    c. 48 – 39 =
    d. 50 – 28 =
    e. 70 – 55 =

4. **Monthly Cost** — Use this word problem to practice problem-solving steps.

    Suppose you wanted to find out the total cost of driving your car per month. You know you spend about $25 on gas, $100 on insurance, and set aside $120 to cover wear and tear on the car. How much monthly does it cost to own a car?

    a. Define:

    b. Plan:

    c. Execute:

    d. Check:

CHAPTER 3: Mental Math and More Operations
LESSON 2. Rounding — Approximate Answers
Worksheet 3.2

> When solving word problems, be sure to write down at least the "executing" step so someone can tell how you arrived at your answer. You can perform the other steps mentally, unless otherwise requested.

1. **Rounding** — Round these numbers to the nearest 10.

    a. 568 = ___
    b. 987 = ___
    c. 254 = ___
    d. 795 = ___

2. **Rounding** — Round these numbers to the nearest 100.

    a. 568 = ___
    b. 987 = ___
    c. 254 = ___
    d. 795 = ___

3. **Rounding to Find an Approximate Answer** — Round these numbers to the nearest 10 to find an approximate answer mentally.

    a. 499 + 63 ≈ ___   b. 897 + 52 ≈ ___

    c. 540 + 96 ≈ ___   d. 395 – 27 ≈ ___

    e. 63 – 24 ≈ ___

4. **Rounding to Find an Exact Answer** — Mentally add the problems below by rounding one number to the nearest 10 and then adding or subtracting the amount you added or subtracted to round it.

    a. 23 + 19 = ___   b. 45 + 17 = ___

    c. 18 + 13 = ___   d. 55 + 77 = ___

    e. 85 + 12 = ___

5. **Mental Subtraction** — Mentally subtract the problems below. Be sure to try adding from the subtrahend.

    a. 48 – 18 = ___

    b. 67 – 15 = ___

6. **Applying It** — One example of how mental addition and subtraction apply is in giving or recieving change. If your total was 59 cents, and you paid three quarters, how much change should you get back? Find the answer mentally.

7. **Reaching a Goal** — If you're trying to save $253 for a new computer, and so far you've managed to set aside $89, how much more do you have left to save?

8. **Question** — How does our need to round numbers remind us of our limited abilities?

PRINCIPLES OF MATHEMATICS

CHAPTER 3. Mental Math and More Operations
LESSON 3. Multiplication

Worksheet 3.3

1. **Multiplication** — As you review your muliplication facts with these problems, notice that we've represented multiplication three different ways: x, •, and ( ).

    9 x 7 = ____

    2 • 6 = ____

    3(9) = ____

    8 • 7 = ____

    5(6) = ____

2. **Daily Pay** — If you make $8 per hour, and work 8 hours per day, how much will you make in one day?

3. **Quarterly Cost of a Subscription** — If an Internet subscription to a site costs $9 per month, how much will it cost for one quarter? *Hint*: A quarter of a year is 3 months.

4. **Comparisons** — Simplify each side and then insert the correct sign of comparison (<, >, or =). *Hint*: Don't let the Roman numerals fool you! Just replace them with their decimal equivalent and solve.

    a.   5(5)   6 + 17

    b.   2(3)   3 + 2

    c.   VIII(III)   VI(IV)

5. **Complements** — Write the complement (the number you have to add to reach 10) for each number listed below.

    a. 7 + ____ = 10
    b. 5 + ____ = 10
    c. 2 + ____ = 10
    d. 1 + ____ = 10
    e. 6 + ____ = 10
    f. 4 + ____ = 10
    g. 3 + ____ = 10

PAGE 55

6. **Mental Math Practice** — You can use any mental technique you like to solve these problems.
   a. Your total is 43 cents, and you give the cashier two quarters. How much change should you get?

   b. Your total is 58 cents and you give the cashier three quarters. How much change should you get?

   c. Your total is 14 cents and you give the cashier one quarter. How much change should you get?

   d. You are buying an item that costs 24 cents, one that costs 19 cents, and another that costs 5 cents. What is your total, assuming there is no tax?

7. **Terms**
   a. In 9 x 4 = 36, what do we call the 9 and the 4?

   b. In 9 x 4 = 36, what do we call 36?

8. **Rounding** — Round 345 to the nearest ten.

9. **Question** — How do the multiplication table and Napier's rods point us to God's power?

PRINCIPLES OF MATHEMATICS

CHAPTER 3: Mental Math and More Operations
LESSON 4. Division

Worksheet 3.4

1. **Practicing Division**
    a. 88 ÷ 8 = _____
    b. 54 ÷ 9 = _____
    c. 8 • _____ = 64
    d. 63 ÷ _____ = 7

2. **Pages Needed** — If you want to finish a 72-page book in 8 days, how many pages do you need to read each day?

3. **Party Favors** — If you buy a box with 56 party favors and divide it amongst 7 guests, how much will each guest get?

4. **Dividing Cookies** — If you buy a box of cookies with 64 cookies in it and share it evenly amongst 8 people, how many cookies will each person get?

5. **Complements** — Write the complement (the number you have to add to reach 10) for each number listed below.
    a. 9 + _____ = 10
    b. 3 + _____ = 10
    c. 5 + _____ = 10
    d. 7 + _____ = 10
    e. 2 + _____ = 10

6. **Mental Math Practice** — Solve the following problems mentally.
   a. If someone was born in 1964, how old were they in 2000?

   b. If someone was born in 1964, how old were they in 2012?

   c. If someone was born in 2012, how old will they be in 2049 (assuming, of course, that the Lord does not return before then and that they are still alive)?

   d. If someone was born in 1987, how old will they be in 2020?

   e. Round 1,897 to the nearest hundred.

7. **Terms** — List the term used to describe each number in 24 ÷ 6 = 4.

8. **Question** — How did the Egyptians divide?

CHAPTER 3. Mental Math and More Operations
LESSON 5. Properties
Worksheet 3.5

1. **Applying the Properties**

    a. Insert either = or ≠ in between these groups of numbers:

    $8 + 9 + 2 \quad\quad 9 + 8 + 2$

    b. What property describes the fact that the above groups of numbers were or were not equal?

    c. Insert either = or ≠ in between these groups of numbers:

    $(2 + 3) + 6 \quad\quad 2 + (3 + 6)$

    The parentheses show which numbers are added together first.

    d. What property describes the fact that the above groups of numbers were or were not equal?

    e. Insert either = or ≠ in between these groups of numbers:

    $6 \div 3 \div 1 \quad\quad 3 \div 1 \div 6$

    f. Insert either = or ≠ in between these groups of numbers:

    $8 \times 1 \quad\quad 8$

    g. What is the property that describes the fact that multiplying by 1 doesn't change the value of a number?

    h. Insert either = or ≠ in between these groups of numbers:

    $8 + 0 \quad\quad 8$

    i. What is the property that describes the fact that adding 0 doesn't change the value of a number?

    j. Insert either = or ≠ in between these groups of numbers:

    $8 \div 8 \quad\quad 1$

    k. Insert either = or ≠ in between these groups of numbers:

    $8 \div 1 \quad\quad 8$

2. **More Property Questions**

    a. Is division commutative?

    b. Is subtraction commutative?

    c. Which of the following operations are both commutative and associative?

    addition, subtraction, multiplication, division

    d. What is the multiplicative identity?

PAGE 59

e. What is the additive identity?

f. What is a property in math?

3. **Complements** — Write the complement (the number you have to add to reach 10) for each number listed below.

   a. 5 + _____ = 10
   b. 2 + _____ = 10
   c. 3 + _____ = 10
   d. 9 + _____ = 10
   e. 7 + _____ = 10

4. **Mental Math Practice** — Solve the following problems mentally. Remember to use the associative property for addition to make it easier, but know that you have to solve the subtraction problems in order because subtraction is not associative.

   a. $60 + $120 + $20

   b. $7 + $8 + $3 + $2

   c. $50 − $10 − $4 − $3

   d. You go to the store with $40. You spend $5 on a book, $10 on a shirt, and $6 on a hat. How much do you have left?

   e. Round 783 to the nearest hundred.

   f. Round 783 to the nearest ten.

5. **Notebook Time** — Write the different properties from Lesson 3.5 in your math notebook.

PRINCIPLES OF MATHEMATICS

CHAPTER 3. Mental Math and More Operations
LESSON 6. Conventions — Order of Operations

Worksheet 3.6

1. **Learning the Language** — Use the conventions and properties you've learned to help you solve these problems. Don't forget that a number next to a parentheses means multiplication: 4(3) means the same thing as 4 x 3. (See the "Combining the Conventions" box in Lesson 3.6.) So 2(4 + 5 − 8) means the same thing as 2 x (4 + 5 − 8).

   a. 4 • 2(4 + 5 − 3)

   b. 5 + 8 x 5 ÷ 5

   c. 8 • 3 ÷ 6 • 2

   d. 50 ÷ (5 + 5) x 2

   e. (7 − 2 x 3) + 8 − 2(2)

2. **Understanding Check** — In problem 1e above, could you have performed the subtraction of 7 − 2 before the multiplication of 2 x 3? Why or why not?

3. **Explaining Parentheses** — Parentheses are quite useful grouping tools. Notice the example meanings.

   a. 4(7 + 8 + 3)

   Example Meaning: If an event costs $7 for admission, $8 for food, and $3 for a program, how much will the event cost for a group of 4 people?

   b. (4 + 2)6

   Example Meaning: Each person in a group of 6 collects an average of $4 worth of items for a yard sale one week and an average of $2 worth of items for the yard sale the next week. Approximately how many dollars worth of items are gathered all together?

   Hint: It doesn't matter that the 6 comes after the parentheses instead of before…it means the *same* thing as 6(4 + 2).

4. **Parentheses in Action** — Use the examples from problem 3 to help you write this problem using parentheses. Then solve.

   If each team member has to order a $4 cap and a $6 T-shirt, and there are 20 team members, how much should the total order for caps and T-shirts be, ignoring tax?

5. **Complements** — Write the complement (the number you have to add to reach 10) for each number listed below.

   a. 1 + _____ = 10

   b. 4 + _____ = 10

   c. 7 + _____ = 10

6. **Problem-Solving Practice** — Use this word problem to practice problem-solving steps.

   Suppose you're reading a book with 315 pages. If you're on page 47, how many pages do you have left?

   a. Define:

   b. Plan:

   c. Execute:

   d. Check:

7. **Notebook Time** — Write the order of operations out in your math notebook so you can easily refer to it. While we won't often mention the math notebook, continue to write anything you think you might need to reference later in it.

PAGE 62

CHAPTER 4. Multi-digit Multiplication and Division
LESSON 1. Multi-digit Multiplication
Worksheet 4.1

1. **Gelosia** — Solve these problems using the Gelosia method.

   a. 71 x 5

   b. 45 x 60

   c. 102 x 463

2. **Traditional Method** — Solve these problems using the traditional method.

   a. 305 x 50

   b. 415 x 6

   c. 202 x 9

   d. 25 x 25

3. **Problem-Solving Practice** — Use this word problem to practice problem-solving steps.

   A school librarian purchased 24 dictionaries at $18 each. What was the total cost of the dictionaries?
   a. Define:

   b. Plan:

   c. Execute:

   d. Check:

   While you can do all but the solving step mentally on these problems, be sure to still think through what you're doing and check to make sure your answer makes sense!

4. **Salary** — At $875 a month, what will a man's yearly salary amount to?

5. **Weight** — If a freight train has 32 cars, each containing 82 barrels of flour, each barrel weighing 196 pounds, how many pounds of flour are in the trainload?

6. **Complements** — Write the complement (the number you have to add to reach 10) for each number listed below.
   a. 4 + ____ = 10
   b. 2 + ____ = 10
   c. 3 + ____ = 10

7. **Skill Sharpening**
   a. 48(52 + 96 − 7)   b. 108(45 − 18 + 15)

8. **Questions**
   a. What is an algorithm?

   b. Why do multiplication algorithms work?

CHAPTER 4. Multi-digit Multiplication and Division
LESSON 2. The Distributive Property — Understanding Why
Worksheet 4.2

1. **Distributive Property in Action** — Show that you solved using the requested method. Notice how you get the same answer either way.

    a. Solve using the order of operations: 3(2 + 6)

    Example Meaning: If it costs $2 in gas plus $6 in wear and tear on the machine to mow a large yard, how much does it cost to mow 3 yards?

    b. Solve using the distributive property: 3(2 + 6)

    c. Solve using the order of operations: (17 + 3)5

    Example Meaning: If one package contains 17 balls and another 3, and I decide to buy 5 of each package, how many balls will I have altogether?

    d. Solve using the distributive property: (17 + 3)5

2. **Property Questions** — What property describes the fact that 4,780 x 28 = 28 x 4,780?

3. **Breaking Up Large Numbers** — Use the distributive property to break apart these multiplications into a series of smaller multiplications. Show your work.

    Example: 5(708)
    5(700 + 8)
    5(700) + 5(8) *Note*: You do not need to show this step.
    3,500 + 40 = 3,540

    a. 7(24)

    b. 3(45)

c. 4(304)

4. **Breaking Up Large Numbers Mentally** — Solve these problems mentally, using the distributive property to help you.

   a. 5(85)

   b. 4(62)

   c. 3(56)

5. **Problem-Solving Practice** — Use this word problem to practice problem-solving steps.

   If a monthly magazine subscription costs $108 a year, how much does it cost per month?

   a. Define:

   b. Plan:

   c. Execute:

   d. Check:

   Hint: You can check your division using multiplication—multiplying the quotient by the divisor should yield the dividend.
   Example: 100 ÷ 5 = 20
   20 x 5 = 100

CHAPTER 4. Multi-digit Multiplication and Division
LESSON 3. Multi-digit Division

Worksheet 4.3

1. **Dividing** — Solve these problems.

    a. $5\overline{)225}$

    b. $7\overline{)623}$

    c. $8\overline{)504}$

2. **Miscellaneous Problems**

    a. If a book is 80 pages long, how many pages should you read each day to finish in 4 days?

    b. Once, my parents noticed we were not going very many miles before we needed to fill up with gas. We immediately took the car into the mechanic and discovered our car had a major problem! You can use division to help you find the approximate miles you are able to drive on a gallon of gas. If you went 297 miles and used 9 gallons, about how far were you able to go on each gallon?

3. **Skill Sharpening** — Solve these problems.

    a. (5 x 6 + 9)25

    b. 75 x 6 ÷ 5

    c. Solve using the distributive property (show your work): 8(789)

4. **Mental Math** — Solve these problems mentally.

   a. Your total is 62 cents and you give the cashier three quarters. How much change should you get?

   b. Your total is 17 cents and you give the cashier one quarter. How much change should you get?

   c. You are buying an item that costs 26 cents, one that costs 18 cents, and another that costs 4 cents. What is your total?

5. **Question** — What did the verse in Lesson 4.3 tell us about work?

1. **Dividing** — Solve these problems.

   a. $36 \overline{)\$3{,}060}$

   b. $28 \overline{)448}$

   c. $15 \overline{)866}$

   d. $80 \overline{)900}$

2. **Working with Nothing** — Solve these problems.

   a. $8 \div 0$
   b. $0 \div 8$
   c. 4×0
   d. $5 \cdot 0$
   e. $852 - 0$
   f. $987 + 0$

3. **Problem-solving Practice** — Use this word problem to practice problem-solving steps.

   If a company spent $13,500 to print 4,500 books, how much did each book cost them?

   a. Define:

   b. Plan:

   c. Execute:

   d. Check:

4. **Miscellaneous Problems**
   a. If a book is 160 pages long, how many pages should you read each day to finish in 20 days?

   b. In order to cover the additional costs of advertising and maintaining the company and still make money, the company decides they need to charge 9 times the actual printing costs when they sell the book in problem 3. How much would they charge for the book?

5. **Skill Sharpening**
   a. $48(29 + 15 \div 3 \times 1)$

   b. $45 \div 15 + 8 + 0$

   c. Solve using the distributive property (show your work): $12(50 + 6)$
   Example: $2(3 + 4) = 2(3) + 2(4) = 6 + 8 = 14$

| PRINCIPLES OF MATHEMATICS | CHAPTER 4. Multi-digit Multiplication and Division
LESSON 5. More Problem-solving Practice / Enter the Calculator | Worksheet 4.5 |

You may use a calculator on this worksheet whenever you see this symbol (🖩).

1. **Distances** (🖩) — Two vessels are 8,888 miles apart and are sailing in opposite directions, one at the rate of 144 mi per day (i.e., traveling 144 miles every day), the other at the rate of 166 mi per day. How far apart they will be at the end of a week?

    a. Define:

    b. Plan:

    c. Execute:

    d. Check:

You don't have to write out every step in the following problems, but it would be wise to use the general process of defining, planning, solving, and checking to solve the remaining problems. Remember to write out enough of your work that someone can see what you added/subtracted/multiplied/divided to find the answer.

2. **Wedding Planning** (🖩)

    a. Say you are planning a wedding, and you are expecting 250 guests. How much will it cost you per guest if you buy the following food items: 1 cake at $1,024 a cake; 3 jars of brochette at $8 a jar; 3 bags of chips at $4 a bag; 4 trays of luncheon meet at $17 a tray; 3 pounds of cheese at $4 a pound; 5 boxes of crackers at $4 a box; 10 pounds of strawberries at $4 a pound; and 5 pounds of dipping chocolate at $10 a pound?

    b. If you increase the guests to 300 and increase the food costs over what you found in problem 2a by adding rolls that come in packs of 60 for $50 (and you buy enough of them to give each guest a roll), how much will it cost you per guest?

    c. If you reduce the food costs from problem 2a by $500 and keep the guests to 250, how much will it cost you per guest?

3. **Apartment Rental** (🖩) — Say you are trying to decide what apartment to rent while you wait for a house to be built. You have narrowed it down to two options: 1) a furnished apartment (which costs $2,000 per month for rent, plus $600 per month for furniture storage) or 2) an unfurnished apartment with delivered furniture (which costs $800 per month for rent, plus $2,600 for a one-time delivery fee).

   a. Which option would be the most economical if you end up staying 1 month?

   b. What is the most economical option for 7 months? How much cheaper is it than the other option?

   c. If the landlord of the unfurnished apartment offered to give you a six-month lease for $3,600 total, how much would you save over renting it month-by-month? *Hint*: You can ignore the cost of the furniture delivery, as it would not change.

4. **Mental Math** — Solve these problems mentally.
   a. 87 − 16 = _____
   b. 23 + 15 = _____
   c. 45 + 18 = _____
   d. 35 − 18 = _____

5. **Skill Check**
   a. 15(860 ÷ 43 + 15)

   b. 89 × 1 + 0

   c. Solve using the distributive property (show your work): 4(12 + 3)

PAGE 72

| | PRINCIPLES OF MATHEMATICS | CHAPTER 4. Multi-digit Multiplication and Division
LESSON 6. Chapter Synopsis | Worksheet 4.6 |

1. **Car** — A car cost $4,565 to buy used. If you expect to get 5 years of wear out of it, what would be the effective cost per year, ignoring maintenance cost?

2. **Pricing** — You want to start a home ice cream business. You've priced out your options, and it will cost you $255 for the necessary equipment.

    a. If you sell each cone for $4, how many cones will you have to sell to make back your equipment costs, assuming your ingredient cost per cone is $1? *Hint*: Don't let this problem fool you—just apply the problem-solving steps you've been learning. Write out the information you know and then see if there are any relationships you do know that can be used to find what you don't know.

    b. What if you sell each cone for $5 instead?

    c. How much profit will you make if you sell 230 cones at the $5 price each?

3. **Learning the Language** — Find these values.

    a. Solve using the order of operations: 7(3 + 2)

    b. Solve using the distributive property (show your work): 7(3 + 2)

    c. Solve using the order of operations: 100 + 2(9 + 5)

    d. Solve using the distributive property (show your work): 100 + 2(9 + 5)

    e. 2 x 3 + 4 x 3

    f. 7 x 0

    g. 8 ÷ 0

PAGE 73

h. 8)525

i. 51 x 25

4. **Out of the Box: Gas Prices** — Find out the average mileage your family's car(s) is getting per gallon. You may need to take a few weeks to complete this problem, as you will need to make two trips to the gas station.

Extra Credit — Make your own Napier's rods and use them to solve 34 x 896.
(See the example in Lesson 4.6.)

To start with, make three copies of the strips on the next page. This will give you 9 blank strips and 3 printed strips. You only need 1 printed strip—you can throw the other 2 away.

Your finished strips should look like the picture—notice how the "2" strip has the answer to 1 x 2 in the top box, the answer to 2 x 2 in the second box from the top, the answer to 3 x 2 in the third box from the top, and so forth. All the other strips follow the same principle, just using 3, 4, etc., as the multiplicand instead of 2.

	2	3	4	5	6	7	8	9
1 x 2 →	2	3	4	5	6	7	8	9
2 x 2 →	0/4	0/6	0/8	1/0	1/2	1/4	1/6	1/8
3 x 2 →	0/6	0/9	1/2	1/5	1/8	2/1	2/4	2/7
4 x 2 →	0/8	1/2	1/6	2/0	2/4	2/8	3/2	3/6
5 x 2 →	1/0	1/5	2/0	2/5	3/0	3/5	4/0	4/5
6 x 2 →	1/2	1/8	2/4	3/0	3/6	4/2	4/8	5/4
7 x 2 →	1/4	2/1	2/8	3/5	4/2	4/9	5/6	6/3
8 x 2 →	1/6	2/4	3/2	4/0	4/8	5/6	6/4	7/2
9 x 2 →	1/8	2/7	3/6	4/5	5/4	6/3	7/2	8/1

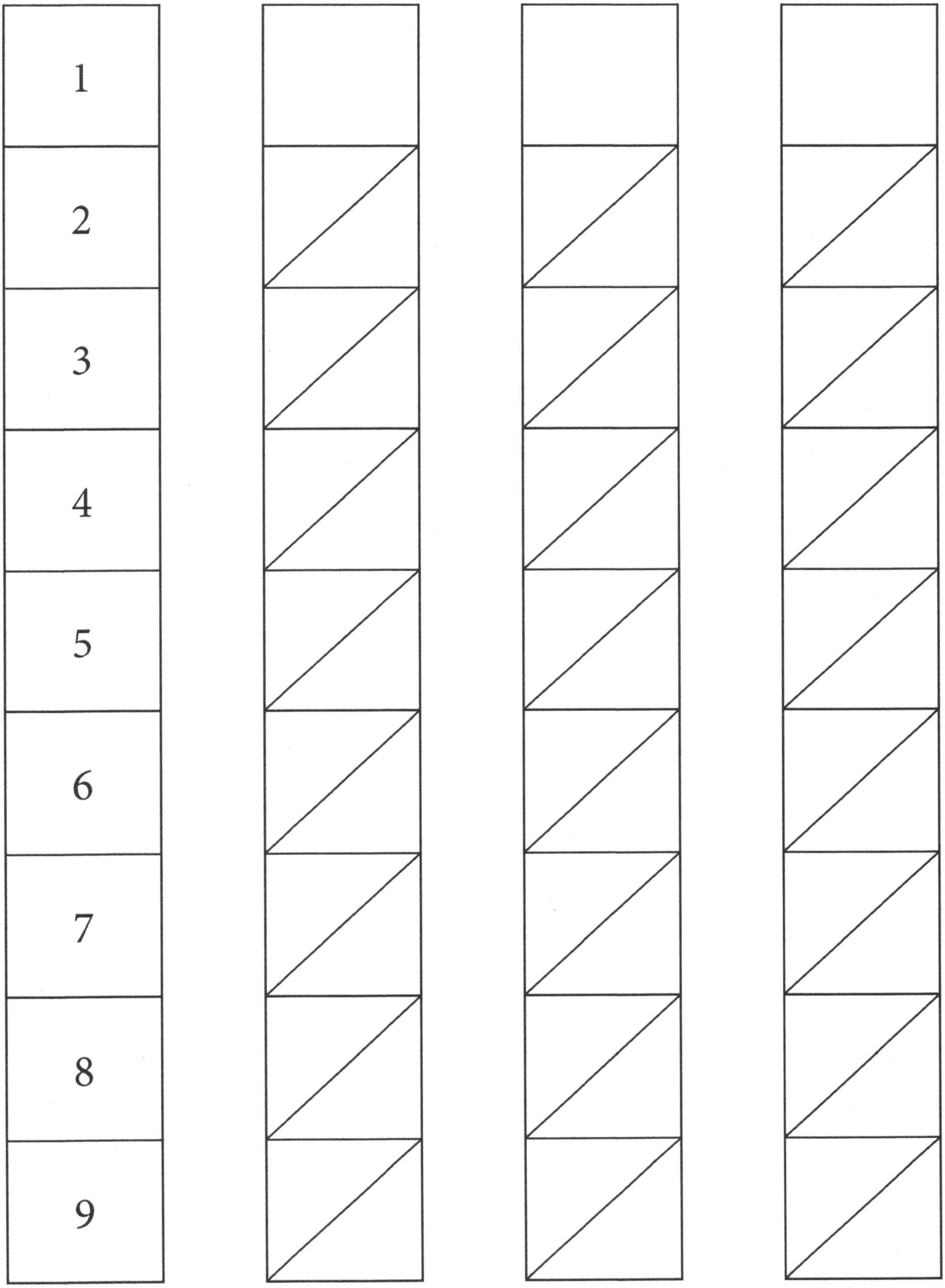
1
2
3
4
5
6
7
8
9

PRINCIPLES OF MATHEMATICS

CHAPTER 5. Fractions and Factoring
LESSON 1. Understanding Fractions

Worksheet 5.1

1. **Fraction Rewriting** — Express the following as fractions.

    a. Finishing 5 out of 20 math problems

    b. Having $30 toward a fundraising goal of $50

    c. Selling 80 out of 100 tickets

    d. 76 buttons *Hint*: Remember, a whole number can be written as a fraction too!

    e. Freezing 34 out of 48 cookies

    f. Dividing a yearly salary of $10,680 to find the monthly salary

    g. 130 people

    h. Dividing the 856 total pages in a book by the 428 pages that you've read

    i. 4 gallons

    j. Dividing 4 pieces of pie evenly amongst 4 people

2. **Division and Fractions** — Rewrite these division problems as fractions.

    a. $3{,}072 \div 256$

    b. $300 \overline{)150}$

    c. $60 \overline{)30}$

    d. $3 \overline{)1}$

3. **Complete the Division**

    a. Complete the division in problem 1f to find the monthly salary.

b. Complete the division in problem 1h.

c. Complete the division in problem 1j.

d. Complete the division in problem 2a.

4. **Term Time**
    a. Circle the numerators in your answers to problems 1a and 1b.
    b. Look at problems in number 1a–1j and put a star next to all of them that we would call improper fractions.
    c. Write in words three ways to read the fraction you wrote in problem 1c.

    d. Write in words three ways to read the fraction you wrote in problem 1e.

    e. In this course, is 0 a whole number?

5. **Questions**
    a. What is a notation?

    b. What are three ways of looking at fractions?

6. **Skill Sharpening** — Solve using the distributive property (show your work).
    a. 42(5 + 92)
    b. 12(6 + 85)

PRINCIPLES OF MATHEMATICS

CHAPTER 5. Fractions and Factoring
LESSON 2. Mixed Numbers
Worksheet 5.2

1. **Fractions and Mixed Numbers**

   $\frac{4}{5}$ $\frac{25}{7}$ $\frac{3}{2}$ $1\frac{8}{9}$ $\frac{12}{13}$ $5\frac{2}{3}$ $\frac{46}{5}$ $\frac{5}{5}$

   a. Circle all the improper fractions.
   b. Convert the improper fractions to mixed or whole numbers.

   c. Convert the mixed numbers to improper fractions.

2. **Remainders as Fractions** — Express the remainders in these division problems as fractional amounts, forming a mixed number. Notice that the mixed number solves part of the division, but leaves the remainder as a fraction.

   a. $684 \div 7$

   b. $25 \overline{)879}$

3. **Completing the Division** — Complete the division in these fractions in order to express them as a whole number instead.

   a. $\frac{88}{8}$

   b. $\frac{9}{9}$

   c. $\frac{54}{6}$

4. **Names** — List all the numerators in problem 3.

5. **Fractions as Division**
   a. Write a division problem to express this: If car insurance costs $700 a year, what does it cost per month?

   b. Rewrite the division from 5a as a fraction.

   c. Rewrite the fraction from 5b as a mixed number. Notice again that the mixed number solves part of the division, but leaves the remainder as a fraction.

6. **Book Boxes** — 2,250 books are to be boxed. If the boxes will hold an average of 120 books each and if each box costs $2, how much will you need to spend on boxes? *Hint*: If you end up with a partial number of boxes you need to buy, give your answer as the next whole number, as you can't buy a portion of a box.
   a. Define:

   b. Plan:

   c. Execute:

   d. Check:

7. **Question** — Why do we need so many mathematical "tools"?

CHAPTER 5. Fractions and Factoring
LESSON 3. Equivalent Fractions and Simplifying Fractions

Worksheet 5.3

Since simplified fractions are easier to read, remember to simplify fractional answers to their lowest terms and to convert improper fractions to mixed or whole numbers, unless otherwise stated.

1. **Equivalent Fractions** — Circle all the fractions below that are equivalent to $\frac{2}{3}$.

   $\frac{8}{9}$ $\quad$ $\frac{18}{27}$ $\quad$ $\frac{8}{12}$ $\quad$ $\frac{12}{24}$ $\quad$ $\frac{16}{19}$

2. **Forming Equivalent Fractions** — Rewrite these fractions by multiplying or dividing as indicated. Do not simplify.

   a. Divide $\frac{28}{40}$ by $\frac{4}{4}$

   b. Multiply $\frac{3}{4}$ by $\frac{12}{12}$

3. **Term Check** — What is the identity property of multiplication?

4. **Question** — Why does multiplying or dividing both the numerator and denominator of a fraction by the *same number* (i.e., a fraction worth 1) result in an equivalent fraction?

5. **Simplifying in Fractional Format** — Use what you know about equivalent fractions to express these quantities from Worksheet 5.1 as fractions, but this time in simplest terms. Compare your answers with what you got on Worksheet 5.1. Notice how much simpler it is to compare fractions when they are in their simplest terms.

   a. Finishing 5 out of 20 math problems

   b. Having $30 toward a fundraising goal of $50

   c. Selling 80 out of 100 tickets

   d. Freezing 34 out of 48 cookies

6. **Simplifying Improper Fractions** — Simplify these improper fractions by expressing them as a whole or mixed number.

   a. $\dfrac{17}{2}$

   b. $\dfrac{8}{4}$

   c. $\dfrac{20}{11}$

7. **Portions of a Dollar** — If each quarter equals 25 cents and 1 dollar equals 100 cents, express the change listed as a fraction of a dollar. Be sure to simplify your answers.

   a. three quarters

   b. two quarters

   c. one quarter

8. **Question** — Why does it make sense for a quarter to be called a quarter?

9. **Buying Pizza** — You're ordering pizza for a party. Each large pizza costs $12 with 1 topping, and $16 if loaded with toppings. If you get 5 1-topping pizzas and 7 loaded pizzas, plus pay a $5 delivery fee, how much will it cost altogether?

PRINCIPLES OF MATHEMATICS

CHAPTER 5. Fractions and Factoring
LESSON 4. Understanding Factoring

Worksheet 5.4

1. **Factoring** — Use a factor tree to factor the following numbers; circle all the prime factors. If a number is already prime, just state so.

   a. 55

   b. 78

   c. 112

   d. 85

   e. 29

   f. 1,000

2. **Prime Numbers** — Copy the list of prime numbers from Lesson 5.4 into your math notebook for easy reference.

3. **Term Time** — What is a prime number, and what is a prime factor?

4. **Rewrite as Division** — Rewrite each fraction or mixed number below as a division problem. *Hint*: First convert to an improper fraction if needed, and then it will be easy to see how to write it as a division problem.

   Important Note: Sometimes, the numerator and the denominator will be written with a diagonal line in between rather than on top of each other. 3/4 is another way of writing $\frac{3}{4}$.

   a. $\frac{7}{9}$

   b. 56 / 7

   c. $1\frac{1}{2}$

   d. $4\frac{8}{9}$

   e. 468/26

   f. $\frac{2,175}{25}$

5. **Question** — Why can we take a concrete concept, work with it abstractly and then bring a concrete application?

6. **Skill Sharpening**

   a. Complete the division in 4e.

   b. Complete the division in 4f.

   c. Solve using the distributive property (show your work): 36(4 + 9)

7. **Business** — Suppose you have an online business reselling old books. Ignoring shipping costs and other supplies and licenses, it costs you $30 dollars a month for a storage unit, $40 a month for a post office box, and $10 a month for a website. How much do you spend in 3 years? *Hint*: There are 12 months in 1 year.

   a. Define:

   b. Plan:

   c. Execute:

   d. Check:

PRINCIPLES OF MATHEMATICS
CHAPTER 5. Fractions and Factoring
LESSON 5. Simplifying Fractions, Common Factors, and the Greatest Common Factor
Worksheet 5.5

You may use a calculator on this worksheet whenever you see this symbol (🖩).

1. **Common Factors** (🖩)
   a. List a common factor of 55 and 110.

   b. List a common factor of 32 and 128.

2. **Exploring** $\frac{39}{52}$ (🖩)
   a. Rewrite this fraction as a fraction of just its prime factors.

   Example: $\frac{6}{12}$ would be rewritten as $\frac{2 \times 3}{2 \times 2 \times 3}$.

   Hint: You can use a factor tree to help if you need to. You'll find it easiest to spot common factors if you list them all in order from least to greatest.

   b. Circle the multiplications the numerator and denominator share.
   c. What is the greatest common factor of the numerator and denominator?

   d. Simplify the fraction to its lowest terms by dividing the numerator and the denominator by the greatest common factor.

3. **Exploring** $\frac{56}{84}$ (🖩)
   a. Rewrite this fraction as a fraction of just its prime factors.

   b. Circle the multiplications the numerator and denominator share.
   c. What is the greatest common factor of the numerator and denominator?

PAGE 85

d. Simplify the fraction to its lowest terms by dividing the numerator and the denominator by the greatest common factor.

4. Exploring $\frac{74}{168}$ (🖩)

   a. Rewrite this fraction as a fraction of just its prime factors.

   b. Circle the multiplications the numerator and denominator share.
   c. What is the greatest common factor of the numerator and denominator?

   d. Simplify the fraction to its lowest terms by dividing the numerator and the denominator by the greatest common factor.

5. **Question** — Fill in the blank. A number can be thought of as a product of its _____.

CHAPTER 5. Fractions and Factoring
LESSON 6. Adding and Subtracting Fractions
Worksheet 5.6

You may use a calculator on this worksheet whenever you see this symbol (🖩).

1. **Adding and Subtracting Fractions** — Add and subtract the following fractions, using other tools (factoring, equivalent fractions, etc.) as needed. Notice that the first two are two of the examples from the lesson text—you now have all the skills to solve them!

   a. $\dfrac{11}{12} - \dfrac{1}{2}$

   Example Meaning: Say you go to the hardware store and find a piece of wood on sale that's $\dfrac{11}{12}$ feet long. You only need $\dfrac{1}{2}$ a foot for your project. How much will you have left?

   b. $\dfrac{1}{2} + \dfrac{1}{4}$

   Example Meaning: Say you added $\dfrac{1}{2}$ cup of flour, then $\dfrac{1}{4}$ cup of flour. How much flour did you use?

   c. $\dfrac{2}{3} + \dfrac{4}{5}$

   d. $\dfrac{9}{10} + \dfrac{5}{16} + \dfrac{46}{80}$

2. **Acreage** — If you own a $\dfrac{1}{2}$ acre lot and buy a $\dfrac{1}{8}$ acre lot adjacent to yours, how much land will you own altogether?

3. **Construction** — A truck needs to deliver $\dfrac{1}{2}$ ton, $\dfrac{1}{4}$ ton, and $\dfrac{2}{3}$ ton of mulch to three different customers. How many tons of mulch does he need to have in his truck to make these deliveries?

4. **Factoring** — What is the greatest common factor of 60 and 90? (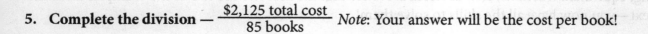)

5. **Complete the division** — $\dfrac{\$2{,}125 \text{ total cost}}{85 \text{ books}}$ *Note*: Your answer will be the cost per book!

6. **Skill Sharpening**

   a. $5 \div 5 + \dfrac{2}{3} + \dfrac{5}{9}$

   b. $5 \times 3 - \dfrac{8}{9}$

PRINCIPLES OF MATHEMATICS

CHAPTER 5. Fractions and Factoring
LESSON 7. Least Common Denominator / Multiple

Worksheet 5.7

You may use a calculator on this worksheet whenever you see this symbol (🖩).

In addition to the prime number list through 100 you've been given, you'll need to know that 293 is a prime number.

1. **Least Common Multiple** — Find the least common multiple of these numbers.

    a. 21 and 84

    b. 15 and 27

Exploring Fractions and Factoring

Example: Exploring $\frac{5}{8} + \frac{17}{20}$.

   a. Rewrite each denominator as a product of its prime factors.

   $$\frac{5}{8} + \frac{17}{20}$$
   $$2 \times 2 \times 2 \quad\quad 2 \times 2 \times 5$$

   b. What is the least common denominator?

   $$\frac{5}{8} + \frac{17}{20}$$
   $$②\times②\times② \quad\quad 2 \times 2 \times ⑤$$

   These two 2s were accounted for in the denominator 8's factors.

   Least Common Denominator = $2 \times 2 \times 2 \times 5$, or 40

   c. Rewrite the fractions so they have the same denominator.

   $$\frac{25}{40} + \frac{34}{40}$$

   d. Add. Simplify your answer if needed.

   $$\frac{25}{40} + \frac{34}{40} = \frac{59}{40} = 1\frac{19}{40}$$

   e. What is the greatest common factor of 12 and 16?

   2×2, or 4

2. Exploring $\frac{5}{12} - \frac{3}{16}$ (🔢)

   a. Rewrite each denominator as a product of its prime factors.

   b. What is the least common denominator?

   c. Rewrite the fractions so they have the same denominator.

   d. Subtract. Simplify your answer if needed.

   e. What is the greatest common factor of 12 and 16?

3. Exploring $\frac{5}{68} + \frac{7}{22}$ (🔢)

   a. Rewrite each denominator as a product of its prime factors.

   b. What is the least common denominator?

   c. Rewrite the fractions so they have the same denominator.

   d. Add. Simplify your answer if needed.

   e. What is the greatest common factor of 68 and 22?

4. **Exploring** $\dfrac{5}{176} - \dfrac{3}{312}$ (🖩)

   a. Rewrite each denominator as a product of its prime factors.

   b. What is the least common denominator?

   c. Rewrite the fractions so they have the same denominator.

   d. Subtract. Simplify your answer if needed.

   e. What is the greatest common factor of 176 and 312?

5. **Record and Simplify** — Write as a fraction and simplify as much as possible.

   a. Completing 56 out of 120 problems on a test

   b. Eating 350 calories out of a 2,000-calorie diet

   c. $48 a year divided by 12 months

   d. 120 pages divided by 8 days to read them

e. 25)overline{36}

6. **Progress** — You are now completing the fifth chapter. Express your progress in a way that compares it with

   a. the total chapters in this quarter (6) and

   b. the total chapters in the course (21).

7. **Term Time** — Define the following terms and explain how you would find them:

   a. Greatest Common Factor

   b. Least Common Multiple

   c. Least Common Denominator

8. **Storage** — The storage charge on certain goods is $30 a month. What do the charges amount to in 3 years? *Hint*: There are 12 months in a year.

| PRINCIPLES OF MATHEMATICS | CHAPTER 6. More with Fractions | Worksheet 6.1 |
| LESSON 1. Multiplying Fractions |

You may use a calculator on this worksheet whenever you see this symbol (🖩).

In addition to the prime number list through 100 you've been given, you'll need to know that 227 is a prime number.

1. **Practicing the Skill** — Multiply the following numbers.

    a. $\dfrac{2}{3} \times \dfrac{4}{5}$

    b. $9 \times \dfrac{1}{2}$

    c. $\dfrac{2}{3} \times \dfrac{5}{9}$

2. **Understanding Check** — Write a sentence describing what each of these problems means. Solve.

    a. $2 \times \dfrac{1}{3}$

    b. $\dfrac{1}{3} \times 2$

3. **Music**

    a. How many beats would a half note (i.e., $\dfrac{1}{2}$ of a whole note) be worth when the whole note is worth 2 beats?

    b. What about a quarter note (i.e., $\dfrac{1}{4}$ of a whole note)?

    c. An eighth note (i.e., $\dfrac{1}{8}$ of a whole note)?

4. **Cooking**

    a. You want to triple a batch of chocolate chip cookies. The recipe calls for $\dfrac{2}{3}$ cups sugar. How much should you put in?

    b. You want to make half a batch of chocolate chip cookies. The recipe calls for $\dfrac{2}{3}$ cups sugar. How much should you put in?

PAGE 93

5. **Discounts** — A department store is offering a buy-one-get-one-$\frac{1}{2}$-off sale. If you want to buy two shirts that cost $12 each, how much will it cost you? *Note*: We haven't covered how to divide by a fraction to check your multiplication, but be sure to see if your answer makes sense!

   a. Define:

   b. Plan:

   c. Execute:

   d. Check:

6. **Exploring 15, 7, and 12** (🖩)

   a. What is the least common multiple of 15, 7, and 12?

   b. Add these fractions: $\frac{4}{15} + \frac{6}{7} + \frac{5}{12}$

   c. What is the greatest common factor of 15, 7, and 12, or is there one?

7. **Skill Sharpening**

   a. $\frac{5}{6} \left(\frac{1}{8} + \frac{2}{7} \right)$

   b. Use the distributive property to solve mentally (i.e., multiply 50 by 3, and 2 by 3, and then add the products together):

   52(3)

CHAPTER 6. More with Fractions
LESSON 2. Working with Mixed Numbers

Worksheet 6.2

You may use a calculator on this worksheet whenever you see this symbol (🖩).

1. **Mixed Numbers** — Solve. (🖩)

   a. $2\frac{2}{3} + 1\frac{7}{8}$

   b. $9\frac{1}{3} - 2\frac{6}{7}$

   c. $6\frac{2}{3} \times 5\frac{1}{4}$

2. **Cooking**

   a. You want to triple a batch of chocolate chip cookies. The recipe calls for $1\frac{1}{2}$ cups flour. How much flour do you need?

   b. You're baking a loaf of bread. You put $1\frac{2}{3}$ cups of flour in the dough, and then gradually add $\frac{1}{2}$ cup more. How much flour did you use?

3. **Miscellaneous**

   a. If you own $10\frac{1}{4}$ acres and decide to sell off $2\frac{1}{2}$ acres, how many acres (abbreviated *ac*) will you have left?

PAGE 95

b. If your 1 cup measurer is missing (maybe someone put it away in the wrong spot or perhaps it is in the dishwasher) and you need to measure $1\frac{1}{2}$ cups of flour, what is one way you could measure the amount using only your $\frac{1}{4}$-cup measurer, and one way you could measure the amount using only your $\frac{1}{2}$-cup measurer?

c. You have been told you need to leave an $\frac{1}{8}$-inch margin around a picture you're printing, as that amount will be cropped. If you send a file that's $4\frac{1}{3}$ inches wide by $7\frac{1}{2}$ inches long, what will the finished size be after the $\frac{1}{8}$ inch all the way around is cropped? *Hint*: Remember that the margin is being taken from both sides of the length and from both sides of the width.

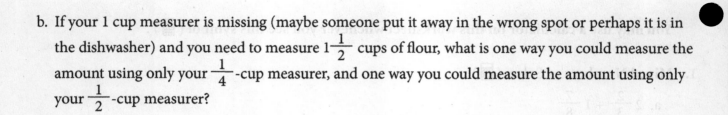

4. Exploring $\frac{5}{26} - \frac{5}{52}$ (🖩)

   a. Rewrite each denominator as a product of its prime factors.

   b. What is the least common denominator?

   c. Rewrite the fractions so they have the same denominator.

   d. Subtract. Simplify your answer if needed.

   e. What is the greatest common factor of 26 and 52?

PRINCIPLES OF MATHEMATICS

CHAPTER 6. More with Fractions
LESSON 3. Simplifying While Multiplying

Worksheet 6.3

1. **Simplifying** — Simplify the following multiplications while you multiply.

   a. $\dfrac{4}{12} \times \dfrac{6}{24}$

   Example Meaning: If $\dfrac{6}{24}$ of the people in a group are women, and $\dfrac{4}{12}$ of the women in the group have black hair, what fraction of the group are women with black hair?

   b. $\dfrac{3}{4} \times \dfrac{12}{36}$

   Example Meaning: If 12 out of 36 problems on the test are about fractions, and we get $\dfrac{3}{4}$ of the fraction problems right, what portion of the total problems did we get right just in the fraction section?

   c. $\dfrac{4}{3} \times \dfrac{27}{12}$

   Example Meaning: If you're increasing the dimensions of a building project by $\dfrac{4}{3}$, and you originally needed $\dfrac{27}{12}$ inches, how many inches do you now need?

2. **Write a Word Problem** — Write a word problem that involves multiplying two fractions. Look at the example meanings from problem 1 for inspiration. Solve.

PAGE 97

3. **Commission** — Some salesmen are on what we call commission — that is, they get a portion of every sale they make. If a salesman gets a $\frac{1}{10}$ commission, that means he gets $\frac{1}{10}$ of every sale.
   a. How much does he make if he sells $40?

   b. How much does he make if he sells $50 $\frac{1}{2}$?

   c. If he needs to make $100 in commissions, how much does he have to sell? *Hint*: If he gets $\frac{1}{10}$ of the sales, then he has to sell 10 times what he wants to make.

4. **Height** — If you were 5 feet $3\frac{1}{4}$ inches last year and are 5 feet $4\frac{12}{16}$ inches this year, how much did you grow over the year? *Hint*: Ignore the 5 feet, as it stayed the same (only the inches changed).

5. **Skill Sharpening**
   a. $\frac{1}{15} + \frac{1}{2}$

   b. What are the prime factors of 100?

PRINCIPLES OF MATHEMATICS

CHAPTER 6. More with Fractions
LESSON 4. Reciprocal / Multiplicative Inverse and More

Worksheet 6.4

1. **Inverse** — Write the reciprocal/multiplicative inverse of these numbers. *Hint*: Convert to an improper fraction if necessary and then find the reciprocal.

   a. $\dfrac{8}{9}$

   b. 4

   c. $3\dfrac{6}{7}$

2. **Fractions and Division** — Rewrite these divisions as fractions. Remember, you can have a fraction on top of a fraction; the division sign means to divide.

   a. $\dfrac{2}{3} \div \dfrac{8}{9}$

   b. $\dfrac{5}{6} \div \dfrac{45}{88}$

3. **Drawing** — Suppose you're trying to duplicate a drawing that is $2\dfrac{4}{5}$ inches wide by $7\dfrac{1}{8}$ inches tall. If you want to shrink the drawing by $\dfrac{1}{2}$, what width and height should you make the drawing?

4. **Making Cards** — If you have a blue paper square that is $4\dfrac{1}{2}$ inches in every direction and want to back it with a green paper square, leaving $\dfrac{1}{4}$-inch border of the green all the way around, how wide should you make the green square?

5. **Skill Sharpening** — Simplify while multiplying.

   a. $\dfrac{7}{16} \times \dfrac{24}{56}$

   b. $\dfrac{3}{14} \times \dfrac{28}{30}$

PAGE 99

CHAPTER 6. More with Fractions
LESSON 5. Dividing Fractions
Worksheet 6.5

You may use a calculator on this worksheet whenever you see this symbol (🖩).

1. **Learning the Skill** — Solve these problems.

   a. $\dfrac{3}{4} \div \dfrac{2}{3}$

   b. $\dfrac{\frac{1}{2}}{\frac{4}{7}}$ *Hint*: Remember that the fraction line means to divide, so this is another way of writing $\dfrac{1}{2} \div \dfrac{4}{7}$.

   c. $\dfrac{2}{3} \div \dfrac{5}{6}$

   d. $\dfrac{1}{2}$ divided by 4

   e. $\dfrac{\frac{9}{8}}{2}$

   Note: Remember, these word problems are designed to mimic real-life settings. You may have to use different tools than those we covered today. Be sure to check your work and make sure your answer makes sense.

2. **Making Cards** — Suppose you want to make a card. You start with a piece of $8\dfrac{1}{2}$ inches wide by 11 inches long. You want your finished card to be half that size, so you fold the longer side in half and cut it, then fold the longer side in half again. How large is your card now? *Hint*: You can grab a piece of paper and actually fold it if needed, but don't use a ruler!

PAGE 101

3. **Cloth** — If $3\frac{3}{4}$ yards of cloth are required to make a garment, how many such garments can be made from a piece containing 56 yards?

4. **Lumber** — Suppose you are trying to build a birdhouse. You have four 8-ft (foot) pieces of wood, and you need $2\frac{1}{2}$ ft of wood for each birdhouse. You want to figure out how many birdhouses you can build with the lumber you have. For this problem, assume that you can glue parts of different pieces together to form the $2\frac{1}{2}$ ft—they do not all need to come from the same piece of wood.

   a. Define:

   b. Plan:

   c. Execute:

   d. Check:

5. **Home Repair** — A certain chemical calls for $\frac{2}{3}$ cup per gallon of water. You only want to make $\frac{1}{4}$ gallon of the solution. How much of the chemical should you use?

6. **Exploring** $\frac{180}{650} - \frac{13}{130}$ (🖩)

   a. Rewrite each denominator as a product of its prime factors.

   b. What is the least common denominator?

   c. Rewrite the fractions so they have the same denominator.

   d. Subtract. Simplify your answer if needed.

   e. What is the greatest common factor of 650 and 130?

7. **Question** — What are two ways to look at division?

PRINCIPLES OF MATHEMATICS

CHAPTER 6. More with Fractions
LESSON 6. Chapter Synopsis

Worksheet 6.6

1. **Learning the Skills** — Solve the following problems.

    a. $\dfrac{\frac{1}{3}}{\frac{1}{4}}$

    b. $\dfrac{8}{9} + 2\dfrac{3}{4}$

    c. $1\dfrac{1}{7} \times \dfrac{4}{5}$

    d. $\dfrac{2}{3} - \dfrac{1}{5}$

2. **Factoring**

    a. Express as a product of its prime factors: 24

    b. Express as a product of its prime factors: 89

    c. What is the greatest common factor of 24 and 89?

    d. What is the least common multiple of 24 and 89?

3. **Multiply while Simplifying** — Be sure to show your work.

    a. $\dfrac{24}{16}$ ounces $\times \dfrac{46}{16}$ ounces

    b. $\dfrac{4}{16} \times \dfrac{16}{1}$

    c. $\dfrac{20}{16} \times \dfrac{16}{1}$

PAGE 103

4. **Fractions as Division** — Use a fraction to represent what needs to be done to solve this word problem. Do not solve—just write the fraction.

   If you pay $650 every six months in car insurance, how much does it cost you each month for car insurance?

   The following problems are ones you might encounter at the grocery store.

5. **Produce** — You buy $3\frac{1}{6}$ pounds of apples at $3 a pound. How much does it cost?

6. **Cheese** — American cheese is on sale for $4 a pound instead of the normal $6 a pound. If you normally buy $\frac{1}{2}$ a pound, how much will you save by buying your normal amount on sale?
   a. Define:

   b. Plan:

   c. Execute:

   d. Check:

7. **Size** — A recipe calls for $\frac{3}{4}$ pound (lb) of cheese. You want to make $1\frac{1}{2}$ times the recipe.
   a. How much cheese do you need?

   b. The cheese comes in $\frac{1}{2}$ pound blocks. How many blocks do you need?

8. **Out of the Box** — Pull out a recipe (cookies work well!) and either double or half it. If the recipe calls for something you can't double/half (like 1 egg—it's kind of hard to half it exactly), just estimate and do your best (put in about half the egg). Notice all the fractions involved in cooking!

PRINCIPLES OF MATHEMATICS　　　Review of Chapters 1–6　Worksheet 6.7

> Use these problems to help you review. If you have questions, review the concept in the *Student Textbook*. The lesson numbers in which the various concepts were taught are in parentheses.

1. **Binary Place-value System (Lesson 1.6)**

    a. What would it mean if a number were written in a binary (base-2) system?

    b. How many digits (including 0) do you need in a binary system?

2. **Addition**

    a. True or False: People have always used + to represent addition. (Lesson 2.1)

    b. Explain in your own words what the purpose is of "carrying" digits in addition. (Lesson 2.3)

3. **Keeping a Checkbook (Lesson 2.4)**

    Input the transactions given, keeping track of the current balance in the balance column.

    10/1 – Opening Balance – $181
    10/5 – Pay Shoe Store with Check 159 – $26
    10/6 – Pay Grocery Store with Check 160 – $33
    10/11 – Deposit Paycheck – $89
    10/20 – Deposit Birthday Money – $64
    11/5 – Withdrawal (Take Money Out of the Bank) – $35

Check Number	Date	Memo	Payment Amount	Deposit Amount	$ Balance

4. **Define the following:**

    a. multiplication (Lesson 3.3)

    b. division (Lesson 3.4)

    c. dividend (Lesson 3.4)

    d. numerator (Lesson 5.1)

PAGE 105

    e. multiplicative inverse (Lesson 6.4)

    f. identity property of multiplication (Lesson 3.5)

5. **Understanding Check** — Is subtraction commutative? (Lesson 3.5)

6. **Mental Math (Lesson 3.2)**

    a. Find an approximate total mentally by rounding each number to the nearest ten and adding: 86 + 29

    b. Find an exact total mentally: 86 + 29

7. **Learning the Language** — Combine what you've learned about properties and conventions (Lessons 3.5 and 3.6), fractions, and division (including Lesson 4.4) to give a decimal answer. We've not ever seen problems exactly like these, so think them through and see if you can figure them out on your own!

    a. $\dfrac{4 + 5(7 + 2)}{10}$

    b. $\dfrac{0}{8}$

    c. $\dfrac{8}{0}$

8. **Distributive Property** — Solve $5(\dfrac{1}{15} + \dfrac{2}{30})$ using the distributive property. Show your work. *Hint*: While you've never seen a problem quite like this one, remember that the distributive property means you can distribute the multiplication . . . saving you from having to add these fractions before multiplying. (Lesson 4.2)

9. **Fractions as Division** — Rewrite this division problem as a fraction in simplest terms. (Lesson 5.1)
$25\overline{)5}$

10. **Factoring/Least Common Multiple/Greatest Common Factor (Lessons 5.4, 5.5, and 5.7)**

    49, 70

    a. Express 49 and 70 as products of their prime factors.

    b. What is the least common multiple of 49 and 70?

    c. What is the greatest common factor of 49 and 70?

11. **Factoring / Least Common Multiple / Greatest Common Factor (Lessons 5.4, 5.5, and 5.7)**

    45, 20

    a. Express 45 and 20 as products of their prime factors.

    b. What is the least common multiple of 45 and 20?

    c. What is the greatest common factor of 45 and 20?

12. **Operations with Fractions (Lessons 5.6, 6.1, and 6.5)**

    a. $1\frac{1}{4} + 2\frac{1}{3}$

    b. $8\frac{1}{2} - 2\frac{5}{8}$

    c. $\frac{1}{3} \times \frac{9}{8}$

    d. $\frac{2}{5} \div \frac{7}{5}$

13. **Applying It** — One of the goals of learning math is to learn it so you can apply it in real life! Test yourself with these word problems. (Lessons 2.6 and 4.5)

   a. If you put $\frac{2}{3}$ cup of flour in a loaf of bread, and then another $3\frac{1}{4}$ cup, how much flour is in the bread altogether?

   b. If you want to make 8 dresses, and each one requires $1\frac{3}{4}$ yard of fabric, how much fabric do you need?

   c. Say you're heading to the fair. You start with $85 in your pocket. You need to have at least $10 left at the end to pay for the train ride home. If you spent $15 on ride tickets, $12 on lunch, $13 on souvenirs, and $2 on a postcard, how much do you have left to spend and still keep the $10 for the ride home?

CHAPTER 7. Decimals
LESSON 1. Introducing Decimals
Worksheet 7.1

> In this worksheet, you'll be converting fractions to decimals. However, it's important to continue getting practice working with fractions. So unless otherwise indicated or unless the problem contains both fractions and decimals, continue to give answers to problems given in fractions as fractions.
>
> You may use a calculator on this worksheet whenever you see this symbol (🖩).

1. **Understanding Decimals** — Rewrite these decimals as fractions of 10, 100, or 1,000.

    a. 0.56

    b. 0.3

    c. 0.30

    d. 0.486

    e. 0.874

2. **Simplifying** — Simplify each fraction you wrote in problem 1, unless the fraction is already in simplest terms. (🖩)

3. **Reading Decimals**

    a. Write "three and fourteen hundredths" using decimal notation.

    b. Express 7.891 in words, using the words you would use if you were reading the decimal portion as a fraction. *Example*: 4.25 would be "four and twenty-five hundredths."

4. **Understanding Decimals** — Rewrite these fractions as decimals.

    a. $\dfrac{5}{10}$

    b. $\dfrac{86}{100}$

    c. $\dfrac{1}{2}$

    d. $\dfrac{12}{25}$

5. **Division Reminder** — Rewrite these division problems as both fractions (you do not need to simplify) and decimals.

   a. $100\overline{)5}$

   b. $10,000\overline{)65}$

6. **Writing Checks** — Fill in the dollars' line with words that describe the amount. Use a fraction to represent the 68 cents.

   [Check image showing amount $1,023.68]

7. **Question** — Why can we explore God's creation with math?

PRINCIPLES OF MATHEMATICS

CHAPTER 7. Decimals
LESSON 2. Adding and Subtracting Decimals

Worksheet 7.2

1. **Adding and Subtracting Decimals** — Rewrite each problem vertically and solve.

    a. 875.01 + 4.78

    b. 78.4 + 109.64

    c. 5,678.32 − 2,593.9

    d. 480 ÷ 8 + 78.96

2. **Adding and Subtracting Fractions**

    a. Add $\frac{14}{25}$ and $\frac{2}{5}$

    b. Now add 0.56 + 0.4

    c. Express 0.56 and 0.4 as fractions in simplest terms. How do they compare with the fractions in problem 2a?

    d. Was it easier to add fractions or decimals…and why?

3. **Understanding Check** — Why does it work to add and subtract decimals using our regular addition and subtraction methods?

PAGE 111

4. **Abacus Fun** — Pull back out your abacus. Cut a circle out of paper and tape it above the third row up from the bottom to represent the decimal dot. Then use your abacus to perform the following addition and subtractions. If you don't have an abacus, you can use an online one—just pretend there's a dot three rows up from the bottom. *Hint*: Rewrite vertically so you can easily see what to add or subtract. Start by forming the first number on the abacus; then add or subtract the second number from the right to left.

   a. 5.46 + 9.368

   b. 2.34 − 1.89

   c. 7.52 − 5.63

5. **Fees** — Let's say you have a plumber come to your house. He charges you an $120 trip and labor charge, plus $4.99, $5.60, and $32.22 for parts. Assuming you do not have to pay sales tax, how much is your total bill?

6. **Checkbook** — Now that we've covered decimals, it's time to revisit a checkbook register and learn how to put in cents. The cents go in the small columns under the Payment, Deposit, and Balance headings. Notice how in the Opening Balance, we've used the small column to record the 45 cents.

   Input the following transactions into the checkbook register, updating your balance as you go. *Hint*: If you need help with how to keep a checkbook register, see the explanation in Lesson 2.4.

   10/3 Paid the telephone bill for $39.99 with check 300

   10/6 Deposited a rebate for $29.12

Check Number	Date	Memo	Payment Amount		Deposit Amount		$ Balance	
	10/1	Opening Balance					1,200	45

PAGE 112

CHAPTER 7. Decimals
LESSON 3. Multiplying Decimals
Worksheet 7.3

1. **Multiplying Decimals**

    a. 42 • 0.15

    b. 0.2 • 24

    c. 1.78 • 9.63

    d. 2.5(48.6 − 9.87)

    e. 1.56 • 0.3

2. **Skill Sharpening**

    a. $1\frac{3}{4} \cdot 98$

    b. $\frac{2}{3} \cdot \frac{6}{5}$

    c. $\frac{4}{7} + \frac{3}{5}$

    d. $\frac{5}{2} - \frac{3}{9}$

3. **Business** — Suppose you sell 50 bags of mulch at $4.55 a bag. You want to find what your gross sales (the amount you recieved without including any of your expenses or deductions) will be.

    a. Define:

    b. Plan:

    c. Execute:

    d. Check:

4. **Storage** — If it costs $59.99 a month for a storage unit, how much will it cost to rent that unit for 2 years?

5. **Oil** — If oil costs $3.56 a gallon, and you typically use 80 gallons a month to heat the house 6 months a year but none the other 6 months, how much should you budget for per year for oil?

PRINCIPLES OF MATHEMATICS

CHAPTER 7. Decimals
LESSON 4. Dividing and Rounding with Decimals
Worksheet 7.4

> From now on, round all decimal answers to the nearest hundredth, unless otherwise specified.
>
> You may use a calculator on this worksheet whenever you see this symbol (🖩).

1. **Rounding** — Round these numbers to the nearest hundredth.
   a. 45.8976
   b. 21.754
   c. 7,892.2138597

2. **Remainders as Decimals** — Divide these numbers, expressing the remainders as decimals (round to the nearest hundredth).

   a. $4\overline{)25}$
   b. $7\overline{)89}$

   c. $3\overline{)101}$

3. **Fractions Review**
   a. $\frac{8}{9} \div \frac{2}{3}$
   b. $1\frac{2}{5} \div \frac{4}{5}$

   c. $1\frac{5}{13} - \frac{6}{13}$

4. **Salary** — Assume 52 weeks in a year, and 40 hours a week. Round your answers.
   If you make $50,000 a year,
   (a) what is your monthly pay,

   (b) what is your weekly pay, and

   (c) what is your hourly pay?

PAGE 115

5. **Finding the Cost** — Round to the nearest dollar and then solve mentally.

    a. If peas cost $1.75 a pound, about how much will 4 pounds cost?

    b. If you only have $10 to spend, about how much more will you have left to spend if you buy the 4 pounds?

    c. Approximately what is $2.99 + $3.99?

6. **Projects** — If you need 2 ft of wood per project, how many projects can you make out of 19 ft? Give your answer in whole projects, assuming you can't make a partial one.

7. **Pricing** (🖩) — Say you're trying to pick up some temporary work to raise $500 for a mission's trip. You discover you can rent a power washer for $69.99 per day. You'll also have the expense of your gas to get to the store and the different neighborhoods (about 2 gallons each day at $3.19 a gallon). You charge $129.99 per house you power wash. Assuming you can wash 3 houses per day,

    a. how much will you make per day?

    b. how much would you make per day if you offer a $79.99 special?

8. **Math and Our Bodies** (🖩) — Math can help us appreciate the incredible way God created our bodies. Although no one knows for sure and the number varies based on body size, the general estimate is that an adult has 60,000 miles of blood vessels. You are going to use math to put this in perspective.

    The earth's circumference at the equator is approximately 24,900 miles. If we could stack the blood vessels in an average adult end to end, how many times could they go around the earth?

    Take a look at your answer and think about the magnitude of blood vessels you have inside you, all working together to transfer oxygen and substance to your body! It's too much for us to even comprehend. Yet God not only comprehends it, but knit us together. Truly we are fearfully and wonderfully made.

    I will praise thee; for I am fearfully and wonderfully made: marvellous are thy works; and that my soul knoweth right well (Psalm 139:14).

    Such knowledge is too wonderful for me; it is high, I cannot attain unto it (Psalm 139:6).

9. **Out of the Box** — When shopping, it's helpful to know approximately how much you're spending *before* you get to the check out (especially when you're on a tight budget). Next time you're at the store, see if you can estimate the total by rounding each item's cost to the nearest dollar as you or your parent puts it into the cart. For example, if you put a can of beans that costs $0.79 into your cart, round it to $1. See how close you can estimate the total!

PRINCIPLES OF MATHEMATICS

CHAPTER 7. Decimals
LESSON 5. Conversion and More with Decimals

Worksheet 7.5

You may use a calculator on this worksheet whenever you see this symbol (🖩).

1. **Conversion Between Fractions and Decimals** (🖩) — Express the following fractions as decimals. Remember to round to the hundredth's place.

   a. $\dfrac{1}{3}$

   b. $\dfrac{11}{12}$

   c. $\dfrac{5}{6}$

2. **Computer Graphics**

   a. Let's say you wanted to change the width of text on a computer so it will print out $3\dfrac{1}{7}$ inches wide. The computer program can set the width of the text if you can give it the dimensions you want using decimals. What number should you type in?

   b. If you want to leave $\dfrac{3}{4}$ inch margins (blank space) all the way around an $8\dfrac{1}{2}$ inch by 11 inch piece of paper, what width and height (in decimals) should you tell the computer to crop a picture you want to fit just inside the margins? *Hint*: Break the problem down into what has to happen to your width and to the height. Your width is $8\dfrac{1}{2}$; you want a $\dfrac{3}{4}$ inch margin on the left *and* a $\dfrac{3}{4}$ inch margin on the right. What should your width be? Your height is 11 inches. You want a $\dfrac{3}{4}$ inch margin on the top *and* a $\dfrac{3}{4}$ inch margin on the bottom. What should your height be? Feel free to draw the problem if needed.

3. **Fractions and Decimals Combined** (🖩) — It's important to become proficient in both decimals and fractions and to know how to work with problems that contain both. For example, when you buy $\frac{1}{4}$ pound of apples for $1.25 a pound, you have both a fraction and a decimal to deal with. In this case, you'd want to express the answer as a decimal, as we use decimals to represent portions of a dollar rather than fractions.

Don't let fractions and decimals in the same problem stump you. You know how to work with both. For these problems, express your answer as a decimal.

Example: $\frac{1}{4} \cdot 1.25 = \frac{1.25}{4} = 0.31$

Notice that we multiplied decimals like we would whole numbers, and then we completed the division (1.25 ÷ 4) to simplify.

a. $\frac{2}{11} \cdot 8.97$

b. $\frac{1}{4} \cdot 5.2$

c. $10.78 - \frac{7}{15}$

d. If you buy $\frac{2}{3}$ a pound of applies at $2.99 a pound, how much will you end up paying?

e. If an $8\frac{1}{2}$-ounce package of noodles costs $2.19 and a 13-ounce package costs $2.99, which package costs less per ounce and by how much?

From now on, if a problem has both a decimal and a fraction, you can pick which notation to use, unless otherwise specified.

4. **Pricing** — Let's say a clothing store buys 200 shirts for $636.56.

   a. If they sell each shirt for $8.99, how many shirts will they have to sell to recuperate their cost?

   b. What if they sell each shirt for $6.99 instead?

5. **Temperature** — If your body temperature is normally 98.6 degrees Fahrenheit and you're current temperature is 101.2 Fahrenheit, how much warmer than usual are you?

6. **Cost per Acre** — Someone is selling 3 acres for $24,790. What is the cost per acre?

7. **Understanding Check** — Explain how to divide a decimal number by a decimal number and what we're really doing when we do. You can use wording from the text if you like.

PRINCIPLES OF MATHEMATICS

CHAPTER 7. Decimals
LESSON 6. Chapter Synopsis
Worksheet 7.6

You may use a calculator on this worksheet whenever you see this symbol (📱).

1. **Purchases** — John bought a hat for $5.55 and a shirt for $3.25. Realizing he'd gotten a great bargain, he decided to buy three more hats (in addition to the ones he bought for himself) and shirts for friends. He then remembered he also needed to pick up sunglasses, and got a pair for $7.99.

    a. How much did John spend altogether?

    b. If he used a $5.50-off coupon, how much did he spend?

2. **You Write It** — Write and solve a problem involving multiplying a decimal number.

3. **Halving a Recipe** — Write down the amounts you should use if halving this recipe.

    a. $1\frac{1}{2}$ cups flour

    b. $\frac{2}{3}$ cup flour

4. **Profit** — Let's say you went into business with two friends or siblings. You all combine resources evenly to purchase baking supplies that cost $10.50, $2.99, $7.68, and $4.99. You bake cookies with the ingredients and sell 24 four-packs for $1.50 each, except for one, which you sell at half price to a friend. If you share the profits evenly, how much will each one of you make after you recuperate the cost of the supplies? *Hint*: Remember to break the problem down by defining, planning (look for how the numbers relate to one another), solving, and checking.

PAGE 121

5. **Conversion** (🖩) — Convert these fractions or mixed numbers to decimals.

   a. $7\frac{1}{16}$ inches

   b. $\frac{2}{3}$ of a cup

   c. $\frac{4}{16}$ of an inch

6. **Cost** — Assume a package of paintballs costs $30 and contains 1,000 paintballs.

   a. How much does each paintball cost?

   b. If you use 30 of these paintballs in a game, how much will it have cost you to play the game if your costs are the paintballs and a $29.99 rental fee?

7. **Fractions and Decimals Combined** (🖩) — Give your answer as a decimal.

   $14.5(\frac{2}{3} + \frac{6}{7} - \frac{2}{9})$

8. **Out of the Box** — Design a letter or report on the computer. Adjust the margins as needed and insert pictures. You can use your own project if you have one (any project with a picture and some text will do); if you don't have anything specific to work on, find a hymn online and copy the words into a document. Change the margins to be $1\frac{1}{2}$ inches on all sides. Insert a couple of pictures (search online for "insert picture (*name of software*)" if you don't know how). Make the pictures $2\frac{3}{4}$ inches wide. Now change the margins to $3\frac{1}{4}$ inches wide. Keep playing around with margins and sizes. When you're happy, print out your hymn!

PRINCIPLES OF MATHEMATICS

CHAPTER 8. Ratios and Proportions
LESSON 1. Ratios
Worksheet 8.1

1. **Recording Ratios** — Record each ratio both as a fraction and with a colon. When representing the ratio as a fraction, include units in the ratio. *Example*: Ratio of problems right to problems wrong if 8 problems were right and 2 problems were wrong = $\frac{8 \text{ problems}}{2 \text{ problems}}$ and 8:2

   a. Ratio of apple pie to rhubarb pie sold if 7 pieces of apple and 1 piece of rhubarb were sold.

   b. Ratio of sunny days to non-sunny days in the month if there were 14 sunny days in a 31-day month. *Hint*: There are two steps to this problem! First find the number of non-sunny days. If there were 14 sunny days and 31 days total, how many days were not sunny?

   c. Rate traveling if you cover 55 miles in 1 hour. *Hint*: You're comparing miles with hours. We would read this rate "55 miles per hour."

2. **Reading Ratios** — What are two ways to read this ratio: 7:8?

3. **Order in Sunflowers** — The seeds in a sunflower are arranged in two spirals. The data below lists the five different spiral combinations we find in sunflowers.

Sunflower A:	Sunflower B:	Sunflower C:	Sunflower D:	Sunflower E:
Spiral 1:	Spiral 1:	Spiral 1:	Spiral 1:	Spiral 1:
8 seeds	21 seeds	34 seeds	55 seeds	89 seeds
Spiral 2:	Spiral 2:	Spiral 2:	Spiral 2:	Spiral 2:
13 seeds	34 seeds	55 seeds	89 seeds	144 seeds

Explore the relationship, or ratio, between the spirals in each sunflower using these steps.

a. Express the relationship between each sunflower's spiral 2 and spiral 1 as a ratio in fractional notation, listing spiral 2 first ($\frac{Spiral\ 2}{Spiral\ 1}$). For example, Sunflower A's ratio would be $\frac{13}{8}$.

b. Use division to represent each ratio as a decimal. Round your answer to the nearest hundredth.

What do you notice from exploring the spirals in sunflowers? Each of the sunflowers had a different number of seeds. Yet, when we divided the number of seeds in the first spiral by the number of seeds in the second spiral, we came up with approximately the same answer every time. Regardless of the number of seeds in a sunflower, the relationship, or ratio, between the two seed spirals did not significantly change.

Guess what? This relationship or proportion between the spirals allows each sunflower to hold the greatest number of seeds possible![1]

Math just uncovered the special relationship God put into sunflowers. God designed them to reproduce in the most efficient way possible. Remember, the God who watches over such a tiny detail in sunflowers is the same God watching over each detail of your life.

And why take ye thought for raiment? Consider the lilies of the field, how they grow; they toil not, neither do they spin: And yet I say unto you, That even Solomon in all his glory was not arrayed like one of these. Wherefore, if God so clothe the grass of the field, which to day is, and to morrow is cast into the oven, shall he not much more clothe you, O ye of little faith? Matthew 6:28-30

---

[1] The number of seeds in each spiral are all neighboring numbers in the Fibonacci sequence, and the ratio between them approaches a common ratio called the golden ratio (≈1.62). See James D. Nickel, *Mathematics: Is God Silent?* rev. ed. (Vallecito, CA: Ross House Books, 2001), p. 241.

4. **Ratios in Art** — Explore the ratio or relationship between the lengths and widths of the rectangles below by

   a. writing a ratio for each rectangle in fractional notation, using the longer side of the rectangle over the shorter side of the rectangle.

   b. representing the ratio as a decimal.

Notice how once again, each ratio was approximately the same. Even though these rectangles were different sizes, the relationship between each length and each width was about the same. And notice that the ratio between each length and width was about the same as the ratio between the spirals in sunflowers.

Since we encounter this ratio so much (it can also be found in the spirals of a nautilus' sea shell, the spirals in a pineapple, and other aspects of creation), we have given it a name: the **golden ratio**. Many people think rectangles with this ratio, nicknamed **golden rectangles**, are visually appealing; hence, you will often find these ratios in art and architecture.

Isn't it interesting that various and completely unrelated parts of creation express a ratio appealing to the human eye? What a wonderful reminder that the same God created both creation and man! It is He who placed this design everywhere around us and created our minds to appreciate the beauty of His creation.

   b. representing the ratio as a decimal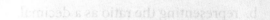

PRINCIPLES OF MATHEMATICS

CHAPTER 8. Ratios and Proportions
LESSON 2. Proportions

Worksheet 8.2

You may use a calculator on this worksheet whenever you see this symbol (🧮).

1. **Missing Number** — Find the missing number in these proportions.

   a. $\dfrac{4}{5} = \dfrac{36}{?}$

   b. $\dfrac{9}{60} = \dfrac{?}{20}$

   c. $\dfrac{7}{?} = \dfrac{14}{58}$

   d. $\dfrac{26}{40} = \dfrac{13}{?}$

2. **Representing Proportions** — Represent the following as proportions. Include the units.

   a. 38 passengers per 1 rollercoaster ride; 380 passengers per 10 rollercoaster rides

   b. 400 bushels of potatoes from 1 acre; 4,000 bushels of potatoes from 10 acres

3. **Finding the Missing Number in a Proportion** (🧮) — Use proportions to find the answer to these questions.

   Supplies

   If every 20 men need 800 rounds of ammunition,

   a. how many rounds will supply 250 men?

   b. how many men will 15,000 rounds supply?

   PAGE 127

Trains

c. If a train of 30 cars carries 1,100 tons, how many cars must be added to the 30-car train so that it can carry 3,300 tons? *Hint*: First figure out using a proportion how many cars you need to carry 3,300 tons, and then figure out how many more than the 30 cars you already have that is.

d. If a train gets 25 miles per 1 gallon of gasoline, how many gallons will it need to go 1,000 miles?

4. **Equal Ratios** — Use = or ≠ to show if these ratios from a proportion.

   a. $\dfrac{17 \text{ ft}}{10 \text{ ft}}$ $\dfrac{8 \text{ ft}}{5 \text{ ft}}$

   b. $\dfrac{4 \text{ eggs}}{2 \text{ batches}}$ $\dfrac{8 \text{ eggs}}{4 \text{ batches}}$

   c. $\dfrac{7 \text{ gallons}}{\$20}$ $\dfrac{5 \text{ gallons}}{\$40}$

5. **Skill Sharpening**

   a. $2\dfrac{4}{5} + 7\dfrac{22}{25}$ b. $\dfrac{7}{8} - \dfrac{9}{64}$

   c. Find the greatest common factor of 28 and 44.

   d. Find the least common multiple of 28 and 44.

   e. $7.8(2.69 + 4.58 - 4\dfrac{1}{2})$ *Note*: Give your answer as a decimal.

PRINCIPLES OF MATHEMATICS

CHAPTER 8. Ratios and Proportions
LESSON 3. Ratios and Proportions Containing Decimals

Worksheet 8.3

You may use a calculator on this worksheet whenever you see this symbol (🖩).

1. **Dividing Ratios** — Divide these ratios to give a single quotient. Give the quotient as a mixed number if written as a mixed number, and as a decimal otherwise.

   a. $\dfrac{8\frac{1}{9}}{7\frac{2}{3}}$

   b. $\dfrac{4.6}{0.25}$

2. **Finding Missing Numbers** — Find the missing number in these proportions.

   a. $\dfrac{4.5}{8.25} = \dfrac{?}{24.75}$

   b. $\dfrac{\frac{3}{6}}{?} = \dfrac{\frac{6}{12}}{\frac{8}{16}}$

3. **Comparing Ratios** — Put = or ≠ in between these ratios. (Remember, you can divide to compare.)

   a. $\dfrac{2.3}{1.3}$ $\dfrac{4.6}{2.6}$

   b. $\dfrac{7}{20}$ $\dfrac{6}{20}$

4. **Purchases** (🖩) — Say you want to asphalt an area of your yard to make a mini basketball court or carport. You get a quote for $8\frac{1}{2}$ square feet for $850. If you decided to change the size to 17 square feet, and the price per square foot stayed the same, how much would it cost you?

5. **All in a Day's Pay** (🖩) — At $3.25 per day, how many days' work will $100 pay for?

6. **Seed Time** (🖩) — If 104 pounds of seed are used on a 2-acre lot, how many pounds would be required for $15\frac{1}{2}$ acres?

7. **Skill Sharpening** (🖩)

   a. Find the least common multiple of 45 and 78.

   b. Find the greatest common factor of 15 and 105.

   c. Use the distributive property to solve: 4(4.78 + 5.67) Show your work.

   d. $\frac{1}{4} \cdot 1.87$

| PRINCIPLES OF MATHEMATICS | CHAPTER 8. Ratios and Proportions | Worksheet 8.4 |
| LESSON 4. Scale Drawings and Models |

You may use a calculator on this worksheet whenever you see this symbol (🖩).

1. **Exploring Scale Ratios** (🖩)

    a. If a treasure map is drawn using a scale ratio of 1:3 (1 inch representing 3 feet), how far in reality would a 5-inch line on the map represent?

    b. If a chart drawn on graph paper uses 1 square to represent $30, how many dollars would be represented by 5 squares?

    c. The Bowing 737 has a wingspan (width) of 34.3 meters and an overall length of 33.6 meters. If you were to make a scale model that's $\frac{1}{300^{th}}$ the size of the original, what wingspan (width) and length should the model have? Round your answer to the nearest **thousandth**, rather than hundredth.

2. **Proportional** (🖩) — If in real life a tie is 24 inches long and the man wearing it is 65 inches tall, and in a drawing of the man the tie is 4 inches long and the man is 10 inches tall, is the drawing proportional? Why or why not?

3. **Skill Sharpening**

    a. $\frac{4}{7} \cdot \frac{5}{16}$ Simplify as you multiply.

    b. $\frac{8}{25} - \frac{1}{5}$

    c. $\frac{4}{7} \div \frac{20}{34}$

4. **Out of the Box** (🖩)

    a. Measure a bookcase in your house and make a $\frac{1}{20^{th}}$, two-dimensional scale drawing of it (view it as a rectangle and just draw the front).

b. View a house blueprint. (Look online under "house blueprints", ask your parents if they have a house blueprint, visit a new home development or your county courthouse, etc.) Notice that a blueprint is really a scale drawing.

5. **The Way We Grow** — If we used the centimeter side of a ruler to measure different lengths in the picture, we would find that the approximate length of the teen's head in the picture is 2 centimeters (abbreviated *cm*), while the baby's head is approximately 1 centimeter. The distance in centimeters from the teen's neck to the bottom of his shoes (i.e., the rest of his body) is approximately 12.5 centimeters, while the distance from the baby's neck to the bottom of his foot (i.e., the rest of his body) is approximately 3 centimeters.

   a. Set up a ratio between the teen's head and the rest of his body and another one between the baby's head and the rest of his body.

   b. Express the ratios as decimal numbers (round to the hundredths place).

   c. Do the two ratios form a proportion?

   Notice that the ratio, or relationship, between the baby's head and the rest of his body was a lot different than that of the teen's head and his body. A baby's head is a lot larger in comparison to the rest of his body than a teen's.

   When we grow, we do not grow proportionally. Our head does not continue to expand at the same rate as our arms and legs. God, in His infinite wisdom and care, designed us to grow in just the right way. He designed our bodies to change proportions, giving us exactly what we need for different stages of lives. Babies fall a lot, so as babies, we needed extra padding on our heads. But as we grow and become more mature, we no longer need such a large head. In fact, if our head continued to grow proportionally to the rest of our bodies, we would look rather strange and probably be unable to walk. Instead of a larger head, we needed larger feet and muscles to support our growing weight. And that is exactly what God gives us!

   By looking at the ratios between a baby and adult, we see that God, in His infinite wisdom and care, designed us to grow in just the right way. As our Creator, He knew our need for different proportions as we grow.

   ...for your Father knoweth what things ye have need of, before ye ask him (Matthew 6:8).

PRINCIPLES OF MATHEMATICS

CHAPTER 8. Ratios and Proportions
LESSON 5. Mental Math and Decimals

Worksheet 8.5

You may use a calculator on this worksheet whenever you see this symbol (🖩).

1. **Making Change** — Figure out mentally how much change (both the amount and the specific bills and coins) you should receive in each of the following situations. Use the least number of coins you can; for example: use a quarter instead of two dimes and a nickel. If you can, grab some money and get someone to act out each exchange with you.

   Example:
   Total = $5.56   Give = $6   Change = 44 cents   Bills and Coins = 1 quarter, 1 dime, 1 nickel, 4 pennies
   a. Total = $4.76   Give = $5.00   Change =   Bills and Coins =
   b. Total = $4.76   Give = $5.01   Change =   Bills and Coins =
   c. Total = $31.09   Give = $31.25   Change =   Bills and Coins =
   d. Total = $31.09   Give = $31.10   Change =   Bills and Coins =
   e. Total = $46.78   Give = $50   Change =   Bills and Coins =
   f. Total = $46.78   Give = $50.03   Change =   Bills and Coins =
   g. Total = $31.23   Give = $32   Change =   Bills and Coins =
   h. Total = $31.23   Give = $32.03   Change =   Bills and Coins =
   i. Total = $25.86   Give = $26   Change =   Bills and Coins =
   j. Total = $25.86   Give = $26.06   Change =   Bills and Coins =

   Notice how sometimes changing up what you give results in fewer coins! For example, if the total is $0.97 cents, giving $1.02 will give you 5 cents back (a nickel) rather than the 3 pennies you'd get back if you just gave $1. The idea is to overpay to the point that you can get fewer, higher-value coins back.

2. **Mental Multiplication and Division** — Figure out the actual totals mentally.

   a. 4 pounds at $3.99 a pound

   b. 10 dimes

   c. 5 nickels

   d. 8 quarters

   e. A roll of quarters is worth $10. How many quarters does it contain?

   f. You have 100 quarters in change after an event. How many dollars and cents is that?

PAGE 133

3. **Cooking** — If you usually use 10 apples to make enough apple pie for 5 people, how many apples will you need to use to make pie for 36 people?

4. **Gallon Fun** — If you're supposed to put $\frac{3}{4}$ cup of a solution in 1 gallon of water, how much should you put in if you want to make 78 gallons? Give your answer as a fraction.

5. **Skill Sharpening** (🖩)

   a. $\frac{1}{2} + \frac{8}{25} \div \frac{4}{50}$

   b. Find the greatest common factor of 34 and 62.

   c. Find the least common multiple of 34 and 62.

   d. Solve, rounding the answer to the nearest whole number: $\frac{78}{100} \cdot 5.5$

CHAPTER 8. Ratios and Proportions
LESSON 6. Chapter Synopsis
Worksheet 8.6

You may use a calculator on this worksheet whenever you see this symbol (🖩).

1. **Making Change** — Figure out mentally how much change you should receive in each of the following situations (both the amount and the specific bills and coins). Use the fewest number of coins you can.

   a. Total = $5.87   Give = $6   Change =   Bills and Coins =

   b. Total = $5.87   Give = $6.02   Change =   Bills and Coins =

   c. Total = $57.83   Give = $60   Change =   Bills and Coins =

   d. Total = $57.83   Give = $60.08   Change =   Bills and Coins =

   e. Total = $1.50 • 3   Give = $5   Change =   Bills and Coins =

2. **Ratios**

   a. Express 1 inch per 8 squares as a ratio.

   b. Convert the ratio in 2a to a decimal number.

3. **Proportions** (🖩) — Write a proportion for each of the following situations and solve. Convert any fractions to decimals.

   a. If a yield of potatoes is 400 bushels per acre, what is the yield for $2\frac{1}{2}$ acres?

   b. If a yield of potatoes is 400.23 bushels per acre, what is the yield for 0.5 of an acre?

   c. If it costs $4.95 for 250 yards of ribbon, what would the cost be for 750 yards?

4. **Making Business Decisions**

   a. If you need 3.5 yards of ribbon to make one wreath and you want to make 26 wreaths to sell at a fundraiser, how many yards of ribbon do you need?

   b. If each yard of ribbon costs $0.99, plus a total of $7.50 shipping for an unlimited number of yards, how much will it cost you to make the wreaths described in 4a?

   c. You have a $5\frac{5}{6}$ acre section of land and would like to plant half of it with apple trees and half with peach trees. If you can fit 40 apple trees per acre, how many apple trees can you plant?

   d. If each tree costs $2.50, how much will it cost you in apple trees if you plant the number you found in 4c?

   e. You're trying to understand a proposed barn blueprint design. The paper uses a 1 inch:6 feet scale. A window is shown in the design 1.5 inches from the door. How far away would that be in real life?

5. **Proportional** — Do these ratios form a proportion?

   $\frac{2 \text{ cm}}{3 \text{ cm}}$ $\frac{20 \text{ cm}}{35 \text{ cm}}$

PRINCIPLES OF MATHEMATICS

CHAPTER 9. Percents
LESSON 1. Introducing Percents

Worksheet 9.1

You may use a calculator on this worksheet whenever you see this symbol (📱).

1. **Fill in the Blank** — Percents are a shorthand way to express ratios as fractions of _____.

2. **Different Ways of Expression** (📱) — Express these rates as fractions, decimal numbers, and percents. You do not need to simplify your fractions.

   a. 4 per 20

   b. 2 per 10

   c. 8 per 80

   d. 70 per 250

   e. 200 per 100

3. **Conversion** — Express the following as percents.

   a. $\frac{4}{25}$ b. 0.89

   c. 0.02   d. 1.25

   e. 0.42   f. $\frac{96}{100}$

PAGE 137

4. **Conversion** — Express the following as fractions in simplest terms.
   a. 25%
   b. 69%

   c. 5%
   d. 72%

   e. 378%
   f. 6%

5. **Conversion** — Express the percents in problem 4 as decimals.

6. **True or False** — If a state charges a flat 10% income tax, the equivalent of 10 out of every 100 taxable dollars made goes to the state.

7. **Coupons** — If you had a coupon for 20% off an item, what would that mean?

8. **Castings** — If 84 pounds of metal are used in making 14 castings, how many pounds of metal will be used for 9 castings? Round your answer to the nearest whole number.

9. **Making Change** — Figure out mentally how much change (both the amount and the specific coins) you should receive in each of the following situations. Use the fewest number of coins you can.
   a. Total = $5.86   Give = $6   Change =   Bills and Coins =
   b. Total = $5.86   Give = $5.91   Change =   Bills and Coins =
   c. Total = $5.86   Give = $6.01   Change =   Bills and Coins =

PAGE 138

	CHAPTER 9. Percents	Worksheet
PRINCIPLES OF MATHEMATICS	LESSON 2. Finding Percentages	9.2

You may use a calculator on this worksheet whenever you see this symbol (🖩).

1. **Finding Percentages** (🖩) — Use a proportion to solve.

   Example: What is 15% of 60?
   $$\frac{15}{100} = \frac{?}{60}$$
   Since 60 ÷ 100 = 0.6, we need to multiply the first ratio by $\frac{0.6}{0.6}$ to find an equivalent ratio with 60 in the denominator.
   $$\frac{15}{100} \cdot \frac{0.6}{0.6} = \frac{9}{60}$$
   9 is 15% of 60.

   a. What is 50% of 250?

   b. What is 25% of 75?

   c. What is 20% of 120?

   d. What is 10% of 420?

   e. What is 3% of 60?

   f. What is 1% of 25?

PAGE 139

2. **Expression** — Express this ratio as a fraction (you do not need to simplify), decimal, and percent: 10 out of 20 people.

3. **Tip** (📱) — Your bill at the restaurant totals $28.45. How much tip should you leave if you want to leave a 20% tip? Use a proportion to find the answer.

4. **Shopping** (📱) — If you use a 30% coupon on a $40 item, how much will you get off of the $40? Use a proportion to find the answer.

5. **Term Time** — In problem 4,

    a. what was your base?

    b. what was your rate?

6. **Conversion** — Convert the following to percents.

    a. 0.32

    b. 0.80

7. **Cooking** — If a recipe calls for $\frac{2}{3}$ cup of sugar, and you put in $\frac{1}{4}$ a cup instead, by how much of a cup did you reduce the sugar?

8. **Making Change** — Figure out mentally how much change you should receive in each of the following situations (both the amount and the specific coins). Use fewest number of coins you can.

    a. Total = $2.52   Give = $3   Change = _____   Bills and Coins = _____
    b. Total = $2.52   Give = $3.02   Change = _____   Bills and Coins = _____
    c. Total = $2.52   Give = $2.62   Change = _____   Bills and Coins = _____

PAGE 140

PRINCIPLES OF MATHEMATICS

CHAPTER 9. Percents
LESSON 3. More Finding Percentages / Multiplying and Dividing Percents

Worksheet 9.3

You may use a calculator on this worksheet whenever you see this symbol (🖩).

1. **More Percents** (🖩) — Solve by multiplying the rate by the base (*Rate • Base = Percentage*). Show your work.

   a. What is 25% of 340?

   b. What is 12% of $500?

   c. What is 15% of 70?

2. **Term Time** — In problem 1b,

   a. what was your base?

   b. what was your rate?

3. **Finding 10%** — Solve by multiplying the rate by the base (*Rate • Base = Percentage*). Show your work.

   a. What is 10% of 12?

   b. What is 10% of 100?

   c. What is 10% of 220?

   d. What is 10% of 8?

   e. What is 10% of 78?

   f. Look at the answers above. Can you think of an easy way to multiply by 10%?

4. **Conversion** — Convert the following to percents. Remember, all you need to do to convert a percent to a decimal is to multiply by 100.

   a. 1.45

   b. 2.75

5. **Commission** — A certain car salesman receives a 5% commission on all sales, meaning he gets paid 5% of everything he sells. How much would he get paid if he sells a car for $12,000?

6. **Nutritional Data** — If you're on a diet that only allows 2,500 calories a day, and you want to intake your calories evenly over three meals, how many calories should you eat at each meal? *Hint*: You want to eat $\frac{1}{3}$ of your calories at each meal, so find $\frac{1}{3}$ of 2,500.

7. **Making Change** — Figure out mentally how much change you should receive in each of the following situations (both the amount and the specific coins). Use the fewest number of coins you can.
   a. Total = $10.77   Give = $11   Change =   Bills and Coins =
   b. Total = $10.77   Give = $10.80   Change =   Bills and Coins =
   c. Total = $10.77   Give = $10.82   Change =   Bills and Coins =

8. **Proportion Time** — If 4 out of 5 customers want extended hours, and there are 569 customers, how many want extended hours? Round your answer to the nearest whole number.

9. **Notebook** — Copy *Rate • Base = Percentage* into your math notebook.

PRINCIPLES OF MATHEMATICS

CHAPTER 9. Percents
LESSON 4. Adding and Subtracting Percents / Finding Totals

Worksheet 9.4

You may use a calculator on this worksheet whenever you see this symbol (🖩).

1. **Adding and Subtracting Percents** — Solve.

    a. 15% + 69%

    b. 2% + 7%

    c. 100% − 13%

2. **At the Store** (🖩) — Answer these questions by finding the discount amount and then subtracting that from or adding it to the original cost. Show your work.

    Example: If you use a 5% off coupon to buy a $50 item, how much will you end up paying?

    Step 1: 0.05 • $50 = $2.50

    Step 2: $50 − $2.50 = $47.50

    a. If you use a 10% coupon to buy a $43.89 item, how much will you end up paying?

    b. If you leave a 20% tip on top of a $55.67 dinner bill, how much did you spend altogether?

    c. If you buy a picture for $52.67 and have a 7% tax added to it, how much will it cost altogether?

3. **At the Store** (🖩) — Answer these questions by first adding together the percentages you need to find.

    Example: If you use a 5% off coupon to buy a $50 item, how much will you end up paying?

    Step 1: 100% (the full price of the item) − 5% (the discount) = 95%

    Step 2: 0.95 • $50 = $47.50

    a. If you use a 20% off coupon to buy a $34.80 item, how much will you end up paying?

PAGE 143

b. If you use a 25% coupon to buy a $14.99 item, how much will you end up paying?

c. If you buy glasses for $42 and have a 10% tax added to it, how much will you end up paying?

d. If you buy dishes for $44.56 and have a 6% tax added to it, how much will you end up paying?

4. **Water** — Water is made up entirely of hydrogen and oxygen. It is approximately 67% hydrogen;* what percent of it is oxygen? *Hint*: Use what you learned about subtracting percents. We could rephrase this question like this: 100% of water is made up of hydrogen and oxygen. So since approximately 67% is hydrogen, how much is oxygen?

5. **Per Mil ( )** — Knowing that *percent* means "per hundred" and that *cent* means "hundred" and *mil* means "thousand", any guesses what *per mil* means? It means "per thousand"! The alcohol content in blood, the salinity in sea salt, and most property taxes,[1] are measured in *per mil* (amount per thousand) rather than in *percent* (amount per hundred). While the symbol for percent has two zeros (% — 100 also has two zeros), the symbol for *per mil* has three zeros (‰ — 1,000 also has three zeros).

See if you can adapt what you know about percents to solve these problems involving per mil. *Hint*: Think about what you would do if it were a percent. Do the same thing, except remember that the denominator of a per mil is 1,000 rather than 100. So when you convert to a decimal, divide the numerator by 1,000 — i.e., move the decimal place over three digits instead of two.

a. What is 8‰ of 2,000?

b. If your property tax is computed at 5‰, how much tax will you owe on a house valued at $325,000?

---

[1] See Wikipedia, s.v. "per mille," for a more complete list of per mil applications. http://en.wikipedia.org/wiki/Per_mil.

| PRINCIPLES OF MATHEMATICS | CHAPTER 9. Percents
LESSON 5. Mental Percents | Worksheet 9.5 |

1. **Approximate Answers** — Pretend these numbers are your total bills at a restaurant. Find the approximate tip you should leave using the mental math process outlined in Lesson 9.5. All steps should be done mentally — only write down the final answer.

    Example: 20% of 130.50

      Step 1: $130.50 rounded to the nearest dollar is $131.

      Step 2: 10% of $131 is $13.10, which rounds to $13.

      Step 3: 20% will be twice 10%; 2 times $13 is $26, so $26 is approximately 20% of $130.50.

    a. 20% of $18.80

    b. 5% of $10.25 *Hint*: 5% is half 10%. (You might want to rethink only leaving a 5% tip!)

    c. 20% of $45.87

    d. 15% of $89.75

2. **Conversion** — Convert the following decimal numbers to percents.

    a. 0.15

    b. 0.05

    c. 1.25

3. **Out of the Box** — Pull out a newspaper (visit a library or look online if you don't have one) and record one use of percents that you find.

4. **Making Change** — Figure out mentally how much change you should receive in each of the following situations (both the amount and the specific coins). Use the fewest number of coins you can.

   a. Total = $19.11   Give = $20   Change =   Bills and Coins =

   b. Total = $19.11   Give = $20.11   Change =   Bills and Coins =

   c. Total = $19.11   Give = $20.01   Change =   Bills and Coins =

5. **Cooking Time** — If you need $2\frac{1}{4}$ c flour for every batch of cookies, how many cups do you need to make $3\frac{1}{2}$ batches?

CHAPTER 10. Negative Numbers
LESSON 1. Understanding Negative Numbers

Worksheet 10.1

PRINCIPLES OF MATHEMATICS

1. **Number Lines** — Find each of these numbers on the number line. Put a dot marking them.

   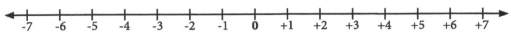

   a. –3
   b. 4
   c. –7
   d. 1.5
   e. –2.5
   f. $-\frac{3}{4}$
   g. -100% *Hint*: Convert the percent to a decimal to see where it fits on the number line.

2. **More Number Lines** — Draw a number line showing –3, –0.25%, and $-3\frac{1}{2}$.

3. **Represent the Following** — Use either positive or negative numbers to represent the following quantities.

   a. $10.50 you owe someone else

   b. $5 you have in your wallet

   c. By analyzing different sound waves, we can learn about and duplicate sounds.

   This graph shows a sound wave. The numbers on the left of the graph tell you the values for that horizontal line. The topmost point the sound reaches to is 40. What is the lowest point the sound wave reaches to?

   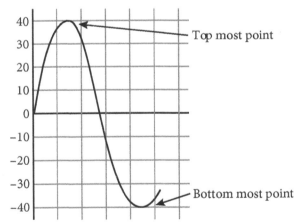

   d. Did you know numbers can help us describe the beating of a heart? Well, they can! If you've ever been to the hospital and seen heart monitors, you may have noticed the lines going up and down as the heart beat. Can you guess what type of numbers (positive or negative) we would use to describe the section of the graph the arrow points to?

   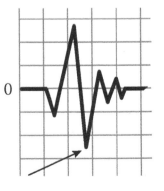

PAGE 147

e. To differentiate money made versus lost, we can use negative numbers (positive numbers would mean a gain; negative numbers would mean a loss). The chart shows the money a company made or lost over time.

To what value is the arrow pointing?

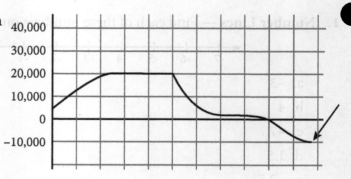

4. **Fertilizer** — A package of fertilizer says it contains 20% nitrogen. If it weighs 5 pounds, how many pounds of nitrogen does it contain?

5. **Nutritional Percents** — If a certain food contains 5% of your daily suggested sodium intake, and your daily suggested sodium intake is the standard recommended 2,400 milligrams,[1] how many milligrams of sodium does the food contain?

6. **Mental Percents** — Find 20% of $78.56 mentally. Round $78.56 to the nearest whole number.

---

1 Department of Health and Human Services, "Code of Federal Regulations Title 21," U.S. Food and Drug Administration, http://www.accessdata.fda.gov/scripts/cdrh/cfdocs/cfcfr/cfrsearch.cfm?fr=101.9 (10/1/14). Quoted on Netrition, "Reference Values for Nutrition Labeling," netrition.com, http://www.netrition.com/rdi_page.html (accessed 10/01/14).

CHAPTER 10. Negative Numbers
LESSON 2. Adding Negative Numbers
Worksheet 10.2A

1. **Number Lines** — Find each of these numbers on the number line. Put a dot marking them.

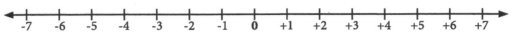

   a. −6.5
   b. 6.5
   c. −4
   d. 5.5
   e. −3
   f. −25%

2. **Subtraction** — Solve each of these problems by viewing the subtraction as adding by a negative number. Draw a number line to help you if necessary.

   a. 7 + −3

   b. 10 + −2

   c. 7 + −18

   d. 10 + −12

   e. (2) + (−7)

   f. −9 + 5

   g. 7 − 10

   h. (−5) + (−2)

3. **Additive Inverse** — Write the additive inverse of each of these numbers (i.e., the number you'd have to add to get to 0).

   a. 7

   b. −6

   c. −0.5

   d. $\frac{3}{4}$

PAGE 149

4. **Amount Owed**

   Suppose you're out at the store and want to buy a $10 item, but only have $7. Your mom agrees to loan you the remaining amount until you get home. How much money do you owe?

   Example: If you want to buy a $12 item but only have $5, how much would you have to borrow from a family member until you got home?

   $5 - $12 = -$7. You would have to borrow $7.

   Note that we typically express the final amount as an amount owed as opposed to -$7; negative numbers help us do the math, but we don't typically use them in everyday speech. So express your final answer as an amount owed, but show how you found it using negative numbers.

PRINCIPLES OF MATHEMATICS

CHAPTER 10. Negative Numbers
LESSON 2. Adding Negative Numbers
Worksheet 10.2B

In this worksheet, you do not need to simply fractions. You should, however, rewrite fractions containing decimals so as to remove the decimal. Example: Rewrite $\frac{4.6}{10}$ as $\frac{46}{100}$.

When given a percent with a decimal in it as part of the problem, do not round it. For example, if told to find a 7.5% tax, use 0.075 to calculate the tax.

You may use a calculator on this worksheet whenever you see this symbol (🖩).

1. **Negative Number Practice**

   a. Represent a negative 5 electrical charge

   b. Solve: 5 mi + −12 mi

   c. Solve: (17 in) + (−20 in)

   d. Solve: 4 − 10

   e. Solve: (−3) + (−5)

   f. What is the additive inverse of 3?

   g. $\frac{3}{5} - \frac{4}{5}$

   h. $\frac{1}{8} + -\frac{3}{8}$

   i. $\frac{1}{4} - \frac{3}{4}$

2. **Traveling** — Suppose you travel 21 miles east before you realize you're going the wrong direction! You turn around and head 51 miles to the west. How much further west are you from your starting point? *Hint*: Treat east as positive and west as negative and find the difference. Show your work, but express your final answer in words (the distance west you went).

3. **Decimals and Fractions Inside Percents** (🖩) — Do not round or simplify.

   Hint: Don't let the decimal in the percent confuse you. Remember that percent means "by a hundred,"[1] so just divide the percent by a hundred to convert it to a fraction or a decimal. 56.5% means $\frac{56.5}{100}$, or 0.565.

   However, to make the fraction easier to read, rewrite the fraction using an equivalent fraction so as to remove the decimal. For example, $\frac{56.5}{100}$ is easier to read as $\frac{565}{1,000}$. They both mean the same thing ($\frac{56.5}{100} \cdot \frac{10}{10} = \frac{565}{1,000}$).

   a. Convert 45.2% to a decimal and a fraction.

   b. Convert $31\frac{1}{4}$% to a decimal and a fraction. (View $\frac{1}{4}$ as 0.25)

   c. Convert 168% to a decimal and a fraction.

   d. Convert 168.42% to a decimal and a fraction.

   e. What is $56\frac{1}{4}$% of 75?

---

[1] See *The American Heritage Dictionary of the English Language*, 1980 New College Edition, s.v. "per cent."

PRINCIPLES OF MATHEMATICS

CHAPTER 10. Negative Numbers
LESSON 3. Subtracting Negative Numbers
Worksheet 10.3

1. **Recognizing the Opposite** — Train yourself to view each negative sign as *the opposite of* by rewriting what it does to the number. The first one makes the number negative, the second positive, the third negative, etc. Then simplify each number to show whether in the end it's positive or negative.

   Example: – – –3

   negative, positive, negative; –3

   a. –6.25

   b. – – –6.25

   c. – – – –6.25

2. **Subtraction** — Solve.

   Example: 10 + – –4

   Since – –4 = +4, this simplifies to 10 + 4, which equals 14.

   a. 8 + – –6
   b. 5 + – –2
   c. 5 + – – –2
   d. 6 + – –$\frac{1}{2}$

3. **Skill Sharpening**

   45.5(1 + 3.2)

4. **Total** — If you buy items that cost $3.50, $4.50, and $7.75 in a state that charges 6% sales tax, what will your total with tax be?

5. **Mental Percents**
   a. Find 20% of $25.67 mentally. Round $25.67 to the nearest dollar.

   b. Solve mentally: What is 40% of 80?

6. **Complete the sentence —** A negative sign means _____.

CHAPTER 10. Negative Numbers
LESSON 4. Temperature and Negative Numbers

Worksheet 10.4

1. **Temperature Time** — List the temperature shown on each thermometer below. Give your answer in both Fahrenheit and Celsius.

    a. 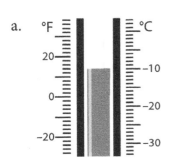   b. 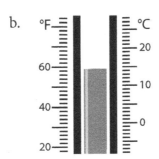   c.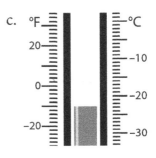

2. **More Temperature Time**

    a. If you heat a solution to 165° F, and then let it cool 60° F, what will the ending temperature be?

    b. If you heat a solution to 32° F, and then cool it 40° F, what will the ending temperature be?

    c. If you cool a solution to 0° C (freezing point of water), and then cool it an additional 20° C, what will its ending temperature be?

3. **Skill Sharpening**

    a. $4 + 10 \div 2$

    b. Convert 0.46 to a percent.

    c. Convert 0.467 to a percent. *Hint*: Remember, you can have decimals in a percent—a percent just means "by a hundred." To convert a decimal to a percent, just multiply by 100 and add a percent sign—even if you still have a decimal. *Example*: 0.789 = 78.9%

    d. Convert 1.325 to a percent.

    e. Convert $32\frac{1}{2}$% to a decimal and a fraction of 1,000. Do not round or simplify.

    f. What is $32\frac{1}{2}$% of 60?

    g. Convert 25.12% to a decimal and a fraction of 10,000. Do not round or simplify.

h. If the blue square is 25.12% shorter than the red square, and the red square is 5 inches long, how long is the blue square? *Hint*: This is a multistep problem.

i. Convert 5.5% to a decimal and a fraction of 1,000. Do not round or simplify.

j. If 5.5% sales tax is added to a $10 purchase, what will the new total be?

k. Solve mentally: 5% of $48.78. Round $48.78 to the nearest dollar.

l. $\frac{1}{6} - \frac{4}{6}$

m. $(\frac{5}{12}) + (-\frac{7}{12})$

n. What is − −5?

4. **Question** — List one thing about Galileo covered in Lesson 10.4.

PRINCIPLES OF MATHEMATICS

CHAPTER 10. Negative Numbers
LESSON 5. Absolute Value

Worksheet 10.5

1. **Absolute Value** — What is the absolute value of these numbers? Remember, absolute value means the distance from 0, regardless of the direction.

   a. $|-3|$

   b. $|-14.5|$

   c. $|20|$

2. **Distance and Temperature**

   a. If a building is 15 miles east of the highway and your home is 10 miles west of that same highway, how many miles apart are your home and the building? *Hint*: Find the answer by subtracting the two distances (using negative to represent west) and then taking the absolute value.

   b. If the temperature for the day starts at –5° F and increases to 25° F, by how many degrees did it increase over the day? *Hint*: You're interested in the absolute value of the change!

   c. If the temperature for a solution starts at –15° F and is heated to 75° F, by how many degrees did its temperature change?

   d. If you travel 10 miles in the positive direction and then turn around and go 20 miles in the negative direction, where do you end up? *Hint*: Don't just assume you solve a problem the same way as the previous ones, really think through what you're trying to find. Draw it out if you need to.

PAGE 157

3. **Skill Sharpening**

   a. Solve mentally: What is 30% of 125?

   b. What is 3.2% of 25.5?

   c. 45(8 + 2) + 20 ÷ 2

   d. $-\frac{1}{3} - \frac{2}{3}$

   e. $(\frac{1}{20}) + (-\frac{3}{20})$

   f. What is – –7?

CHAPTER 10. Negative Numbers
LESSON 6. Multiplying and Dividing Negative Numbers
Worksheet 10.6

PRINCIPLES OF MATHEMATICS

1. **Multiplying Negative Numbers** — Find the answer to these problems, noticing the example meanings.

   a. $5 \cdot -20$

   Example Meaning: If I am traveling in the negative direction (west) at 20 miles an hour, then in 5 hours I will be $5 \cdot -20$ miles from where I am now.

   b. $-5 \cdot -20$

   Example Meaning: If I have been traveling for the past 5 hours in the negative direction (west) at a rate of 20 miles an hour, then 5 hours ago, I was $-5 \cdot -20$ miles from where I am now.

   c. $\frac{1}{2} \cdot -3$

   Example Meaning: If I am traveling in the negative direction (west) at 3 miles an hour, then in $\frac{1}{2}$ an hour, I will be $\frac{1}{2} \cdot -3$ miles from where I am now.

   d. $-\frac{1}{2} \cdot -3$

   Example Meaning: If I have been traveling for the past $\frac{1}{2}$ hour in the negative direction (west) at a rate of 3 miles an hour, then $\frac{1}{2}$ hour ago, I was $-\frac{1}{2} \cdot -3$ miles from where I am now.

   e. $3 \cdot -2$

   Example Meaning: If we add 3 particles, each of which has a -2 electrical charge, to an object, the resulting charge of the object will change by $3 \cdot -2$.

   h. $-3 \cdot -2$

   Example Meaning: If we take away 3 particles, each of which has a -2 electrical charge, from an object, the resulting charge of the object will change by $-3 \cdot -2$.

2. **Dividing Negative Numbers** — Find the answer to these problems, noticing the example meanings.

   a. $(-4) \div (2)$

   Example Meaning: If we travel 4 miles in the negative direction in 2 minutes, how far are we traveling each minute?

   b. $(-40) \div (-20)$

   Example Meaning: If we set up a ratio comparing an average temperature of -40 degrees with an average temperature of -20 degrees, we'd have -40 divided by -20.

PAGE 159

c. $(-16) \div (4)$

Example Meaning: If a temperature is cooling 16 degrees in 4 minutes, how much is it cooling each minute?

3. **Digging Deeper** — Sometimes we end up with a crazy number of negative signs in a problem. Don't let the number of negative signs fool you! Each just means *the opposite of*. Think in terms of negative, positive, negative, positive, etc., counting the positive and negative signs to determine if the ending number is positive or negative.

   a. $-----5$

   b. $-1 \cdot -1 \cdot -1 \cdot -1 \cdot -1 \cdot 5$

   c. $-2 \cdot -4 \cdot -3$

   d. $-2 \cdot -2 \cdot -2$

4. **Skill Sharpening**
   a. $4 - 7$ *Hint*: Remember, you can think of this as $4 + -7$.

   b. $-12 + 8$

   c. $-89 + 64$

   d. $\frac{1}{15} - \frac{6}{15}$

   e. $\left|-\frac{1}{3}\right|$

5. **Fill in the blank** — Each negative sign means _____.

PRINCIPLES OF MATHEMATICS

CHAPTER 10. Negative Numbers
LESSON 7. Negative Mixed Numbers

Worksheet 10.7

1. **Fun with Force 1** — Be sure to read the example of what the problem could represent—you'll be asked to solve similar problems next!

   a. $(2\frac{3}{4}) + (-5\frac{1}{4})$

   Example Meaning: A machine has $2\frac{3}{4}$ pounds of force in the positive direction and $5\frac{1}{4}$ pounds in the negative. What is the resulting force on the machine?

   b. $5\frac{2}{5} - 7\frac{1}{5}$

   Example Meaning: A machine has $5\frac{2}{5}$ pounds of force in the positive direction and $7\frac{1}{5}$ pounds in the negative. What is the resulting force on the machine?

2. **Fun with Force 2**

   a. If $2\frac{1}{3}$ pounds (LB) of force is applied to an object in the positive direction, and 4 pounds of force in the negative direction, what is the resulting force on the object? *Hint*: Add the two forces together.

   b. Let's say you tie a $3\frac{1}{3}$-ounce weight to a helium balloon. If the helium in the balloon has an upward pull (force) of 4 ounces, what is the resulting force on the balloon? *Hint*: The $3\frac{1}{3}$-ounce weight would be pulling the balloon in the negative direction—it's a downward pull (force) of $3\frac{1}{3}$ ounces.

3. **Skill Sharpening**
   a. $5 \cdot -2$

   b. $-7 \cdot -5$

   c. $-3 \cdot 3 \cdot -3$

   d. $-3 \cdot -3 \cdot -3$

   e. $-20 \div 5$

   f. $-20 \div -5$

   g. $|-5|$

4. **Sales Tax** — If a state charges 4.5% sales tax, what will the total with tax be if the pre-tax total is $45?

PRINCIPLES OF MATHEMATICS

CHAPTER 10. Negative Numbers
LESSON 8. Negative Fractions

Worksheet 10.8

1. **Understanding Division as Fractions** — Write a fraction to describe each of the following situations.

   a. Negative fifty miles divided by 1 hour
   Example Meaning: How far a car traveled in 1 hour?

   b. Negative two hundred miles divided by an 4 hours
   Example Meaning: How far a car traveled in four hours?

   c. One day divided by 30 days
   Example Meaning: What portion of a month is a day?

   d. $780 divided by 4 weeks
   Example Meaning: The cost per week if it costs $780 for 4 weeks?

2. **Mastering the Skill** — Simplify these expressions to a single number or a mixed number.

   a. $\dfrac{-80}{40}$

   b. $70 + -\dfrac{-120}{40}$

   c. $60 + -\dfrac{30}{-10}$

   d. $40 + -\dfrac{-16}{-8}$

PAGE 163

3. **Skill Sharpening**

    a. $-8 \cdot -4$

    b. $-5 \cdot 2$

    c. $-8 + 5$

    d. $\frac{1}{4} - \frac{3}{4}$

    e. $-30 \div 3$

    f. $-30 \div -3$

    g. $\left|-\frac{5}{6}\right|$

4. **Number Lines** — Find each of these numbers on the number line. Put a dot marking them.

    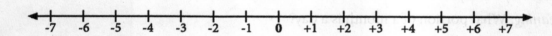

    a. $-2\frac{2}{3}$

    b. $\left|-2\frac{2}{3}\right|$

    c. -5

    d. -50%

5. **Tipping** — If your total bill was $45.67, how much of a tip should you leave if you wanted it to be 20%? Round your total bill to the nearest dollar and find the tip amount mentally.

PAGE 164

PRINCIPLES OF MATHEMATICS

CHAPTER 11. Sets
LESSON 1. Sets and Venn Diagrams
Worksheet 11.1

1. **Understanding Venn Diagrams**

    a. What does this Venn Diagram tell you about how the set "Dinner" relates to the set "Recipes"?

    b. Are any subsets of "Recipes" shown? If so, what are their names?

    c. What does this Venn Diagram tell you about how the set "Dinner" relates to "Crowd Pleasers"?

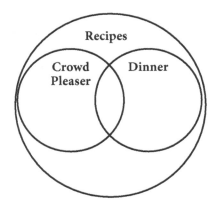

2. **Diagrams and Sets** — Using the diagram, list *all* the sets and subsets each book belongs to.[1] Remember, if a circle is nested within another circle, that set is a subset of the bigger circle.

    Example: *Growing Tomatoes*

    Answer: Books, Nonfiction, Gardening, and Vegetable Gardening

    Note: We could have organized the books by author, by type (paperback or hardback), or by some other characteristic. By topic is the method used for nonfiction at libraries because that's how people are typically trying to find the book. It's easier to find a paperback by the author's name, but you look for a nonfiction book by the topic.

    a. *Pilgrim's Progress*

    b. *How to Use Your Computer's Software*

    c. *Guide to Growing Vegetables*

    d. *Flower Growth*

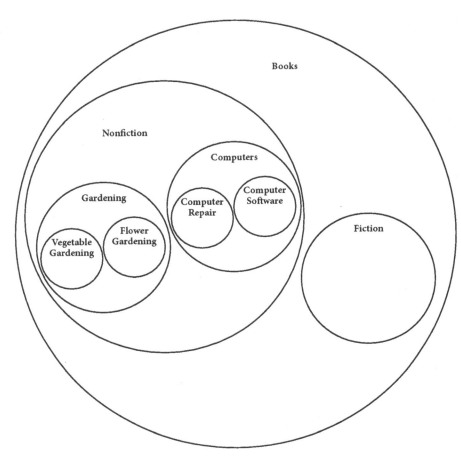

---

[1] All but *Pilgrim's Progress* are made up titles. Any resemblance to actual books is accidental.

PAGE 165

3. **Drawing a Venn Diagram** — Draw a Venn Diagram showing how a set of all foods containing "Chocolate" would relate to a set of all "Ice Cream" flavors.

4. **Grouping** — Let's say you're writing a report on animals and trying to figure out how to organize it. Start by drawing an arrow from each of the following animals to all the sets that describe its characteristics (you can decide what makes an animal a pet, farm animal, etc.). Notice that the same animal can go in more than one set. Which sets you use in your report would depend on the point you were trying to make.

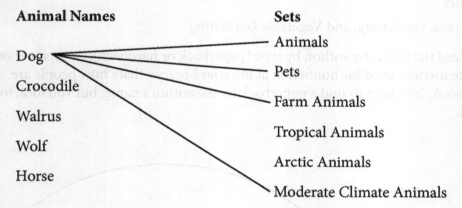

5. **Skill Sharpening**

   a. $\dfrac{-36}{-18}$

   b. $\dfrac{-8}{2} + \dfrac{-9}{3}$

   c. $1\dfrac{2}{8} - 7\dfrac{7}{8}$

   d. $4 \cdot \dfrac{-3}{-1}$

PRINCIPLES OF MATHEMATICS

CHAPTER 11. Sets
LESSON 2. Number Sets

Worksheet 11.2

You may use a calculator on this worksheet whenever you see this symbol (🖩).

1. **Putting Numbers into Sets** — Circle all the sets listed that can be used to describe the given number.

    a. 8

            Whole Numbers   Prime Numbers   Even Integers

    b. –7

            Whole Numbers   Odd Integers   Negative Numbers

    c. 7.5

            Whole Numbers   Positive Numbers   Integers

    d. $-\frac{4}{5}$

            Odd Integers   Whole Numbers   Negative Numbers

2. **Flashcards** — Make flashcards for each of the different number sets mentioned in Lesson 11.2. It's important to get familiar with them so you'll recognize the terminology when you see it.

3. **Skill Sharpening**

    a. $-7 + -11$

    b. $-2 \cdot -5$

    c. $-2 \cdot 5$

    d. $25 \div -5$

    e. $\frac{-50}{-5}$

    f. $2\frac{3}{12} + -5\frac{4}{12}$

4. **Percent Time** (🖩) — If you purchase items that total $87.45 at a 3.5% tax rate, what will your total with tax be?

PAGE 167

PRINCIPLES OF MATHEMATICS

CHAPTER 11. Sets
LESSON 3. More on Sets
Worksheet 11.3

1. **Expressing Specific Sets** — Represent the following as a set using >, <, =, or ≠.

    Example: Buying a gift for less than $15
    {< $15}

    a. Spending less than $20 at the store

    b. Any number *but* 9

2. **Understanding the Signs**

    a. Is 5 a part of this set?
    {1, 2, 3 … 10}

    b. Is 9 a part of this set?
    {2, -9, 6, 1.5}

    c. Can sets be random?

3. **Organizing the Silverware** — What sets, or groupings, does your family use to organize your silverware? (big forks, small forks, etc.) List all the sets, and then draw a Venn Diagram showing them all as subsets of the set of silverware.

4. **Skill Sharpening**

    a. $(-2)(-2)$

    b. $\dfrac{6}{-2}$

    c. $20 + - -3$

    d. $-15 + 8$

5. **Flashcards** — Review your set flashcards.

PAGE 169

6. **Percent Time**

   a. If you buy a scarf for $5.99 plus 7.5% tax, how much will you end up paying?

   b. On a $28.89 dinner bill, how much tip should you leave to leave approximately 20%? Solve mentally; round $28.89 to the nearest dollar before finding.

7. **Question** — What is one way sets are used in programming?

CHAPTER 11. Sets
LESSON 4. Ordered Sets (i.e., Sequences)
Worksheet 11.4

You may use a calculator on this worksheet whenever you see this symbol (🖩).

1. **Finish the Sequence** — Find the next two numbers in these sequences, assuming the pattern continues. If you get stumped, write down the difference between each number and look for a pattern.

    a. {3, 6, 9, ___, ___, ...}
    b. {9, 7, 5, 3, 1, –1, –3, ___, ___, ...}
    c. {1, 2, 4, 7, 11, ___, ___, ...}
    d. {7, 9, 11, 16, 18, 20, 25, ___, ___, ...}
    e. {8, 16, 32, 64, _____, _____, ...}
    f. {–2, –4, –6, –8, _____, _____, ...}

2. **Common Differences and Common Ratios**

    a. In problem 1, which sequence(s) has a common difference? What is that difference?

    b. In problem 1, which sequence(s) has a common ratio? What is that ratio?

3. **The Golden Ratio** (🖩)

    {0, 1, 1, 2, 3, 5, 8, 13, 21, 34, 55, 89...}

    Starting with 2, use a calculator to divide each number in the sequence above (the Fibonacci Sequence) by the number before it. As the numbers get larger, what number do you start to consistently get (round to the nearest hundredth)?

    You should have noticed that the ratio, or relationship expressed as division, between each of the numbers in this sequence is approximately the same, especially as the numbers get higher. We call this common ratio the **Golden Ratio**.

    Back in 8.1, you explored sunflowers and discovered that the different spirals in a sunflower relate together according to this ratio (and that this ratio allows for them to fit the most number of seeds in any given sunflower). We also discussed then that rectangles built with this ratio are used extensively in art and architecture due to how visually appealing they are.

    The Fibonacci Sequence is just a useful way of describing these numbers that we find all over creation—numbers that testify to God's caring hand. He thought of everything—down to the way spirals relate in a sunflower and pineapple. All throughout creation we see evidence of God's care.

4. **Out of the Box: Counting Change** — Round up all the change you can find and put it in piles of pennies, nickels, dimes, and quarters. Then find the total, counting the nickels by 5s (5, 10, 15, ...), the dimes by 10s, and the quarters by 25. When you count by 5s, 10s, and 25s, you are really naming numbers in sequences!

PAGE 171

5. **Skill Sharpening**

   a. $3 \div -1$

   b. $\dfrac{-8}{2} + \dfrac{-6}{-3}$

6. **Flashcards** — Review your set flashcards.

PRINCIPLES OF MATHEMATICS

Review of Chapters 7–11 — Worksheet 11.5

Use these problems to help you review. If you have questions, review the concept in the *Student Textbook*. The lesson numbers in which the various concepts were taught are in parentheses.

You may use a calculator on this worksheet whenever you see this symbol (🖩).

1. **Shopping Time** (🖩)

    a. If you buy 5 packages of noodles at $1.48 each, what's the total? (Lesson 7.3)

    b. If a 5.5% tax is added to the total from problem 1a, what would be the new total? (Lesson 9.4)

    c. If you're interested in buying a scarf that costs $14.95 and a shirt that costs $8.99, and you have a 15% off coupon, what will your adjusted total be? (Lesson 9.4)

    d. If you give $21 for the purchase in problem 1c, how much change should you get? (Lesson 8.5)

    e. If you can buy 5 pieces of candy for $2.50, how much will it cost for 18 pieces of candy, assuming the candy is sold by the piece? (Lesson 8.2)

    f. If you owe your mom $20 you borrowed at a store, how could you represent what you owe her using negative numbers? (Lesson 10.1)

    g. If one store is 20 miles in the positive direction and another is $10\frac{1}{3}$ miles in the positive direction, how far apart are the two stores? (Lesson 10.5)

    h. If one store is 20 miles in the positive direction and another is $10\frac{1}{3}$ miles in the negative direction, how far apart are the two stores? (Lesson 10.5)

2. **More Shopping**

    a. If you buy 20 pounds of wheat for $5.59, how much does each pound cost? (Lesson 7.4)

    b. If you start your shopping trip with $21.31 and end with $5.87, how much did you spend? (Lesson 7.2)

3. **Make It to Scale (Lesson 8.4)** (🖩) — If a bookshelf is 4.4 feet wide by 8.5 feet high, and you want to draw a scale drawing so that 1 foot equals 0.5 inch, how many inches should you make the width and height?

4. **Miscellaneous**

   a. Find the next two numbers in this sequence assuming the pattern continues: {4, 11, 18…} (Lesson 11.4)

   b. What is the common difference in the sequence in problem 4a.? (Lesson 11.4)

   c. List all the number sets you can think of (Integer, etc.) that describe the common difference in problem 4a. *Hint*: Look back at Lesson 11.2 to remind yourself of the different sets. (Lesson 11.2)

   d. Solve: $-6 \cdot 8$ (Lesson 10.6)

   e. Solve: $8\frac{1}{2} + - -\frac{1}{2}$ (Lesson 10.3)

   f. Solve: $\frac{-9}{3}$ (Lesson 10.8)

   g. Find the missing number: $\frac{5.5}{25} = \frac{?}{5}$ (Lesson 8.3)

   h. If a substance cools from 10° C to −5° C, by how many degrees did it change? (Lesson 10.4)

   i. Represent $8 per 3 pounds as a ratio. (Lesson 8.1)

   j. What is the decimal equivalent of the ratio in problem 3i? (Lesson 8.1)

   k. Find 20% of $56.89 mentally. Round $56.89 to the nearest dollar. (Lesson 9.5)

5. **Term Time**

   a. Is "People Photos" a subset of "Good Photos"? If not, redraw the diagram so that it is. (Lesson 11.1)

   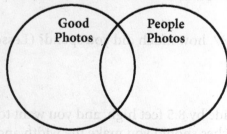

   b. What is a base in a percent problem? (Lesson 9.2)

   c. What is a set? (Lesson 11.1)

PRINCIPLES OF MATHEMATICS

CHAPTER 12. Statistics and Graphing
LESSON 1. Introduction to Statistics

Worksheet 12.1

You may use a calculator on this worksheet whenever you see this symbol (🖩).

1. **Frequency Distributions** (🖩)

    a. Fill out the frequency column in the table below using this data (representing the ages of students enrolled in a class): 12, 13, 12, 13, 12, 12, 15, 16, 12, 12

Age	Frequency	Relative Frequency
12		
13		
15		
16		

    b. Fill in the relative frequency column above with a percentage that shows what percent of the whole each age represents. *Hint*: Example: If you have 10 students age 12 and 40 students total, then the relative frequency is $\frac{10}{40}$, or 25%.

2. **Leaving a Tip** — Find an approximate 20% tip mentally for each of the following totals. Round to the nearest dollar before finding the 20%.

    a. $35.89

    b. $14.50

    c. $19.05

3. **Skill Sharpening**

    a. $(-2)(-5)$

    b. $\frac{60}{-5}$

    c. $4\frac{1}{7}(\frac{1}{3} + \frac{1}{4})$

    d. If 34 out of 65 employees opted for health care plan 1 at one facility, if the same ratio opts for it at another facility with 234 employees, how many will opt for it there?

4. **Accurate Stats?** — Read the following claim and identify at least one possible problem with the conclusion presented.

    The unemployment rate went down by 20% in the last 4 years; therefore, reelect the current governor.

5. **Definition** — Statistics is "The mathematics of the collection, organization, and _____ of numerical data."[1]

---

[1] *The American Heritage Dictionary of the English Language.* New College Edition, 1980, s.v. "statistics."

CHAPTER 12. Statistics and Graphing
LESSON 2. Collecting Data — Sampling

Worksheet 12.2

1. **Taking a Sample** — Let's say you want to find out the opinion of people in your whole neighborhood about a proposed addition of a light. Suppose you conduct a random survey of a sufficient portion of the people in the neighborhood.

    a. In this situation, which of the following constitutes what we would call the **population**?
        i. The people in favor of the light
        ii. The people not in favor of the light
        iii. Everyone living in the neighborhood
        iv. Everyone who received a survey
        v. Everyone who returned a survey

    b. Which of the following constitutes what we would call the **sample**?
        i. The people in favor of the light
        ii. The people not in favor of the light
        iii. Everyone living in the neighborhood
        iv. Everyone who received a survey
        v. Everyone who returned a survey

2. **Understanding Sampling** — List two things we need to make sure a sample is.

3. **Margin of Error** — For each of these stats listing the results of hypothetical surveys, look at the margin of error (MOE) and list the range the results really showed.

    Example: One candidate has the lead in the polls with 51% of people surveyed saying they are planning to vote for him. (MOE = 5%)
        Answer: Range = 46%–56%

    a. A survey shows 51% of the people favor a certain bill. (MOE = 2%)

    b. A survey shows 55% of the people favor one candidate. (MOE = 2%)

    c. A survey shows 55% of the people favor one candidate. (MOE = 10%)

4. **Confidence Level** — If a survey gives a 42% result with level of confidence of 93% and an MOE of 4%, that means we're 93% confident the results of the whole would fall within what percents?

PAGE 177

5. **Accurate Stats?** — Read the following claims and identify at least one thing that's either wrong about how the data was collected or could possibly be wrong (i.e., what more information you'd need to know).

   a. Our new spelling curriculum will teach you how to spell! 90% of people using the curriculum scored in the 80th percentile or above for spelling on standardized tests.

   b. A bookstore wants to learn what book is most popular with the teens in the city, as they want to advertise that book. They decide to interview all the teens on a football team.

6. **Percentage Off** — If you apply a 25% coupon to a $56.78 item, how much will you end up paying for that item?

7. **Skill Sharpening**
   a. 7 + –13

   b. $\frac{9}{-3}$

   c. Find the missing number in this proportion: $\frac{14}{7} = \frac{?}{2}$.

   d. What is the next number in this sequence assuming the pattern continues: {1, -1, 2, -2, 3, -3, 4, -4, ___, ...}

Extra Credit — Read *How to Lie with Statistics* by Darrell Huff[1] (your library will likely carry it). You'll be fascinated by the many ways stats are twisted — and better aware of what to look for when presented with a stat. And don't worry — the book is quite thin and written in a light-hearted, easy-to-read manner.

---

1 Darrell Huff, *How to Lie with Statistics* (New York: W. W. Norton, 1993).

1. **Elections**

    a. What percent of the people polled said they are planning to vote for Candidate A?

    b. If 1,000 people were polled, how many people are planning to vote for Candidate A?

    c. If the margin of error was 3% with a confidence level of 95%, we're 95% confident that the percent planning to vote for Candidate A fall in what range?

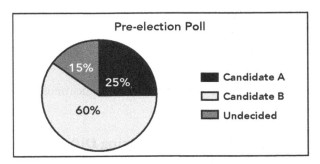

2. **Restaurant Choices**

    a. According to this survey, what was the most popular dish and by what percent did it beat out the next most popular dish?

    b. If the restaurant found out the staff only handed the survey to families who came in together, how could that skew the results if they were trying to find the most popular dish overall, not just among families?

    c. By only targeting families, was the staff collecting a *random* sample?

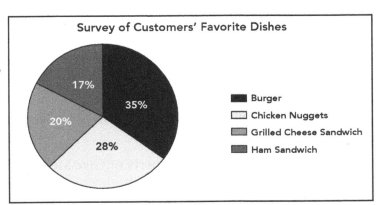

3. **War Strength**

    a. About how many men did Russia have in its army in 1907?

    b. About how many men did Japan have in its army in 1907?

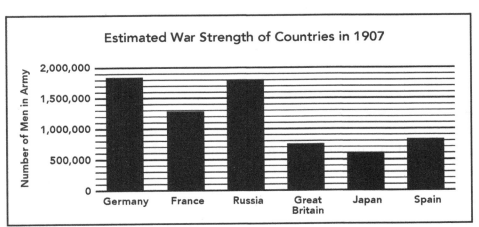

4. **Interpreting Bar Graphs** — Use the "Dr. Brian Ray" box in Lesson 12.3 to answer these questions.

   a. How many students were home educated in 1973?

   b. How many students were home educated in 2007?

5. **Reviewing Sets: Organizing Different Ways** — Practice the organization skills we looked at in the last chapter by thinking of two headings you could use in a report on these events (i.e., two sets into which you could divide the various events). List the two headings and which events you would cover under each heading. *Hint*: Think of different attributes the events have in common. There is no single right answer; you just need to have formed logical groupings. Feel free to look up any event online if you need to learn more about it. We can organize things—whether events; closets; ideas; numbers; or, as we'll see in the next chapter, shapes—many different ways.

   Plymouth Plantation Founded

   President Lincoln Elected

   Lincoln's Assassination

   The Declaration of Independence Signed

   The Gettysburg Address

   Patrick Henry's "Give Me Liberty or Give Me Death"

6. **Skill Sharpening**

   a. $-6 \cdot -5$

   b. $\frac{-18}{-2}$

   c. $\frac{1}{2} \cdot \frac{3}{8}$

   d. $\frac{2}{3} + \frac{24}{25}$

   e. If you need $\frac{1}{4}$ yard of ribbon per wreath, and you're making 86 wreaths, how many yards of ribbon do you need?

7. **Understanding Graphs**

   a. What do pie graphs make it easy to see?

   b. What do bar graphs make it easy to see?

   c. True or false: A bar graph with 15 bars would be easy to read.

CHAPTER 12. Statistics and Graphing
LESSON 4. Organizing Data — Drawing Bar/Column Graphs

Worksheet 12.4

1. **Drawing Bar Graphs** — Using graph paper, draw a bar graph to represent the following data.

   Expenses in 2012: $600

   Expenses in 2013: $800

   Expenses in 2014: $1,000

2. **Frequency Distributions**

   a. Fill out the frequency column in the table below using this hypothetical data about a team's wins, losses, and ties: win, win, loss, win, loss, loss, loss, loss, win, tie, win, tie, loss, loss, loss, loss

Score	Frequency	Relative Frequency
Win		
Loss		
Tie		

   b. Fill in the relative frequency column above. You can round your percent to the nearest whole percent (i.e., round 21.45% to 21%). *Hint*: First figure out your whole—how many game results (or data points) you have. Then find the percent of that whole each frequency represents by dividing the frequency by the whole.

3. **Accurate Stat?** — Take a look at both the bar graphs below.

   a. What is the only difference between them?

   b. While neither one is inaccurate, which one better shows how the utility costs compare with each other and why?

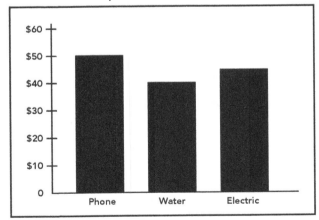
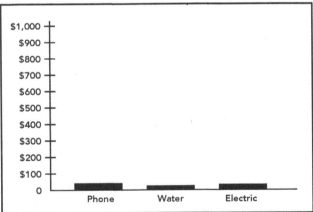

PAGE 181

4. **Skill Sharpening**

   a. $89.56 \div 17$

   b. $-7 \cdot -3$

   c. $5 + 5(\frac{5}{15})$

   d. If $20 out of $80 made during a month went to savings, if you keep the same ratio but make $240, how much would go to savings?

Extra Credit: Do It the Tech Way — Following the instructions below, use Microsoft Excel (or another computer program) to generate a pie graph for the data in problem 2.
- Open the program and type in the score and relative frequency. Make sure you put each piece of data in a separate cell in two columns—one column for the score, and the other for the relative frequency.
- Choose Insert – Chart (if you don't have an Insert menu, look online for "insert chart in your program" to find out how to insert a chart).
- Play around with different settings and charts you can make. Again, you should be able to find further instructions online (or you could play around with the buttons to figure it out yourself).

PRINCIPLES OF MATHEMATICS

CHAPTER 12. Statistics and Graphing
LESSON 5. Coordinates

Worksheet 12.5

1. **Labeling the Space** — Using coordinates, describe the box on the map to which the arrow is pointing. List the horizontal coordinate first, and enclose your coordinates in parentheses.

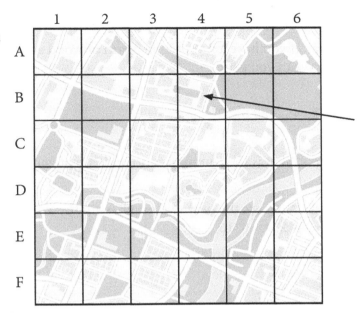

2. **Label the Points** — Label the points on this coordinate graph using the horizontal, vertical convention. For example, a point 2 over horizontally and 3 up vertically would be labeled like this: (2,3).

   a.
   b.
   c.
   d.
   e.
   f.

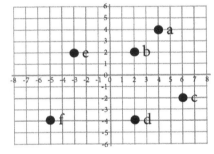

3. **Navigation** — A lighthouse is at the coordinates (7,–2) and a boat is at coordinates (5,1). Graph the location of both the lighthouse and the boat, and then draw a diagonal line connecting the two points. The diagonal line will be the shortest way for the boat to reach the lighthouse!

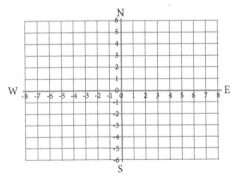

4. **Draw the Graph** — Using graph paper, draw your own coordinate graph (make it look like the one in the previous problem), and label it with the following points.

   a. (3,2)
   b. (4,–2)
   c. (7,3)
   d. (6,–4)

PAGE 183

5. **Skill Sharpening**

   a. If you have a 5-foot long board of wood and need to cut it into $\frac{2}{3}$-foot sections, how many sections can you get?

   b. $45.6(89.6 \div 2 + 56.2)$

   c. What is the common ratio in this sequence? $\{-2, 4, -8, 16...\}$

   d. What are the next two numbers in this sequence, assuming the pattern continues?
   $\{-2, 4, -8, 16, \underline{}, \underline{}\}$

CHAPTER 12. Statistics and Graphing
LESSON 6. Organizing Data — Line Graphs
Worksheet 12.6

You may use a calculator on this worksheet whenever you see this symbol (🖩).

1. **Pension Roll**

   a. What is this graph displaying?

   b. What scale is used to show the dates? In other words, how many years does one box represent?

   c. What scale is used to show the number of people? In other words, how many people does one box represent?

   d. Is the data for every year shown on the chart, or were some years skipped?

   e. In general, what happened between 1898 and 1903?

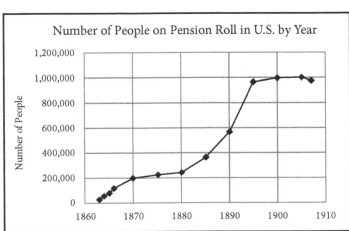

2. **Profit and Loss Over Time** — The line graph below shows the amount of money a company gained or lost over 5 years.

   a. What gain/loss did the company have in 2010?

   b. What gain/loss did the company have in 2014?

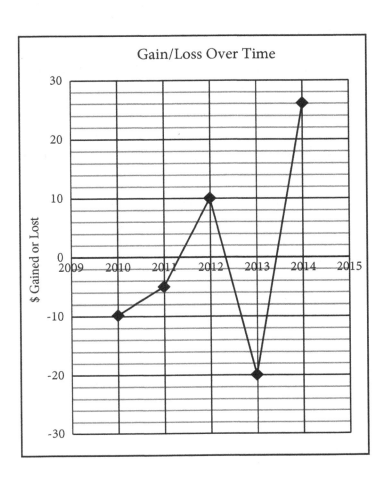

PAGE 185

3. **Spending Habits** (🖩) — The data below shows the average amount spent, along with the amounts of that spent on entertainment and reading, according to a government survey.[1]

A. Year	B. Average Annual Expenditure (spending)	C. Entertainment	D.	E.	F. Reading	G.	H.
2003	40,817	2,060	$\frac{2,060}{40,817}$	5.05%	127	$\frac{127}{40,817}$	0.31%
2004	43,395	2,218	———		130	———	
2005	46,409	2,388	———		126	———	
2006	48,398	2,376	———		117	———	
2007	49,638	2,698	———		118	———	
2008	50,486	2,835	———		116	———	

a. In the columns labeled D and E, compare the amount spent on entertainment with the total amount spent by writing a ratio of column C over column B, and then expressing that ratio as a percent. The first one is done for you. Round your percent to two decimals, as done in the first row.

b. In the columns labeled G and H, compare the amount spent on reading with the total amount spent by writing a ratio of column F over column B, and then expressing that ratio as a percent. The first one is done for you. Round your percent to two decimals, as done in the first row.

c. Based on this data, in general, do people spend as much on reading as on other entertainment?

d. Conclusions: Is it valid to conclude from this data that people don't read a lot?

[1] Source: U.S. Bureau of Labor Statistics, *Consumer Expenditure Survey, Consumer Expenditures in 2009*, October 2010. Found in "1232 - Expenditures Per Consumer Unit for Entertainment and Reading"; http://www.census.gov/compendia/statab/cats/arts_recreation_travel/consumer_expenditures.html, accessed 10/23/13.

4. **Graphing Expenditures** Using the data from the previous problem, finish the line graph by adding points for entertainment in 2006-2009 and connecting the dots. Notice how the line graph makes it easy to compare entertainment and reading by year on the same graph. Don't worry if you can't get your dots exact—just do your best.

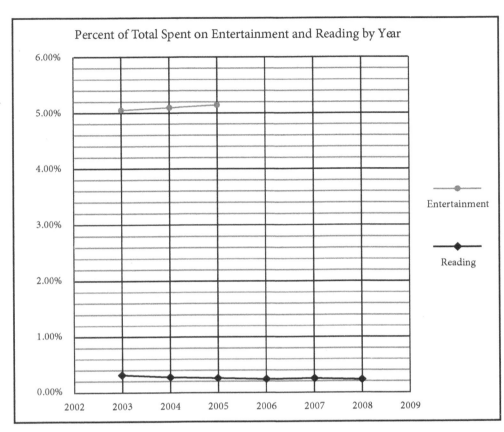

5. **Skill Sharpening**

   a. Find the least common multiple of 16 and 24.

   b. $\frac{5}{16} + \frac{9}{24}$

   c. What is the common difference in this sequence? {75, 74.25, 73.5, …}

   d. What are the next two numbers in the sequence given in 5c, assuming the pattern continues?

6. **True or False** — Line graphs are useful in showing trends.

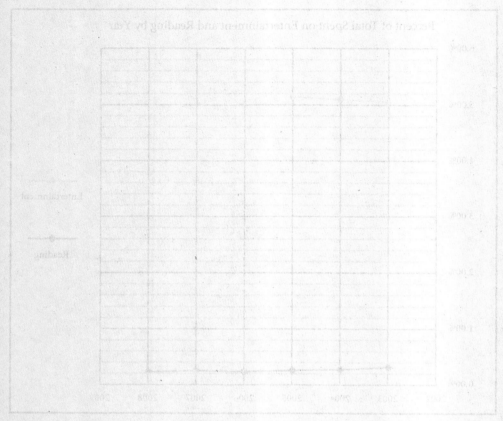

PRINCIPLES OF MATHEMATICS

CHAPTER 12. Statistics and Graphing
LESSON 7. Organizing Data — Averages

Worksheet 12.7

You may use a calculator on this worksheet whenever you see this symbol (🖩).

1. **Average** (🖩)

    a. Below are some hypothetical numbers for the ages of students in a class.

41	35	12	85	43
42	39	20	55	42

    Use what you learned about averages to find the average age of students in the class. Round to the nearest whole number (i.e., whole age).

    b. If one class received these math SAT scores, what was the average math SAT score for the class?

510	450	390	590	475	480
495	517	530	540	525	546

    c. If these are the sale prices of homes that have sold in an area, what's the average sale price for the area?

$150,000	$275,000	$175,000	$349,000	$255,000
$230,000	$285,000	$115,000	$290,000	$280,000

    d. On the line graph below, draw a straight line showing approximately where the average of the sale prices from 1c would fall.

    e. In Worksheet 12.6, we saw that, according to a survey, the average spending on reading for 2008 was $116. Does that mean that every household who submitted a survey spent $116 on reading?

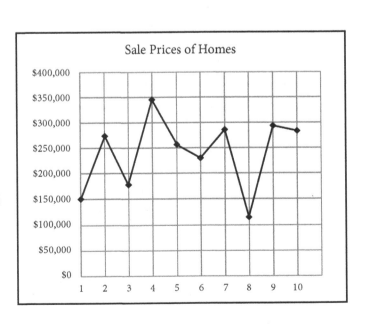

PAGE 189

2. **Baseball Time** (🖩) — Find these player's batting averages given the data listed. Remember, the shortcut for finding the batting average is to divide the total hits by the total at-bats. Round all averages to the thousandths place; include 0s if necessary to give each batting average to the thousandths place. For example, a result of 0.6 should be listed as 0.600.

   a. Jason: 5 hits; 40 at-bats

   b. Jeremy: 10 hits; 50 at-bats

   c. Mark: 15 hits; 65 at-bats

3. **Accurate Stats?** — What's wrong with this statement?
   If the average child learns to walk around age 1, there must be something wrong with Lydia because she is $1\frac{1}{2}$ and still not walking.

4. **Graphs with Averages**— The following chart, which is based upon old life insurance records, shows average weights of men 5 foot 10 inches tall.

Age	Average Weight
15–24	154
25–29	159
30–34	164
35–39	167
40–44	170
45–49	171
50–54	172
55–59	173
50–64	174

   Notice that, rather than listing the actual weight for every individual person, the weights have been summarized by age ranges and an average weight for that range has been given rather than a frequency count of an exact weight.

Finish the histogram (a bar graph with ranges!) below by appropriately shading the remaining columns.

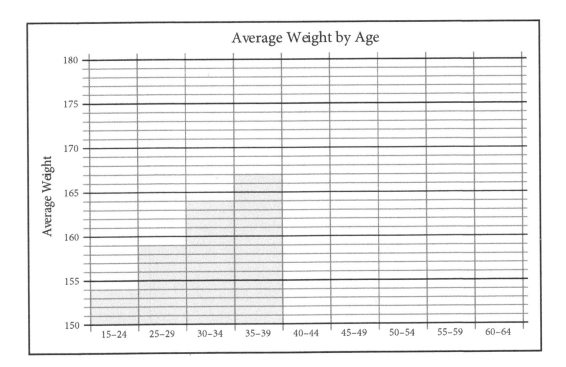

5. **Mental Math** — Add or subtract mentally.

   a. 45 − 28

   b. 87 + 96

6. **Skill Sharpening**

   c. Find the next three numbers in this sequence, assuming the pattern continues:
   $\{\frac{2}{3}, 1\frac{1}{3}, 2, 2\frac{2}{3}, \underline{\quad}, \underline{\quad}, \underline{\quad}\}$

   d. Find the next three numbers in this sequence, assuming the pattern continues:
   $\{2, 7, 5, 10, 8, 13, 11, \underline{\quad}, \underline{\quad}, \underline{\quad}\}$
   Hint: This pattern is a little more in-depth, but it's there! Look at what changed between each number.

Finish the histogram (a bar graph with ranges) below by appropriately shading the remaining columns.

Average Weight by Age

5. **Mental Math** — Add or subtract mentally.

   a. 45 − 28

   b. 87 + 96

6. **Skill Sharpening**

   c. Find the next three numbers in this sequence, assuming the pattern continues.

   $\frac{1}{2}, \frac{1}{3}, \frac{2}{3}$ _____ _____ _____

   d. Find the next three numbers in this sequence, assuming the pattern continues.

   [?] 5, 10, 6, 13, 11, _____ _____ _____

   Hint: This pattern is a little more in-depth, but it's there! Look at what changed between each number.

PRINCIPLES OF MATHEMATICS

CHAPTER 12. Statistics and Graphing
LESSON 8. Organizing Data — More on Averages

Worksheet 12.8

You may use a calculator on this worksheet whenever you see this symbol (🖩).

Understanding Modes
A mode is the most frequent data point. However, not all sets of data have a mode or a single mode.

- **Data with No Mode** — When no data point occurs more frequently than any other, there is no mode.
  Example: 54, 56, 60, 62, 69, 70, 80
  There is no mode, as no number appears more frequently than any other.

- **Data with Multiple Modes** — When there's a tie for which data point appears the most, there is more than one mode.
  Example: 69, 69, 70, 70, 72, 72, 80, 82
  In this case, 69, 70, and 72 are the modes, as they all appear more than any other number.

1. **Find That Middle** (🖩) — For the following sets of hypothetical data, find the average, the median, and the mode. *Hint*: Don't forget to arrange your data in order. Also be sure to read the gray box above for a helpful note.

    a. Number of problems a class of college students got correct on an exam:
       9, 3, 7, 13, 15, 8, 7, 5, 7, 10, 14

    b. House selling prices in a community for the previous 6 months (values represent thousand dollars, i.e., $101 represents $101,000): $101, $130, $125, $134, $120, $122, $128, $120

    c. Final grade for students in a college class: 95, 55, 60, 72, 70, 73, 75, 71, 80, 79, 65, 69, 68, 70

    d. High temperatures for a city: 90, 76, 80, 82, 77, 91, 87

    e. Age of criminal offenders in a local jail: 18, 19, 18, 25, 30, 19

2. **Frequency Chart** — Finish filling out the frequency chart using the data from problem 1c. Notice that this time we've summarized the data into ranges based on the grade the students would have received. Round your relative frequency to a whole percent. *Example*: 12.69% rounds to 13%.

Grade	Frequency	Relative Frequency
0–60 (F)	2	14%
61–70 (D)		
71–80 (C)		
81–90 (B)		
91–100 (A)		

PAGE 193

3. **Median Filters**

   a. A programmer is programming a robot to move on its own. In order to do this, the robot sends out radar signals and interprets the signals to figure out how far away it is from objects. Yet sometimes the signal the robot receives gets corrupted. To avoid this problem, the programmer programs the robot to take the median from five signals and use that as the correct distance. If the robot's last five signals were 5 ft, 5 ft, 0 ft, 5 ft, and 5 ft, what would the median be? *Note*: The programmer here was using what's called a median filter. Median filters apply in lots of different areas. For example, in photo editing, a median filter would use the median of a group of pixels to modify an image.

   b. If the robot had based a decision on the distance based on the average (i.e., mean) instead of the median, what distance would he have used?

4. **Just Give Me the Mode** — Suppose a company gave a choice of 1 through 6, with each number representing a different favorite dish. On one survey, the following results were obtained: 2, 3, 6, 3, 2, 3, 6, 3, 1, 3, 3, 3 *Note*: In real life, there would have been a *lot* more results.

   a. The company wants to know what dish was people's favorite. In other words, they want to know the mode—the actual numerical value that was chosen more frequently than others. Find the mode.

   b. Now find the average (i.e., mean). Notice how this number doesn't mean anything in this situation. Each number stood for a dish—finding the average doesn't make sense, as instead we want to know which *exact* dish people liked best, not what the average number was. Remember, math is a tool—be sure to think through what you're trying to find and use the tool what will help you find it!

5. **Book Length** — An online article published by the Huffington Post titled "Average Book Length: Guess How Many Words Are In A Novel"[1] makes this statement: "According to Amazon's great Text Stats feature, the median length for all books is about 64,000 words." In your own words, explain what this stat means? How would it be different if they used the average instead of the median?

6. **Navigation** — If your ship log shows you to be at the point shown on the map below, how would you describe your current location using coordinates? Remember to list the horizontal position first.

7. **Drawing a Bar Graph** — The value, in millions of dollars, of agricultural implements exported from the United States has been as follows: during 1880, 2.25; during 1885, 2.56; during 1890, 3.86; during 1895, 5.41; during 1900, 16.1. Use graph paper to draw a bar graph expressing this. Round each value to the nearest whole number first. Be sure to label your chart!

---

[1] Gabe Habash, "Average Book Length: Guess How Many Words Are In A Novel" (PWxyz, posted 03/09/12) accessed 10/16/13, http://www.huffingtonpost.com/2012/03/09/book-length_n_1334636.html.

CHAPTER 13. Naming Shapes: Introducing Geometry
LESSON 1. Understanding Geometry
Worksheet 13.1

1. **Skill Sharpening**

    a. $(-5) + (-21)$

    b. $67\% + 23\%$

    c. $12\% \cdot 50$

    d. 5% of 25

    e. 45 out of 100 people surveyed indicated they would vote for a certain candidate. Express this as a fraction (in simplest terms), a decimal number, and a percentage.

    f. Convert 0.465 to a percent.

    g. Complete the division: $\frac{-8}{2}$.

    h. $-9 + -5\frac{2}{3}$

    i. $-8 \cdot -5$

    j. If a business' expenses were $4,500 one year and their revenue was $19,500, how much revenue would they have to make to keep at least the same ratio between expenses and revenue if they increase their expenses to $5,400?

    k. $\frac{32}{40} - \frac{108}{160}$

    l. $\frac{2}{5} - \frac{3}{5}$

2. **Just Your Average** — Find the average test score if these were the actual scores on a test: 78, 90, 75, 81, 82, 50, 99, 64

3. **Flashcards/Notebook** — It's important for you to know all the bolded terms from Lesson 13.1; make flashcards for any terms you were not familiar with and review those cards each day until you have them mastered. Be sure to also add the terms to your math notebook.

4. **Comprehension Check**
    a. What does the word geometry mean?

    b. Who formalized geometry?

    c. What are two different definitions of a line?

    d. What can different geometry systems remind us of?

CHAPTER 13. Naming Shapes: Introducing Geometry
LESSON 2. Lines and Angles
Worksheet 13.2

1. **Angle and Line Identification** — Identify the lines (parallel, perpendicular, or neither) and the angles (acute, obtuse, or right—or not applicable if the lines don't form an angle) in the pictures below.

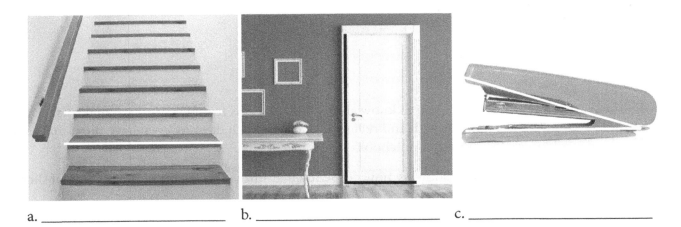

   a. _____ b. _____ c. _____

2. **Drawing Angle** — Show which angle in problem 1 is a right angle. Remember, we mark right angles with a half box, as shown.

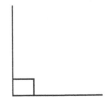

3. **Angles and Lines** — Since perpendicular lines always form a right angle, if we know that two lines are perpendicular, what can we assume about the angle they form?

4. **Labeling** — Remember, the middle letter in an angle label represents its vertex.

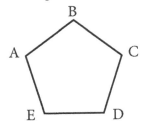

   a. Use a purple crayon/highlighter/marker to trace $\overline{AE}$.
   b. Use a blue crayon/highlighter/marker to trace ∠ABC.
   c. Use a red crayon/highlighter/marker to trace ∠DCB.
   d. Use a green crayon/highlighter/marker to trace ∠CDE.
   e. Use an orange crayon/highlighter/marker to trace ∠DEA.

5. **Labeling** — Add letters and then use them to represent the angle shown.

6. **Flashcards** — It's important for you to know all the bolded terms from Lesson 13.2; make flashcards for any terms you were not familiar with and review those cards each day until you have them mastered. Be sure to also add them to your math notebook.

7. **Sets** — Draw a Venn Diagram showing how perpendicular lines relate to lines. *Hint*: Think of them both as sets.

8. **Just Your Average** — Find the average high temperature for the week if these were the actual highs:
45, 50, 62, 55, 40, 59, 60

9. **Skill Sharpening** — Find the answers mentally.

    a. Total = $95.78 Give = $100   Change =   Bills and Coins =

    b. Total = $95.78 Give = $100.03 Change =   Bills and Coins =

    c. 89 • 5   d. 5 • $3.99

    e. 86 − 75   f. 230 − 58

CHAPTER 13. Naming Shapes: Introducing Geometry
LESSON 3. Polygons

Worksheet 13.3

1. **Name That Shape** —What two-dimensional shape best describes these pictures? Choose the most descriptive shape name you can.

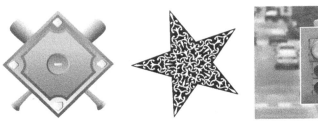 

a. _____ b. _____ c. _____ d. _____ e. _____

2. **More Shape Naming**

   a. Which of the shapes above are polygons?

   b. Which of the shapes above are quadrilaterals?

   c. Which of the shapes above are trapezoids?

   d. Which shape is an irregular polygon?

   e. Is the shape below a polygon? Why or why not?

3. **Labeling** — Circle ∠DEF.

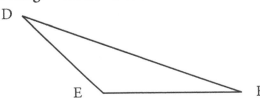

PAGE 199

4. **Label Me** — Label each corner point on this rectangle using the (horizontal, vertical) format.

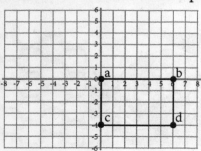

5. **Draw Me** — Draw a triangle on graph paper with following coordinates: (3,0), (2,3), (0,0) *Hint*: Just plot the points, and then connect the lines to form a triangle.

6. **Venn Diagram** — Draw a Venn Diagram illustrating how the sets of all polygons, quadrilaterals, trapezoids, parallelograms, and squares relate. *Hint*: You'll have multiple circles inside of circles!

7. **Flashcards** — Keep reviewing your flashcards and add flashcards for any shape names in Lesson 13.3 that you didn't already know.

8. **Baseball Time** — Find these player's batting averages given the data listed. Remember, the shortcut for finding the batting average is to divide the total hits by the total at-bats. Batting averages are rounded to the third decimal (i.e., the thousandths place).

   a. David: 7 hits; 42 at-bats

   b. Matthew: 11 hits; 52 at-bats

   c. Amir: 65 hits; 205 at-bats

9. **Skill Sharpening**

   a. $\frac{-8}{2} + -\frac{-10}{-2}$

   b. $-9 \cdot -2$

   c. $-1 \cdot 8$

PAGE 200

CHAPTER 13. Naming Shapes: Introducing Geometry
LESSON 4. Circles, Triangles, and Three-Dimensional Shapes

Worksheet 13.4

1. **Drawing** — Many art books suggest reducing everything you want to draw to smaller shapes, and then going back in later and connecting the shapes together to form object. For example, if you were drawing a dog, you might draw an oval for his stomach, a small circle for his head, and a few quadrilaterals for his legs. Then you could go back and erase/draw lines as necessary to transform those shapes into the shape of your dog.

   Pick an object in your house (any object will do) and draw it using this strategy.

2. **Find That Shape** — List six things that can be represented by circles.

3. **Identify That Triangle** — Name each triangle shown based on its angles (right, acute, obtuse) and its sides (isosceles, equilateral, or scalene). For example, label a triangle with all equal sides and all acute angles as an "acute equilateral triangle."

   a.   b.   c.

4. **Sets**
   a. Draw a Venn Diagram showing the relationship between these sets: isosceles triangles and right triangles. **Use a compass to draw the circles!** *Hint*: Are any right triangles also isosceles?

b. Draw a Venn Diagram showing the relationship between acute triangles and equilaterals triangles if you are told all equilateral triangles have all acute angles, but that not all acute triangles have equal-length sides.

5. **Name That Shape** — What term would you use to describe these objects as three-dimensional objects? Be as specific as you can be.

   Example:

   Answer: rectangular prism

   a.   b.   c.   d.

6. **Skill Sharpening**

   a. $8 + 20 \div 2$

   b. Convert 0.62 to a percent.

   c. Convert 0.257 to a percent.

d. Convert 3.429 to a percent.

e. Convert 43.7% to a decimal and a fraction of 1,000.

f. What is $25\frac{1}{2}$% of 60?

g. Convert 47.15% to a decimal and a fraction of 10,000.

h. Convert $7\frac{1}{2}$% to a decimal and a fraction of 1,000.

i. If 8.5% sales tax is added to a $38.99 purchase, what will the new total be?

j. Find the missing number in this proportion: $\frac{7.8}{9.6} = \frac{?}{3.2}$. Round your answer to the nearest tenth.

7. **Naming Angles** — How would you name the marked angle?

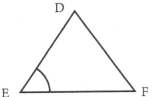

8. **Flashcards** — Keep reviewing your flashcards, and add flashcards for any shape names in Lesson 13.4 that you didn't already know.

i. If 8.5% sales tax is added to a $38.79 purchase, what will the new total be?

CHAPTER 13. Naming Shapes: Introducing Geometry
LESSON 5. Fun with Shapes

Worksheet 13.5

1. **Congruency Fun**
   a. Trace the triangles below and cut them out.
   b. Put them on top of each other. Notice how they have the same length sides and the same degree angles.
   c. Use your triangles to form a rotation.
   d. Use your triangles to form a reflection.
   e. Use your triangles to form a translation.
   f. What kind of triangles are these? Describe the triangles by their angles and by their sides.

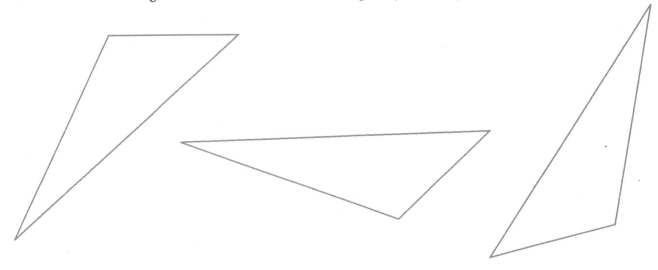

2. **Rotating, Reflecting, and Translating** — Notice how the blades in the windmill are all the same shape, but they've been moved around a fixed point. What is the name to describe this sort of movement?

PAGE 205

3. **Reflect Me** — Finish the reflection below by graphing the remaining two points (just think about where you'd need to draw them to create a mirror image) and connecting the dots.

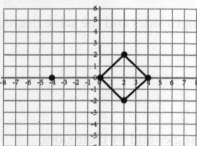

4. **Translate Me** — Move this triangle by increasing each horizontal and vertical value by 1 (i.e., moving each point one box to the right and one box up). The first point has been moved for you. You do not have to label each point.

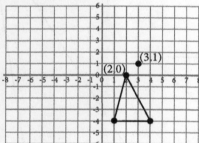

5. **Just Your Average** — If plants in an experiment grew 7 in, 5 in, 10 in, 6.5 in, 7.8 in, and 9 in, what was the average growth? Round your answer to the nearest tenth, since that was the accuracy of the measurements we took of the plants.

6. **Out of the Box** — Pick one of the following to do.

    - Grab some graph paper and draw a shape. Now make a design by reflecting, translating, rotating, etc., that shape.

    - Search online for "wallpaper patterns." Pick one image and identify one form of transformation used. Even though the patterns may not use polygons, you'd be amazed how many include reflections, rotations, translations, etc., of a design!

    - If you've looked at quilts, you may have noticed that many quilt patterns make use of triangles rotated and otherwise rearranged to create an interesting design. Take a trip to the library and check out a book on folk art quilts or use the Internet to look up folk art quilts. Find three different examples of how triangles were rotated, reflected, translated, or scaled to create a pattern.

7. **Flashcards** — Keep reviewing your flashcards, and add flashcards for any bolded terms in Lesson 13.5 that you didn't already know. Be sure to add them to your math notebook too.

CHAPTER 13. Naming Shapes: Introducing Geometry
LESSON 6. Chapter Synopsis
Worksheet 13.6

1. **Check It Out at the Source**

    a. Read Proverbs 9:10. What is the beginning of wisdom?

    b. Read Psalm 119:105 and 130. According to these verses, what gives understanding?

    c. Read Romans 1:19–25. What does verse 25 say men will do?

    d. From what you read in Lesson 13.6, how did the Greek mathematicians/philosophers in general worship and serve "the creature more than the Creator"?

    e. Read Psalm 51:5, Job 38:4–5, and 1 Corinthians 3:19. List at least two reasons why human reasoning and our mind is not a good place to put our trust.

2. **Recording Real-life Shapes** — Write down 10 objects around your house that you could describe using a rectangle.

3. **Name That Shape**

    a. Notice how we've drawn lines around the snowflake to find the shape it best fits inside. Name the shape drawn.

    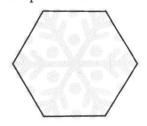

PAGE 207

b. Describe this triangle based on its angles.

c. Describe the triangle from 3b based on its sides.

4. **Labeling** — Add letters to the yield sign in 3b, and use them to describe the top-left angle and the top line (i.e., the horizontal line).

5. **Flashcards** — Have your parent or teacher quiz you on your geometry flashcards. Did you get them all correct? Continue to study any you did not know.

PAGE 208

PRINCIPLES OF MATHEMATICS

CHAPTER 14. Measuring Distance
LESSON 1. Units for Measuring Distance

Worksheet 14.1

> You may use a calculator on this worksheet whenever you see this symbol (🖩).

1. **Does It Fit?** — It's no fun to buy a piece of furniture only to discover it doesn't fit in the car or in the closet you'd intended it for...or to buy a box and realize it won't fit in your car. Measuring the furniture, box, etc., and the space we're putting it in can save us a lot of trouble.

    a. Let's say your closet is 4 feet long. You've found a tiny dresser that's $2\frac{1}{2}$ feet long. How much length will still be left in your closet if you put it there?

    b. Grab a measuring tape and measure the width of your bedroom door in inches. Will a desk 25 inches wide fit through the opening?

2. **Ordering Blinds** — If you want to install new blinds on one of the windows in your home, you need to know the window's measurements! Grab a measuring tape and pick a window. Measure from one end of the window to the other.

    a. What was the measurement in inches?

    b. If you want the blinds to hang two inches beyond the left and right side of the window, how wide should the blinds be?

3. **Greater Than, Less Than, or Equal to** — Use >, <, or = to compare the measurements of these lines.

    a. _____   _____

    b. _____   _____

    c. A line 5 ft long   A line 7 ft long

4. **Percents and Measurement** (🖩) — Include the units of measure in your answers!

    a. What is 20% of 8 feet?

    b. Convert $2\frac{1}{4}$% to a decimal number.

    c. What is $2\frac{1}{4}$% of 60 yards?

    d. If a runner is 35% done a 20-mile race, how far has he ran so far?

    e. How far does the runner in question 4d have left to go? *Hint*: You can think of the total race as 100% of the distance he has to run. He's gone 35%, so he has ___ percent left. Then find that percentage of 20 miles. Or you can subtract the amount he's gone from the total miles in the race.

f. If a person is 40% finished with a 500-mile trip, how many miles has he traveled?

g. How many miles does the person in 4f have left to go?

5. **Choosing a Measurement** — For each measurement listed, list logical unit choices for both the metric and customary systems. Be prepared to explain your answers. For example, if measuring a football field, yards (or feet) and meters would be logical choices, as it would take quite a lot of centimeters or inches to fill a field, and quite a few fields to make a mile or kilometer.

   a. A bookshelf

   b. The length of a swimming pool

   c. The length of a paper

6. **Skill Sharpening**

   a. $8 \cdot \frac{1}{2}$

   b. $\frac{4}{5} + \frac{11}{60}$

   c. Simplify: $\frac{21}{70}$

   d. Find the missing number in this proportion: $\frac{8}{75} = \frac{?}{105}$

   e. Find the average of these lengths (round to the nearest foot): 8 ft, 10 ft, 12 ft, 11 ft, 9 ft, 7 ft

7. **Flashcards** — Make flashcards for all the bolded measurements in Lesson 14.1 you didn't already know. Keep reviewing them, along with any flashcards from last chapter you don't know yet.

8. **Questions**

   a. What is the definition of a meter?

   b. How do we measure up to God's standard of holiness?

   c. Why did the length of a cubit vary?

PRINCIPLES OF MATHEMATICS

CHAPTER 14. Measuring Distance
LESSON 2. Conversions via Proportions

Worksheet 14.2

Remember to include the unit in your answer whenever one is given!

You may use a calculator on this worksheet whenever you see this symbol (🖩).

1. **Unit Conversions (🖩)** — Use proportions to solve these problems.

    a. 6 feet equal how many inches?

    b. 2 miles equal how many feet?

    c. 87 inches equal how many yards?

    d. Your GPS tells you to turn right in 1,000 feet. What fraction of a mile is that? *Hint*: Convert feet to miles!

2. **Sewing (🖩)** — Say you measured the height of a window at 100 inches and its width at 40 inches. You want to drape fabric over both sides and the top of the window as shown. How many yards of fabric would you need? Include 10 inches extra fabric to allow the fabric to hang properly. *Hint*: Watch your units! Find the total number of inches of fabric you need, and then convert it to yards, which is how fabric is typically sold.

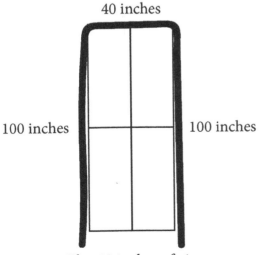

40 inches

100 inches   100 inches

Plus 10 inches of give

PAGE 211

3. **Buying Ribbon** (🖩) — Suppose you want to make Christmas gifts for neighbors. You need 2 feet of ribbon for each gift. The ribbon is sold by the yard.

   a. If you want to make 20 gifts, how many feet of ribbon do you need?

   b. How many yards of ribbon do you need? Give your answer as a fraction.

4. **Percents** (🖩)

   a. If you finish 8% of a 21-mile bike ride before stopping for lunch and then finish another 8% before stopping again, how many miles altogether did you go?

   b. In the previous problem, how many miles would you still have left to go on the bike ride?

5. **Flashcards** — Be sure to keep reviewing your flashcards.

PRINCIPLES OF MATHEMATICS

CHAPTER 14. Measuring Distance
LESSON 3. Different Conversion Methods
Worksheet 14.3

Remember to include the unit in your answer whenever one is given.

You may use a calculator on this worksheet whenever you see this symbol (🖩).

1. **Unit Conversions via the Ratio Shortcut** (🖩) — Use the ratio shortcut method to answer the following questions. Show your work.

    a. 7 feet equal how many inches?

    b. 5 miles equal how many feet?

    c. 105.75 inches equal how many yards?

    d. Your GPS tells you to turn right in 500 feet. What fraction of a mile is that?

2. **Unit Conversions via Mental Math** — Answer the following questions mentally.

    a. 15 feet equals how many yards?

    b. 3 feet equals how many inches?

3. **Thinking Outside of the Box**

    a. Simplify: $\dfrac{5 \text{ ft} \cdot \text{in}}{\text{in}}$

    b. Simplify: $\dfrac{a \cdot b}{a}$

    Hint: The *a* and *b* are just placeholders—they could represent any number. But both the *a*s would be the same number, so you can safely know that their division will equal 1, as any number (other than 0) divided by itself equals 1. You'll learn more about using letters to represent numbers in Book 2.

PAGE 213

4. **Ratio Review** — Write the conversion ratio between the units listed two ways: with the first unit in the numerator and with the second unit in the numerator.

   Example: inches and feet

   $\dfrac{12 \text{ in}}{1 \text{ ft}}$ and $\dfrac{1 \text{ ft}}{12 \text{ in}}$

   a. centimeters and meters

   b. feet and miles

   c. inches and yards

5. **Motorboat Speed** (🔢) — If a motorboat goes 65.57 miles in 60 minutes (which is another way of saying 65.57 miles per hour), how far can it go in 1 minute? *Hint*: Use a proportion to find the answer.

6. **Flashcards** — Be sure to keep reviewing your flashcards.

| | PRINCIPLES OF MATHEMATICS | CHAPTER 14. Measuring Distance LESSON 4. Currency Conversions | Worksheet 14.4 |

Remember to include the unit in your answer whenever one is given!

You may use a calculator on this worksheet whenever you see this symbol (🖩) .

1. **Currency Conversions** (🖩) — Use a conversion rate of 1 U.S. Dollar (represented $) = 0.63 British Pounds (represented £).

    a. Convert to pounds: $405

    b. Convert to pounds: $33.46

    c. Convert to dollars: £56.90

    d. Convert to dollars: £890.76

2. **Visiting Thailand** (🖩) — Use a conversion rate of 1 U.S. dollar = 31 Thai baht.

    a. You brought $80 and an ATM card with you. You exchange the $80 for bahts upon arrival. You then pay 100 bahts for a taxi to your hotel, 50 bahts for a snack from the hotel lobby, and 620 bahts for your first night's hotel. How many bahts do you have left?

    b. You go shopping and see something you'd like for 250 bahts. How many U.S. Dollars is that?

    c. You go out to eat later that day; your meal costs 40 bahts. How many bahts should you pay in total if you want to leave a 20% tip?

PAGE 215

3. **Visiting Spain** (🖩) — Use a conversion rate of 1 euro = 1.3 U.S. dollars.

   a. While visiting Spain, you exchange $80 American dollars for euros. How many euros should you get, assuming the place doing the conversion doesn't charge a fee?

   b. While in Spain, you go out to eat. You order items costing 8 euros and 5 euros. They charge a 7% tax. How much in euros will your total be if you include a 15% tip on your after-tax total? *Hint*: Find the total with tax, and then the tip on top of that.

   c. How much in U.S. dollars did it cost you for the meal in the previous problem?

4. **Just Your Average** — If you spent the following on souvenirs to bring home, what did the average souvenir cost you?

   $4.50, $0.75, $0.20, $5.75, $47.23, $2.30, $56.21, $16.78, $3.45, $2.60

5. **Flashcards** — Be sure to keep reviewing your flashcards.

PRINCIPLES OF MATHEMATICS

CHAPTER 14. Measuring Distance
LESSON 5. Metric Conversions

Worksheet 14.5

Remember to include the unit in your answer whenever one is given!

You may use a calculator on this worksheet whenever you see this symbol (🖩).

1. **Drawing in Metric**

   a. Draw a line 6 centimeters long each. (The lines will also be 60 millimeters long, as there are 10 millimeters per centimeter.)

   b. Draw a line 40 millimeters long.

   c. Draw a triangle with sides 2.5 centimeters long.

2. **Measuring in Metric** — Use a ruler to find the requested measurements. List all measurements in centimeters (express millimeters as portions of a centimeter using a decimal).

   When asked to identify a shape, remember that polygons can be both regular and irregular—count the sides on irregular polygon(s) to identify the shape. Be as specific as you can be.

   a. Shape that would best describe the end of the spatula:

   b. Length of $\overline{GH}$:

   c. Shape of this figure:

   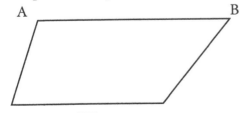

   d. Length of $\overline{AB}$:

PAGE 217

e. Shape of this figure:

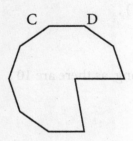

f. Length of $\overline{CD}$:

g. Shape of this figure:

h. Length of $\overline{EF}$:

i. How many centimeters long is line $\overline{AB}$?

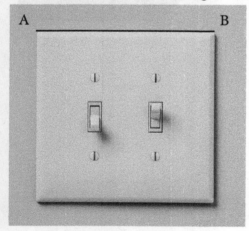

3. **Converting Between Metric** — Solve these problems by mentally multiplying or dividing by 10, 100, 1,000, etc.

   a. 150 centimeters equal how many meters?

   b. 7 centimeters equal how many millimeters?

PAGE 218

c. −7 centimeters equal how many millimeters? *Hint*: Remember, negative just means *the opposite of*. So −7 centimeters means 7 centimeters in the opposite direction.

d. 8 kilometers equal how many meters?

4. **Finding the Middle** — What is the **mode** of these lengths?
   8 m, 10 m, 8 m, 11 m, 7 m, 8 m, 7 m, 18 m, 2 m

5. **Out of the Box: Metric Heights**
   a. Grab a measuring tape and get someone to help you find your height in centimeters.

   b. Now find the other person's height in centimeters.

   c. What is the difference between your heights in centimeters?

6. **Questions**
   a. Share your thoughts about the need for a standard for right and wrong?

   b. Why is it easy to convert mentally between units in the metric system?

CHAPTER 14. Measuring Distance
LESSON 6. Multistep Conversions
Worksheet 14.6

Remember to include the unit in your answer whenever one is given!

You may use a calculator on this worksheet whenever you see this symbol (🖩).

1. **Unit Conversions and More** (🖩) — Use the ratio shortcut method method to solve these problems in one step.

    a. Convert 300,000 inches to miles.

    b. Football fields in the U.S.A. are 120 yards long. Represent that distance in miles using only the knowledge that 5,280 ft equals a mile and that 3 ft equals 1 yd.

    c. How many times would you have to run the length of a 100-yd football field to run a mile? *Hint*: You'll need your answer from the previous problem, along with a different skill besides unit conversion!

    d. There are 18,228.3465 feet in a league. Knowing this, how many yards are in two leagues?

2. **Just Your Average** (🖩) — If you travel the following distances in a week, how many miles do you travel on average in a day? Round your answer to the nearest whole number.

    15 mi, 10 mi, 20 mi, 7 mi, 0 mi, 9 mi, 45 mi

3. **Name and Measure That Shape**

    a. What is the length of $\overline{AB}$ in millimeters?

    b. What is the length of $\overline{CB}$ in centimeters?

    c. What kind of angle is ∠ CDA ?

PAGE 221

d. What two-dimensional shape best describes the front of the suitcase?

e. What three-dimensional shape describes the entire suitcase?

4. **Skill Sharpening** — Solve these problems mentally.

   a. Find the exact cost of 4 packages of $0.99 noodles.

   b. How old was someone born in 1945 in 2008?

   c. Total = $7.76   Give = $8.00   Change =   Bills and Coins =
   d. Total = $7.76   Give = $8.01   Change =   Bills and Coins =

5. **Flashcards** — Don't forget to keep reviewing your flashcards.

6. **How Big Was the Ark?** (🖩)

   "And this is the fashion which thou shalt make it of: The length of the ark shall be three hundred cubits, the breadth of it fifty cubits, and the height of it thirty cubits." Genesis 6:15

   Convert the dimensions of the ark into feet. Assume the cubit equals 18 inches (it may have been 17.5 inches or closer to 22 inches—we don't know exactly).[1] *Hint*: First convert into inches, and then into feet.

Extra Credit: Worldview Building — Watch the "Answers Academy: After Its Kind" videos on the Answers in Genesis Website and write a one-paragraph summary of one of the key points from the videos. The videos, which total about an hour, give an understanding of the variety God built into His creation, which explains how Noah fit the animals on the ark (see the "Plenty of Room for All" box).

http://www.answersingenesis.org/media/video/ondemand/aa-kind

---

[1] See Tim Lovett, "Noah's Cubit," (Worldwideflood.com, 2004), http://www.worldwideflood.com/ark/noahs_cubit/cubit.htm for an explanation of why Noah's cubit was most likely larger than 18 inches.

Noah's Ark: Plenty of Room for All

By looking at the dimensions of the ark and comparing it to that of modern aircraft carriers, we can see that the ark was a very large ship. According to the Navy's website,[2] Nimitz class aircraft carriers are 1,092 feet long—about $2\frac{1}{2}$ times the length of Noah's ark (if we use the smallest measurement for a cubit to compute its size).

The carriers can carry 5,000–5,200 people, 60+ aircraft, supplies, and more!

If the ark was about half the size of the Nimitz class aircraft carrier, there would have been plenty of room on the ark for Noah, the animals, and anyone else willing to come!

Still wondering how all the animals fit, including dinosaurs? Noah didn't have to take two of every single animal we've identified today. Because of the variety of genetic information God coded into animals, Noah only had to bring two of each kind of animal. He would only have needed two of the canine kind in order to produce all the varieties of dogs and wolves we have today.[3] Noah could have fit two of each kind of animal on the ark with plenty of room to spare. As for dinosaurs, they start out small, and not all dinosaurs were enormous. Noah didn't have to bring gigantic dinosaurs—in fact, bringing younger ones would have made sense, as they'd be more capable of replenishing the earth after the flood.[4]

---

[2] Stats taken from Department of the Navy, "Aircraft Carriers = CVN" *United States Navy Fact File* ,http://www.navy.mil/navydata/fact_display.asp?cid=4200&tid=200&ct=4 (accessed 12/09/14).

[3] The fact that two canine creatures could give rise to the variety of dogs and wolves around today has often been misunderstood to mean that animals evolve. However, the opposite of evolution has actually taken place. Every new breed of dogs *loses* genetic information. A dog with short hair has lost the genetic information for long hair. Nothing is evolving at all! Dogs are still dogs. Watch Ken Ham, "Answers Academy: After Its Kind" (Answers in Genesis, 2009), http://www.answersingenesis.org/media/video/ondemand/aa-kind/aa-after-its-kind for more details.

[4] See Answers in Genesis' website for more information, specifically "Were Dinosaurs on Noah's Ark?" (July 26, 2000), http://www.answersingenesis.org/articles/2000/07/26/dinosaurs-on-the-ark.

PRINCIPLES OF MATHEMATICS

CHAPTER 14. Measuring Distance
LESSON 7. Conversions Between U.S. Customary and Metric

Worksheet 14.7

Remember to include the unit in your answer whenever one is given!

You may use a calculator on this worksheet whenever you see this symbol (🖩).

1. **Conversions Between the Systems (🖩)** — Use the ratio shortcut method to solve in one step.

   a. Convert 20 feet into centimeters.

   b. Convert 280 kilometers into miles.

   c. Convert 80.78 centimeters into inches.

2. **Apply It (🖩)** — Suppose you take a trip to France, where they use the metric system. Are you ready to translate the distances?

   a. If a sign says you're 108 kilometers away from your destination, how many miles is that?

   b. If you notice the driver is going 90 kilometers per hour, how many miles per hour is that?

   c. A shop wants to sell you ribbon priced at 3 euros per meter. You'd like to get 3 yards of it for a friend's dress. How many euros will it cost you? *Hint*: Don't let this fool you! You have all the tools you need. Just watch your units!

3. **Measuring**

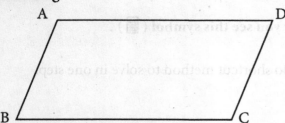

   a. What is the length in millimeters of $\overline{AD}$?

   b. What is the length in millimeters of $\overline{CD}$?

   c. What shape is the shape above?

   d. What kind of angle is $\angle ABC$?

   e. What kind of angle is $\angle BAD$?

4. **Flashcards** — Add any conversion ratios from Lesson 14.7 you don't know, and review your flashcards.

PRINCIPLES OF MATHEMATICS

CHAPTER 14. Measuring Distance
LESSON 8. Time Conversions

Worksheet 14.8

Remember to include the unit in your answer whenever one is given.

You may use a calculator on this worksheet whenever you see this symbol (🖩).

1. **More Conversions** (🖩)

    a. How many hours are in 2 days?

    b. How many hours are in 1 week?

    c. How many minutes are in 1 day?

    d. Convert −30 feet to centimeters.

    e. Convert 7,850 feet to kilometers.

    f. Convert 7.62 centimeters to feet.

    g. Let's say your total travel time to get to Thailand by air (including layovers) is scheduled to be 34 hours. How many days will you be traveling?

2. **Traveling** (🖩)

    a. If you travel at 60 miles per hour, how far will you travel in 4 hours?

    b. If you travel at 60 miles per hour, how far will you go in 35 minutes?

    c. If you travel at −35 miles per hour (i.e., 35 miles per hour in the opposite direction), how far will you go in 35 minutes?

    d. Going 30 miles in 60 minutes (which is another way of saying 30 miles an hour), how many minutes will it take you to go 5 miles? *Hint*: Set up a proportion to find the answer.

3. **Just Your Average** (🖩) — Suppose someone tracked how many hours a day he spent watching TV for a week and got these results:

    120 min, 60 min, 0 min, 30 min, 120 min, 0 min, 30 min

    a. On average, how many minutes a day were spent watching TV?

    b. What percentage of *hours* a day were spent watching TV?

    c. Over the course of a week, how many hours altogether were spent watching TV?

    Suggestion: It's often helpful to take stock of where we're spending our time—and math can help us! Keep a log of how many minutes you spend each day doing some activity (watching TV, reading, texting, exercising, etc.) for a week. How many hours total did you spend on that activity? Were you surprised by the result?

4. **Time for Everyday Time** — Solve mentally.

    a. If a class that lasts 2 hr 45 min starts at 9:30 a.m., at what time will it finish?

    b. If a meeting that lasts 5 hr starts at 8:15 a.m., at what time will it end?

    c. If a meeting that lasts 5 hr starts at 8:15 a.m., at what time will it end if two 30-minute breaks are given?

    d. If you need 20 min in drive time plus 15 min at a store plus another 45 min in drive time, at what time do you need to leave the house to make a 4 p.m. appointment?

5. **Watching a Movie** — Solve mentally.

    a. If a movie is 144 minutes long, how many hours and minutes is that?

    b. If a movie is 120 minutes long and you start it at 7 p.m., when will it finish?

6. **Flashcards** — Add any ratios from Lesson 14.8 you don't know and review your flashcards.

CHAPTER 15. Perimeter and Area of Polygons
LESSON 1. Perimeter
Worksheet 15.1

Remember to include the unit in your answer whenever one is given!

You may use a calculator on this worksheet whenever you see this symbol (🔢).

1. **Perimeter** — Find the perimeter of the following figures. As always, be sure to include the proper unit in your answers.

   a. A regular nonagon with 34-inch sides

   b. A regular (equilateral) triangle with 2-feet sides

   c. The irregular hexagon shown

   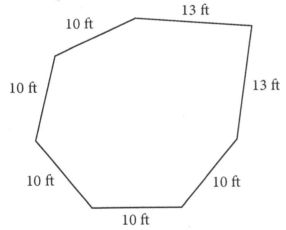

   d. A square with 3-inch sides

   e. A rectangle 20 feet long by 5 feet wide

2. **Fencing a Garden** (🖩) — Say you decided to fence a garden in order to keep deer out. You measure your garden, and it is basically a rectangle 7 feet long by 11 feet wide. *Hint*: Watch your units as you answer the following questions.

   a. How much fencing do you need?

   b. The fencing costs $3.99 a yard. How much will it cost before tax?

   c. If you use a 20%-off coupon, how much will the fencing cost before tax?

   d. If a 5% sales tax is added to the total found in 2c (the total after the 20% discount), how much will the fencing cost?

3. **Skill Sharpening**

   a. $\dfrac{78}{28} - \dfrac{16}{56}$

   b. $8(4 - 5)$

CHAPTER 15. Perimeter and Area of Polygons
LESSON 2. Formulas — Worksheet 15.2

You may use a calculator on this worksheet whenever you see this symbol (🧮).

1. **Formulas and Flashcards** — Express the following relationships as formulas.

    a. perimeter of a polygon

    b. perimeter of a regular polygon

    c. perimeter of a rectangle

    d. Make flashcards to help you remember how to find these perimeters.

2. **Formulas Beyond Geometry** — Use letters to express these relationships.

    a. The total cost for participating in a sports team equals the cost of the jersey plus $25.

    b. The power running through an outlet always equals the voltage times the currency.

    c. The change in our distance equals the opposite of 5 miles times the number of hours.

3. **Tiling a Pool**

    a. Suppose your family is installing a rectangular pool that's 13 feet long by 5 feet wide. You want to put one row of decorative tile around the top of the entire pool. How many feet of tile do you need?

    b. If each tile and its grout lines take up 6 inches, how many tiles will you need?

    c. If each tile costs $2.99, and you also need $5.99 worth of grout, how much will it cost if you also have to pay a 9% sales tax on the tile and grout?

4. **Framing a Picture** (🖩) — Suppose you decide to get a picture you painted custom framed. The picture is a 3-foot square. A custom frame costs $7.99 a foot, plus $10 for the labor, but you have a 40% off coupon you can apply toward just the frame itself. If your state charges a 4.5% sales tax on the frame cost (not labor) after the discount has been applied, how much will it cost you altogether? *Hint*: This problem (like many in real life) involves many steps. Before you begin, take a moment to think through what you need to find and how you'll find it.

5. **Skill Sharpening** — Solve these problems mentally.

   a. 4 • $18

   b. If someone is celebrating a 32-year wedding anniversary and they're 55 years old, how old were they when they got married?

   c. Total = $13.97   Give = $14   Change =   Bills and Coins =
   d. Total = $13.97   Give = $14.02   Change =   Bills and Coins =

PRINCIPLES OF MATHEMATICS

CHAPTER 15. Perimeter and Area of Polygons
LESSON 3. Area: Rectangles and Squares

Worksheet 15.3

You may use a calculator on this worksheet whenever you see this symbol (🖩).

1. **Area** — Use the formulas we discussed today to find the area of the following figures. As always, be sure to include the proper unit in your answers—for area, make sure you mention that it's a *square* unit!

   a. A 7-inch square

   b. A rectangle 6 feet long by 4 feet wide

   c. A 12-inch square

   d. A rectangle 17 yards long and 24 yards wide

2. **Gardening** (🖩)

   a. Suppose you have a square garden with 30-foot edges you need to fertilize. The fertilizer package says you need to apply 1 pound of fertilizer per 450 square feet. How many pounds of fertilizer do you need to apply to cover your garden? *Hint*: This is a multi-step problem! Start by defining what you know, then planning (i.e., look for relationships), and then solving and checking. Apply proportions along with what you learned about area.

   b. If you fertilize 6 times per season, how many pounds will you need per season?

   c. The fertilizer is sold in 1-pound bags for $4.99, plus 5% tax. How much will it cost to fertilize the garden all season?

3. **Traveling** (🖩) — Suppose you drive into Canada and see that your destination is 120 kilometers away. How far is that in miles?

4. **Carpentry** — You want to build a 54-inch long table. If you buy an 8-foot long wood top that's the appropriate width, how many inches will you have left over?

5. **Flashcards** — Add the formulas covered in the corresponding text to your flashcards, unless you already know them. Review any flashcards you still need to learn.

CHAPTER 15. Perimeter and Area of Polygons
LESSON 4. Area: Parallelograms
Worksheet 15.4

You may use a calculator on this worksheet whenever you see this symbol (🖩).

1. **Area** — Find the area of these parallelograms. As always, be sure to include the proper unit in your answers.

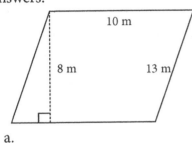
a.

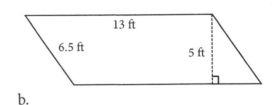
b.

2. **Perimeter Review** — What is the perimeter of the parallelogram in 1b above? Opposite sides of a parallelogram are equal.

3. **Painting a Room** (🖩)

   a. Suppose you've obtained permission to paint your room. Your ceilings are 8 feet tall. You have two walls that are 6 feet long, and two that are 8 feet long. One of the 8 feet walls has a doorway that's 3 feet wide and 6 feet tall, and the other has a closet that is $4\frac{1}{2}$ feet wide and 6 feet tall. How many square feet do you need to paint, assuming you don't paint the door or the closet? *Hint*: When you face a problem like this, it's helpful to draw it out, as we've done here.

   Notice that we drew each wall as a rectangle, with the height of 8 ft (our ceiling height) and the length specified (8 ft and 6 ft). Then we drew in the door and the closet. Remember, we can use shapes to represent different objects—including walls!

   To find the area you're painting, first find the area of each wall and add those together. Then find the area taken up by the door and the closet and subtract those areas from the overall area to find the area you're actually going to paint.

   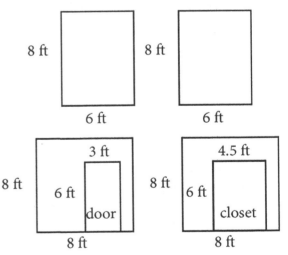

   b. If one quart of paint covers 95 square feet, how many quarts will you need to purchase to paint the area you found in 3a?

4. **Construction** (🖩)— Make the following estimates for a room 16 feet wide and 22 feet long with $9\frac{1}{2}$ feet ceilings. *Note*: The pricing in this problem is all circa 1908. Prices have likely gone up considerably.

   a. Draw rectangles representing each wall of the room. Remember, there are four walls in the room. Two are 16 feet by $9\frac{1}{2}$ feet, and two are 22 feet by $9\frac{1}{2}$ feet. *Hint*: It does not matter the actual lengths that you draw, so long as you label the lengths correctly, as the purpose of drawing in this case is simply to help us find the area.

   b. What is the area of all the walls in feet?

   c. What is the area of all the walls in yards if 1 sq yd equals 9 sq ft? *Hint*: Set up a proportion to solve.

   d. A bunch of wood pieces (called laths) will cover 4 square yards. How many bunches are required to cover the walls of this room? Round your answer up to the next whole number, as you need to make sure that you have enough.

   e. The wood pieces cost $0.30 per bunch. How much will the number of bunches needed cost?

   f. A 100 pounds of hard wall plaster will plaster 5 square yards. How many pounds of plaster will it take to plaster this room?

   g. Hard wall plaster is sold in 100-pound bags. How many bags are needed? Round your answer up to the next whole number, as you need to make sure that you have enough.

   h. Hard wall plaster costs $0.33 per bag. What will the plaster cost?

   i. How much should the workman charge for his labor if he charges 33 cents per square yard?

   j. If the customer also decides to add molding around the ceiling of the room, how much molding would it take?

5. **Flashcards** — Make a flashcard for the area of a parallelogram. Review any flashcards you still need to learn.

PRINCIPLES OF MATHEMATICS

CHAPTER 16. Exponents, Square Roots, and Scientific Notation
LESSON 1. Introducing Exponents

Worksheet 16.1

You may use a calculator on this worksheet whenever you see this symbol (🖩).

Now that we've covered exponents, be sure to use exponents to express square units.

1. **Exponent Time** — Use exponents to express these multiplications more simply.

   Example: $2 \cdot 2 = 2^2$

   a. $8 \cdot 8 \cdot 8 \cdot 8 \cdot 8 \cdot 8 \cdot 8 \cdot 8 \cdot 8 \cdot 8 \cdot 8 \cdot 8$

   b. $50.6 \cdot 50.6 \cdot 50.6$ *Hint*: Yes, you can rewrite decimals as exponents too!

   c. $3 \times 3 \times 3 \times 3 \times 3 \times 3 \times 3 \times 3 \times 3 \times 3 \times 3$

   d. $\frac{1}{4} \cdot \frac{1}{4} \cdot \frac{1}{4} \cdot \frac{1}{4}$ *Hint*: Repeated multiplication of fractions can be written using exponents as well! Just put the fraction inside parentheses to avoid confusion. *Example*: $\frac{2}{3} \cdot \frac{2}{3} = (\frac{2}{3})^2$

   e. $-4 \cdot -4 \cdot -4 \cdot -4 \cdot -4 \cdot -4$ *Hint*: When using exponents to show repeated multiplications of a negative number, put parentheses around the negative number to make it clear you mean that the negative number is multiplied that many times, and not just that you want the opposite of the results. *Example*: $-2 \cdot -2 = (-2)^2 = 4$, whereas -2^2 means the opposite of 2^2, or -4.

   f. cm • cm

   g. in • in

2. **More Exponents** — Rewrite these exponents as repeated multiplications and solve.

   a. 4^3

   b. $(\frac{1}{2})^2$

   c. 0.2^4

   d. $(-10)^5$

3. **Applying Exponents to Geometry** (🖩)

a. Represent an area of 87 square miles using an exponent to show that the miles are squared.

b. Find the area of the parallelogram below.

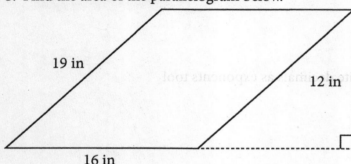

c. Find the area in yards of a square with 120-foot sides. *Hint*: Convert to yards before finding your area.

4. **Order of Operations**

   a. $7 + 5^2$

   b. $4 \text{ ft}^2 + (9 \text{ ft})^2$

5. **Question** — What is 2 raised to the 5th power?

6. **Flashcards** — Add the area of a square to your flashcards and review others as needed.

PAGE 238

PRINCIPLES OF MATHEMATICS

CHAPTER 16. Exponents, Square Roots, and Scientific Notation
LESSON 2. Understanding Square Roots

Worksheet 16.2

You may use a calculator on this worksheet whenever you see this symbol (📱).
Do not use a calculator's square root button.

Unless otherwise indicated or in a word problem where the positive value is clearly meant, put ± in front of each answer to show that the square root could be positive or negative.

1. **Square Roots** — Find the square root of each of these numbers. If you need help, try factoring them and using that information to find the square root.

    Example: $\sqrt{25} = \pm 5$

    a. $\sqrt{49}$

    b. $\sqrt{196}$

    c. $\sqrt{225}$

    d. $\sqrt{400}$

2. **Lengths of Squares** — Find the length of each side of the following squares. You can assume the answer is positive.

    Example: A square with an area of 25 ft²
    Answer: $\sqrt{25 \text{ ft}^2} = 5$ ft

    a. A square with an area of 144 ft²

    b. A square with an area of 81 in²

    c. A square with an area of 25 yd²

3. **Term Check**

    d. What does "two squared" mean, and what does it equal?

    e. What does "the square root of sixteen" mean, and what does it equal?

4. **Exponents** (📱) — Use exponents to express these multiplications more simply! Then solve. You can plug the multiplication into the calculator, however if your calculator has an exponent button, do not use it yet—it's important to get practice with exponents first.

    a. 24 • 24 • 24 • 24 • 24 • 24

    b. −10 • −10 • −10

c. $\frac{1}{5} \cdot \frac{1}{5} \cdot \frac{1}{5}$

d. 60 in • in

5. **Exponents 2 (🖩)** — Solve.

   a. $5^2 + 96$

   b. $2 + 9^2$

6. **Conversion Practice (🖩)**

   a. Convert 30 inches to yards

   b. Convert 20 kilometers to meters

   c. Convert 20 kilometers to yards

7. **Order of Operations with Roots** — For these problems, assume the square roots are positive.
   a. $4 \cdot \sqrt{9} - 5$

   b. $5 - 2 + \sqrt{64}$

   c. $4 \cdot \sqrt{121}$

PAGE 240

PRINCIPLES OF MATHEMATICS

CHAPTER 16. Exponents, Square Roots, and Scientific Notation
LESSON 3. Square Unit Conversions

Worksheet 16.3

> You may use a calculator on this worksheet whenever you see this symbol (🖩).

1. **Unit Conversion** (🖩) — Convert the following units.[1]

   a. 24 ft² into in²

   b. The area of the state of Utah is 84,897 mi². How many km² is that?

   c. The area of the state of Delaware is approximately 6,446 km². How many mi² is that?

   d. The area of Israel is approximately 20,770 km². How many mi² is that?

2. **Exploring the Great Pyramid** (🖩) — The base of the Great Pyramid is approximately a square, with sides measuring about 230 m. Use this information to find the following. Round all your answers to the nearest whole number.

   a. The perimeter of the base of the Great Pyramid in meters.

   b. The perimeter of the base of the Great Pyramid in yards.

   c. The area of the base of the Great Pyramid in meters.

   d. The area of the base of the Great Pyramid in yards.

3. **Installing Trim** (🖩) — How much trim would you need to install floor trim in a room that is 10 ft wide and 13 ft long with 8 ft ceilings if there is one doorway in the room that is 6 ft tall and 3.75 ft wide and you want the trim to go around the floor of the room, but not across the doorway?

---

[1] Facts from Sarah Janssen, sr. ed, M. L. Liu, Shmuel Ross, and Nan Badgett, eds, *The World Almanac and Book of Facts, 2012* (Infobase Learning, NY: 2012), 430, 789.

4. **Challenge Problem** (🖩) — Solve, assuming the square root is positive.
   $26 + \sqrt{64} \ (52 \div 2)$

5. **Simplified Expression** — Use a symbol to represent the following more concretely, and then solve. *Hint*: Don't forget to use ± when an answer could be positive or negative.

   a. $\frac{2}{3} \cdot \frac{2}{3} \cdot \frac{2}{3} \cdot \frac{2}{3}$ *Hint*: Remember to use parentheses around the fraction.

   b. 48(48)

   c. $-7 \cdot -7 \cdot -7$

   d. The number that, times itself, equals 100

   e. The number that, times itself, equals 400.

6. **Building a Chicken Coup**

   a. If you've been told you need 49 ft² for your chickens to roam, what length should you make each side of your square chicken coup?

   b. How many feet of fencing will you need to enclose the coup?

   c. How much will the fencing cost you if it's sold for $2.50 a yard, plus 11.25% tax? Round to the nearest whole dollar.

CHAPTER 16. Exponents, Square Roots, and Scientific Notation
LESSON 4. Scientific Notation
Worksheet 16.4

You may use a calculator on this worksheet whenever you see this symbol (🖩).

1. **Understanding Check**
   a. What does 6^0 equal?
   b. What does 6^1 equal?

2. **Powers of 10** — Complete the multiplication.
   a. $10^1 = 10$ = _____
   b. $10^2 = 10 \cdot 10$ = _____
   c. $10^3 = 10 \cdot 10 \cdot 10$ = _____
   d. $10^4 = 10 \cdot 10 \cdot 10 \cdot 10$ = _____
   e. $10^5 = 10 \cdot 10 \cdot 10 \cdot 10 \cdot 10$ = _____

3. **Multiplying by a Power of 10**
   a. 5×10^4
   b. 5×10^6
   c. $5 \times 5 \times 10^6$

Notice how multiplying by a power of 10 is just a matter of moving decimals to the right the same number of digits as the power of 10.

4. **Learning the Notation** — Represent these numbers using scientific notation.
   a. 489,000,000,000,000

   b. 1,337,000,000 (population of China in 2011; rounded to nearest million and not including Hong Kong or Macao[1])

   c. 3,050,000,000 (one of the bacteria measurements from a fish pond[2])

5. **Learning the Notation 2** — Convert these problems from scientific notation to standard decimal notation.
   a. 1.4×10^9 meters (diameter of the sun)

   b. $\$1.77 \times 10^{13}$ (national debt as of October 8, 2014, rounded to the nearest hundred trillion[3])

   c. 3.72×10^{13} (estimated number of cells in a human body[4])

---

[1] Facts from Sarah Janssen, sr. ed, M. L. Liu, Shmuel Ross, and Nan Badgett, eds, *The World Almanac and Book of Facts, 2012* (Infobase Learning, NY: 2012), 732.
[2] Fisheries and Aquaculture Department, "Preliminary Studies on the Analysis Of Bacterial Types in the Fish Ponds Applied with Four Kind of Animal Manure and the Effects of Manuring on the Ecosystem & Yield," FAO Corporate Document Repository, http://www.fao.org/docrep/field/009/ag149e/ag149e01.htm (accessed 10/08/14).
[3] Based on "United States Debt Clock," National Debt Clocks, http://www.nationaldebtclocks.org/debtclock/unitedstates, (accessed 10/8/14).
[4] Department of Experimental, Diagnostic and Specialty Medicine, University of Bologna, Bologna, Italy, "An Estimation of the Number of Cells in the Human Body," PubMed.gov, http://www.ncbi.nlm.nih.gov/pubmed/23829164 (accessed 10/8/14).

6. **Perimeter and Area Review**

   a. Find the perimeter in feet of a regular decagon with 45-inch sides.

   b. Find the perimeter of the irregular polygon below.

   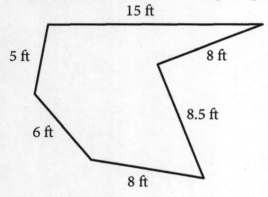

   c. What is the area of a square with 12-foot sides?

   d. What is the area of a rectangle that's 8 m long and 6 m wide?

7. **Term Check**

   a. What does "four squared" mean, and what does it equal?

   b. What does "the square root of four" mean, and what does it equal?

8. **Carpet Cleaning (🖩)** — If it costs $4.50 a square yard for carpet cleaning, how much total will it cost to clean carpets with these dimensions: 8 feet by 10 feet, 5 feet by 7 feet, and 3 feet by 4 feet.

CHAPTER 16. Exponents, Square Roots, and Scientific Notation
LESSON 5. More Scientific Notation
Worksheet 16.5

You may use a calculator on this worksheet whenever you see this symbol (🖩).

You will need a calculator that can handle scientific notation to solve problems 3 and 4. Free scientific notation calculators are available online. See www.christianperspective.net/math/pom1

1. **Comparing Numbers in Scientific Notation** — Use the symbols >, <, or = to compare these numbers:
    a. 5.8×10^{22} 6×10^{22}
    b. 2.5×10^5 3.05×10^6
    c. 1.105×10^9 6.31×10^8

    These numbers are the "average value of total bacteria" of two ponds in an experiment.[1]

2. **Mastering the Notation**
    a. Rewrite the numbers in problem 1c in decimal notation.

    b. Express 875,200,000,000 in scientific notation.

3. **Scientific Notation on a Calculator** (🖩) — Perform these multiplications on a calculator. The answers it gives you should be in scientific notation! Round to 2 decimal places.
    a. 450,265,000 • 250,600,000

    b. 287,963(45,000,000)

4. **Distance to the Stars** (🖩) — The closest star to earth, Proxima Centauri, is about 4.3 light years away. Now that may not sound too far, but if we use math to explore this number, we'll discover just how far away that really is—and get a glimpse of God's greatness. Because you'll be multiplying and dividing such large numbers to find this, you'll find your calculator will switch to scientific notation on you! Also, while there is a way to multiply and divide in scientific notation, since we haven't covered it yet, we'll be converting numbers to decimal notation in order to multiply and divide.

    a. **Seconds in a Year** — Have you ever wondered how we know the distance light travels in a year? We cannot physically measure the distance, but we can measure the distance light travels in a second. Then we can use math to find the distance traveled in a year. Start by figuring out approximately how many seconds there are in a year. To do this, you'll need to convert from 1 year to days (use 365.25 days per year), from days to hours (there are 24 hours in a day), from hours to minutes (there are 60 minutes in an hour), and from minutes to seconds (there are 60 seconds in a minute). *Note*: Remember, a light year is a unit based on the distance we observe light travel in a second here on earth. Keep in mind, though, that this doesn't mean light took this long to reach earth. See Appendix A in the *Student Textbook* for more details.

[1] Fisheries and Aquaculture Department, "Preliminary Studies on the Analysis of Bacterial Types in the Fish Ponds Applied with Four Kind of Animal Manure and the Effects of Manuring on the Ecosystem & Yield," FAO Corporate Document Repository, http://www.fao.org/docrep/field/009/ag149e/ag149e01.htm (accessed 10/08/14).

b. **Distance Light Travels in a Year** — Light travels approximately 186,282 miles per second. Given this, and the number of seconds in a year you found in 4a, approximately how far does light travel in a year? Give your answer in scientific notation. Round to three decimal places. Important: Do *not* clear the full answer from the calculator; you will need it for the next step.

c. **Distance to Proxima Centauri in Miles** — Now that we know how many miles are in a light year (your answer to problem 4b), let's see how many miles away the closest star to earth—Proxima Centauri—is to earth. It is 4.3 light years away. How many miles is this from the earth? Give your answer in scientific notation; round the decimal portion to three digits, but again, do not clear from the calculator.

d. **Putting It in Perspective** — According to MapQuest, it is about a 3,000 mile (40 hour) drive from Massachusetts to California. How many times longer is a trip to Proxima Centauri? Express your answer in scientific notation; round the decimal portion to three digits.

e. **Distance Light Travels in a Year in Decimal Notation** — Re-express your answer to 4b in decimal notation.

f. **Distance to Proxima Centauri in Miles in Decimal Notation** — Re-express your answer to 4c in decimal notation.

g. **Putting It in Perspective in Decimal Notation** — Re-express your answer to 4d in decimal notation.

5. **Comparing the Debts**[2] (🖩) — Use <, >, or = to show how these debts compare. They're all given here in dollars. (Numbers are approximate and as of October 2014.)

   a. U.S. 1.77×10^{13} China 5.04×10^{12}
   b. U.S. 1.77×10^{13} Russia 2.31×10^{11}

6. **Challenge Problem** (🖩) — Always remember not to panic when you see unfamiliar problems—just think them through step by step.

   $2(5 + 2)^2 + 60 \div 3$

> "When I consider thy heavens, the work of thy fingers, the moon and the stars, which thou hast ordained; What is man, that thou art mindful of him? and the son of man, that thou visitest him?" Psalm 8:3–4

---

[2] Data based on "United States Debt Clock," National Debt Clocks, http://www.nationaldebtclocks.org, (accessed 10/8/14) and Google Finance.

Review of Chapters 12–16 — Worksheet 16.6

You may use a calculator on this worksheet whenever you see this symbol (📱).

Use these problems to help you review. If you have questions, review the concept in the *Student Textbook*. The lesson numbers in which the various concepts were taught are in parentheses.

1. **Frequency Table** (📱) — A restaurant sells the following shakes during a lunch rush: vanilla, chocolate, vanilla, chocolate, vanilla, vanilla, vanilla, chocolate, strawberry, vanilla, chocolate, chocolate, chocolate, strawberry, vanilla, chocolate, chocolate, vanilla, vanilla, vanilla, strawberry. Draw a frequency table showing the percent of each flavor sold as compared to the total sold. Round the relative frequency to the nearest whole percent. *Example*: 0.4578 rounds to 46%. (Lesson 12.1)

Type	Frequency	Relative Frequency
Vanilla Shakes		
Chocolate Shakes		
Strawberry Shakes		

2. **Bar Graphs** (📱) — Use graph paper to graph the data below showing the time it took each runner to complete the race. (Lesson 12.4)

    Mary: 30 min

    Sarah: 32 min

    Jenna: 35 min

    Sally: 40 min

3. **Coordinates** — How would you describe the four corners of this rectangle using coordinates? Follow the (horizontal, vertical) convention to list the coordinates. (Lesson 12.5)

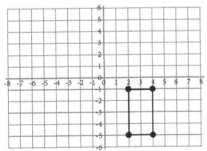

4. **Averages** (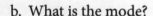) — Use the hypothetical high temperatures (in degrees Fahrenheit) shown on the line graph to answer the questions. *Hint*: Write down the value of each point on the graph, and use the values to find the various measures of central tendency. (Lessons 12.6 and 12.8)

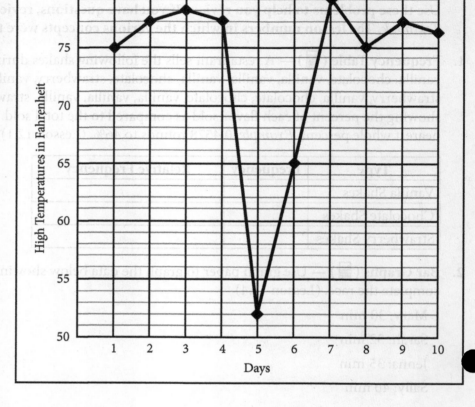

   a. What is the average temperature (i.e., the mean)?

   b. What is the mode?

   c. What is the median?

5. **Shape Time**

   a. Find the area of a rectangle that's 2 in by 4 in. (Lesson 15.3)

   b. What is the area of this shape? (Lesson 15.4)

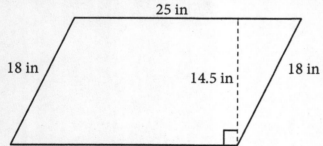

   c. What is the name of the shape from problem 5b (be as specific as possible)? (Lesson 13.3)

PAGE 248

d. What is the area of this square? (Lesson 15.3)

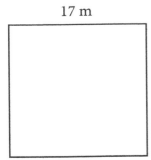
17 m

e. What is the perimeter of this irregular polygon? (Lesson 15.1)

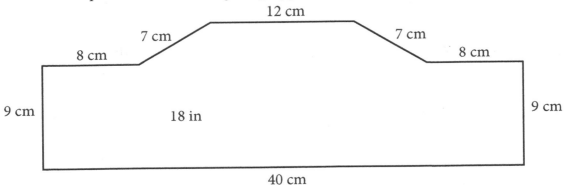

6. **Math in Action** (🖩)

   a. If you're traveling in a cab at the rate of 45 miles per hour, how far can you travel in 35 minutes? (Lesson 14.8)

   b. If you owe the cabdriver in the previous problem $2.50 a mile plus a 15% tip, how much will you owe at the end of those 35 minutes? (Lesson 9.4)

7. **Conversion** (🖩) — Convert the following units.

   a. $-5\frac{1}{4}$ kilometers to yards (Lesson 14.6)

   b. $50.75 to euros, if the exchange rate is $1 to 0.81 euros (Lesson 14.4)

   c. 70 yd² to ft² (Lesson 16.3)

   d. 2 ft² to yd² (Lesson 16.3)

8. **Checking the Language**

    a. $\sqrt{324}$ (Lesson 16.2)

    b. 5^4 (Lesson 16.1)

    c. Rewrite in decimal notation: 1.26956×10^{21} (Lesson 16.4)

    d. Rewrite in scientific notation: 405,897,000,000 (Lesson 16.4)

9. **Term Time**

    a. If the margin of error is 2% and a survey result is listed as being 44%, what is the expected range of the opinion of the entire population? (Lesson 12.2)

    b. What do you know about a triangle if you know it's scalene? (Lesson 13.4)

    c. Is this shape rotated or scaled? *Note*: It is also translated. (Lesson 13.5)

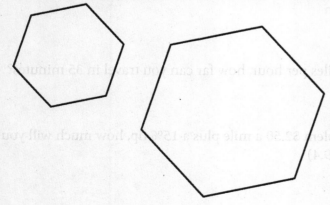

PRINCIPLES OF MATHEMATICS

CHAPTER 17. More Measuring: Triangles, Irregular Polygons, and Circles
LESSON 1. Area: Triangles

Worksheet 17.1

You may use a calculator on this worksheet whenever you see this symbol (🖩).

Don't forget to watch your units.

1. **Area of Triangles** (🖩) — Find the area of these triangles.

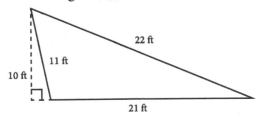
a.

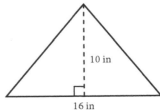
b.

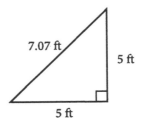
c.   d.

2. **Just Your Average** (🖩) — What was the average area of the triangles in problem 1 in square feet?
   Hint: First convert the areas of the triangles in 1b and 1c to square feet.

3. **Roofing** (🖩)

   a. How many square yards of roofing would you need to roof the following two sections of a roof? The section on the right is a rectangle.

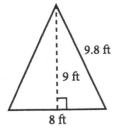
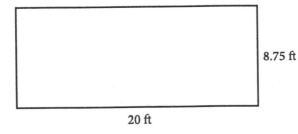

PAGE 251

b. If the roofing supplies cost $12.99 a square yard plus 5.5% tax, how much would the total supply cost be?

c. If it costs $50 an hour for installation, and it takes 5 hours to install the roof, how much will it cost for both the labor and the supplies?

d. If you could use roofing supplies that costs $1.29 a square foot plus 5.5% tax instead, would you save money over using the roofing supplies in 3b if installation stayed the same? If so, how much would you save?

4. **Grassing a Football Field** (🖩) — Including the end zones, a football field is 120 yards long and 53.33 yards wide. How many bags of grass seed would you need to seed it, if each bag of grass seed covers 200 square feet? Remember when you choose your answer that you can't buy a portion of a bag.

5. **Skill Sharpening**

   a. Simplify: $(-76)^2$

   b. Simplify: $-(76)^2$

   c. Simplify: $\frac{4}{7^5}$

   d. Find $8 \cdot \sqrt{16} - 5$ assuming a positive square root.

   e. If a square has an area of 441 ft², what are the lengths of each of its sides? *Hint*: Find the $\sqrt{441}$

   f. $4(8 + 5\frac{2}{5})$

   g. $7^0 + -2(3)$

6. **Flashcards** — Add a flashcard with the formula for the area of a triangle if you do not already know it; review all your flashcards.

PRINCIPLES OF MATHEMATICS

CHAPTER 17. More Measuring: Triangles, Irregular Polygons, and Circles
LESSON 2. Area: More Polygons

Worksheet 17.2A

You may use a calculator on this worksheet whenever you see this symbol (🖩).

1. **Finding the Perimeter** — Find the perimeter of each of the following shapes. Keep in mind that these shapes could represent the boundaries of property, bridges, road signs, etc.

   Remember, when finding the perimeter of a figure, use all you know about the figure to help you find the missing sides. For example, you know a rectangle has two pairs of equal-length sides, so only one of the pairs will be labeled.

   Figure 1b is a parallelogram with opposite sides equal, and figure c is a regular hexagon.

   a.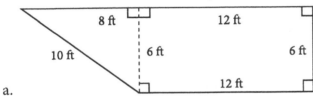

   b.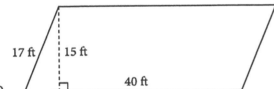

   c.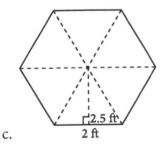

2. **Finding the Area**

   a. Find the area of problem 1a.

   b. Find the area of problem 1b.

   c. Find the area of problem 1c.

3. **Surveying** ( ) — Use the diagram to the right to solve. *Hint*: You know from the right angle marks that a portion of the land is a rectangle.

    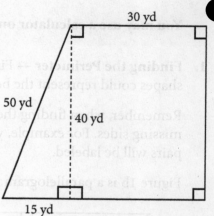

    a. Find the area of the land shown.

    b. If the homeowner decided to fence his property, how many feet of fencing would he need to enclose the entire property?

    c. How much would that fencing cost if the homeowner could buy the materials for $5 a foot, minus a $50-off coupon applied before a $6\frac{1}{2}$% tax?

4. **Skill Sharpening**

    a. Convert the area you found in 2a to square centimeters.

    b. Convert the area you found in 2c to square inches.

    c. What is the perimeter of the shape in problem 1b in meters?

    d. What is the perimeter of the shape in problem 1c in inches?

    e. Rewrite using exponents and solve: 3 • 3 • 3 • 3

    f. $\sqrt{256}$

PAGE 254

PRINCIPLES OF MATHEMATICS

CHAPTER 17. More Measuring: Triangles, Irregular Polygons, and Circles
LESSON 2. Area: More Polygons

Worksheet 17.2B

You may use a calculator on this worksheet whenever you see this symbol (📱).

Shape Time — Use the figures shown to solve problems 1 and 2.

Keep in mind that these shapes could represent the boundaries of property, bridges, road signs, etc.

Figure a is a rectangle and two triangles, figure b is a parallelogram (so its opposite sides are equal) and figure c is a regular heptagon.

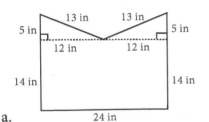
a.

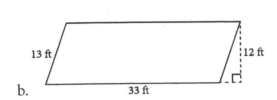
b.

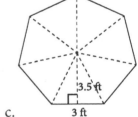
c.

1. **Finding the Perimeter** (📱) — Find the perimeter of each of the shapes shown above. Remember, when finding the perimeter of a figure, use all you know about the figure to help you find the missing sides. For example, you know a rectangle has two pairs of equal-length sides, so only one of the pairs will be labeled.

    a.

    b.

    c.

2. **Finding the Area** (📱) — Find the area of each of the shapes shown above.

    a.

    b.

    c.

3. **Average Area** (📱) — What was the average area of the 3 areas you found in problem 2? *Hint:* Be sure to make the units the same before you find the average!

4. **Hexagons and Bees** (🖩) — By categorizing and noticing the shapes of plants and animals, we often discover some fascinating things about God's design. For example, take a look at this picture of a honeycomb.

   a. What shape is the honeycomb made out of?

   Notice how the regular hexagons in a honeycomb perfectly tile—that is, they match up without leaving any gaps. A square and regular triangle also tile perfectly. However, if we compare these shapes, we'll see the wisdom in building the honeycomb out of regular hexagons—a wisdom God built into every bee.

   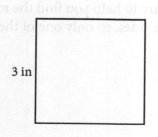 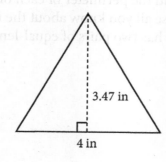 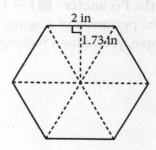

   b. Find the perimeter of the square.

   c. Find the perimeter of the regular triangle.

   d. Find the perimeter of the regular hexagon.

   e. Find the area of the square.

   f. Find the area of the regular triangle.

   g. Find the area of regular hexagon.

   h. Compare the area and perimeter of each shape to itself by setting up a ratio between the area and the perimeter. Put the area in the numerator.

i. Divide the ratios you formed in 4h to express them as a decimal number in order to make them easier to compare.

j. The bee has to use supplies to make the perimeter of the honeycomb, so the largest area per perimeter would allow it to store its honey most efficiently. Which shape allows it to do this? Math just helped us see the wisdom God placed in the honeybee!

5. **Skill Sharpening**

   a. $(\frac{15}{30})^2$

   b. $(-25)^2$

   c. If a square has an area of 900 ft², what are the length of each of its sides?

   d. Find $\sqrt{144} + 7^2 \cdot 2$ if the square root is positive.

   e. $8 \div 2 + (3 - 6^0)$

   f. $24(\frac{3}{25} - \frac{1}{75})$

PRINCIPLES OF MATHEMATICS

CHAPTER 17. More Measuring: Triangles, Irregular Polygons, and Circles
LESSON 3. Measuring Circles

Worksheet 17.3

You may use a calculator on this worksheet whenever you see this symbol (🖩).

Use 3.14 for π throughout this course.

1. **Circumference** (🖩) — Find the circumference of the circles given.

   a. A circle with a diameter of 100 yards.

   b. A circle with a radius of 32 inches.

   c. A circle with a diameter of 25 feet.

2. **Area** (🖩) — Find the area of the following. Remember, you can find the radius by dividing a diameter by 2, and the diameter by multiplying a radius by 2.

   a. A circle with a radius of 32 inches.

   b. A circle with a diameter of 100 yards.

   c. A circle with a diameter of 25 feet.

3. **A Dog's Run** (🖩)

   a. How much area will a dog have to run around in if he's tied to a pole with a 5 ft chain? Remember, he can run in a circle around the pole. *Hint*: Draw it out!

   b. How many feet of fencing would it take to fence the circular dog pen described above?

   c. How much would it cost to fence the circular dog pen if it costs $4.50 a foot for the fencing, plus 5% tax, and $150 for labor? There is no tax on the labor portion.

4. **Skill Sharpening**

   a. If a square has an area of 324 yd², what are the length of each of its sides?

   b. $-4(-3)(\frac{5}{65})$

   c. If a triangle has a base of 6 m and a height of 3 m, what's its area?

   d. 7^4 (🖩)

   e. Convert 18 yd² to in². (🖩)

   f. Convert 3,000 ft to km. (🖩)

5. **Flashcards** — Add flashcards to help you learn the circumference and perimeter of a circle, as well as the value of pi. Review your other flashcards.

PAGE 260

CHAPTER 17. More Measuring: Triangles, Irregular Polygons, and Circles
LESSON 4. Irrational Numbers

Worksheet 17.4

You may use a calculator on this worksheet whenever you see this symbol (🖩).

1. **Pi** — Give the value of 5 • π, rounding pi to 3.14.

2. **Number Sets**

   a. What is an irrational number?

   b. Is the result of 1 ÷ 6 an irrational number?

3. **Circle Time** (🖩)

   a. Find the circumference of a circle with a radius of 10 yards.

   b. Find the area of a circle with a radius of 10 yards.

   c. Find both the perimeter and the area of the figure shown, which is a semicircle (i.e., half a circle) with a line across the diameter. *Hint*: You can find the perimeter by finding the measurement of the semicircle (which would be *half* the circumference of a circle with the same diameter) and adding it to the length of the straight line. You can find the area by finding *half* the area of a circle with the same diameter.

   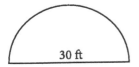
   30 ft

4. **Building** (🖩)

   a. Using the diagram to the right, find the area of this plot of land. *Hint*: The base of the triangle will be the diameter of the semicircle, or 50 ft.

   b. If 10% of the land must be left without a structure or driveway on it, what percent of land is available for structures/driveways?

   c. Using your answer to problems 4a and 4b, how many square feet of land is available for structures/driveway in this situation?

   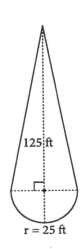
   125 ft
   r = 25 ft

PAGE 261

5. **Recording Real-life Shapes** (🖩) — Find a circular object around your house (the top of a bowl or a lid to a circular container, a CD/DVD, etc.) and find its circumference and its area.

6. **Skill Sharpening** — Solve these problems mentally.

   a. 4($2.50)

   b. If someone graduated high school in 1986, how old were they in 2014? Assume they graduated at 18 years old.

   c. Total = $17.11   Give = $18   Change =   Bills and Coins =
   d. Total = $17.11   Give = $18.11   Change =   Bills and Coins =

PAGE 262

PRINCIPLES OF MATHEMATICS

CHAPTER 18. Solid Objects and Volume
LESSON 1. Surface Area

Worksheet 18.1A

You may use a calculator on this worksheet whenever you see this symbol (🖩).

1. **Surface Area** (🖩) — Answer the questions, given that this is a rectangular prism.

   a. What is the area of the dotted rectangle?

   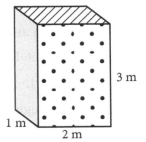

   b. What is the area of the grey rectangle? *Hint*: The height of the grey rectangle is the same as the height of the dotted rectangle.

   c. What is the area of the patterned rectangle? *Hint*: Even though the patterned rectangle doesn't have any dimensions next to it, you know its length and width from other parts of the figure.

   d. What is the total surface area of the entire shape? *Hint*: The entire shape has 6 sides.

2. **More Surface Area** (🖩) — Find the total surface area of the rectangular prism below. *Hint*: Follow a similar process to what we did in the previous problem: first find the individual areas of each surface, and then add them altogether.

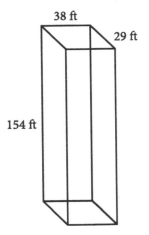

PAGE 263

3. **Unit Conversion (🖩)** — Find the total surface area of the rectangular prism in problem 2 in square meters instead of square feet.

4. **Surface Area and Triangular Prisms (🖩)** — It's time to apply surface area to triangular prisms by finding the total surface area of this triangular prism. *Hint*: In a triangular prism, there are not two of every side as there are in a rectangular prism. Try to think through what the sides are. Remember that the area of a triangle equals $\frac{1}{2}$ • base • height. All three sides of this triangle are 12 inches long.

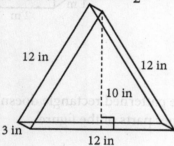

5. **Exponents** — Rewrite these multiplications as exponents and solve.

   a. $\frac{15}{60} \cdot \frac{15}{60}$

   b. $-2 \cdot -2 \cdot -2 \cdot -2$

   c. $4.36 \cdot 4.36 \cdot 4.36$

PRINCIPLES OF MATHEMATICS

CHAPTER 18. Solid Objects and Volume
LESSON 1. Surface Area
Worksheet 18.1B

You may use a calculator on this worksheet whenever you see this symbol (🖩).

1. **Surface Area Around the House** (🖩) — Find the following objects around your house. Using a measuring tape or ruler, measure the sides of the object. Then draw a picture of the objects and find their total surface area.

    a. Dictionary

    b. Tissue Box

2. **Fabric for a Tent** (🖩) — Suppose you need to waterproof the tent below. You found a solution you can spray; each bottle of the solution covers 400 ft². How many bottles of the solution do you need to spray the entire outside of the tent?

    Let's break this problem down. First, notice how in the drawings we've viewed the tent as rectangles and triangles.

    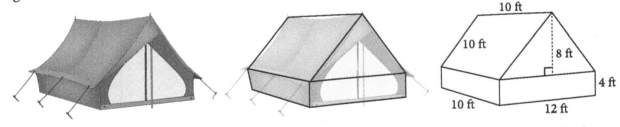

    a. Find the total surface area of the outside of the tent. *Hint*: While there are more individual areas than in previous problems, the same principles apply. Find the areas of each individual shape (the triangle and three different rectangles), and remember that there are **two** of every one on the tent: the shape you can see and the one on the backside that you can't see.

    b. Now use the total surface area from problem 2a and the knowledge that one bottle covers 400 square feet to figure out how many bottles of the spray you need to waterproof the tent.

3. **Pi** — What is the value of pi to two decimal places?

4. **Miscellaneous Mechanical Applications** (🖩)

   a. What is the area of a cross section in square millimeters of a water pipe (what you'd get if you cut straight through the pipe at one point) that is 3 inches in diameter? *Hint*: Watch your units.

   b. A circle has a diameter of 8 inches, with a hole that is 2 inches in diameter cut out of it. Find the area of the part of the circle still remaining (i.e., don't include the area of the hole).

5. **Skill Sharpening** — Solve these problems mentally.
   a. 5 + 25 + 66
   b. 91 − 67
   c. Total = $15.13   Give = $16   Change =   Bills and Coins =
   d. Total = $15.13   Give = $16.03   Change =   Bills and Coins =

CHAPTER 18. Solid Objects and Volume
LESSON 2. Volume
Worksheet 18.2A

You may use a calculator on this worksheet whenever you see this symbol (🧮).

1. **Find the Volume** — *Hint*: Be sure to list your answers in cubic units. Remember, to find the volume, first find the area of the base, and then multiply that by the depth (height).

   Find the volume of...

   a. ...the drawer if it is 10 inches long, 8.5 inches wide, and 2.5 inches thick.

   b. ...the freezer chest if the top is 4 feet long by 2 feet wide, and the height of the chest is 3 feet.

   c. ...the box of cereal if it is 8 inches tall, 6 inches wide, and 2 inches deep.

   d. ...the pan if it is 18 inches in diameter with a depth of 3 inches.

   e. ...the can of peaches if the diameter of the lid is 2.5 inches, and the height is 3.5 inches. Round to the nearest whole inch.

PAGE 267

2. **Find the Total Surface Area** (🖩)—Use the dimensions given in the previous problem to help you find the total surface area of these solid objects. *Hint*: Draw the problems out if you need to.

   a. the freezer chest in problem 1b

   b. the box of cereal in problem 1c

3. **Skill Sharpening** — Solve these problems mentally.

   a. 8(99)

   b. Find an approximate total (round each number to the nearest dollar): $2.60 + $5.99 + $1.99

   c. Total = $25.69   Give = $26   Change =   Bills and Coins =
   d. Total = $25.69   Give = $26.04   Change =   Bills and Coins =

4. **Flashcards** —Make flashcards for finding volume of prism and cylinders. Keep reviewing your flashcards, as you will be quizzed on them at the end of the chapter.

PRINCIPLES OF MATHEMATICS

CHAPTER 18. Solid Objects and Volume
LESSON 2. Volume
Worksheet 18.2B

You may use a calculator on this worksheet whenever you see this symbol (🖩).

1. **Around the House** — Find the following objects around your house. Use a measuring tape or ruler to find its dimensions. Then find its volume. You can draw a picture of it if you need to.

    a. Filing Cabinet or Box

    b. Tissue Box

2. **Miscellaneous Problems** (🖩)

    a. The circular basin of a fountain is 20 feet in diameter. How many square yards of floor tile would be required to tile it?

    b. What will it cost to tile the floor of the fountain in problem 2a at $9.50 a square yard, plus 5.5% tax, plus $100 (no tax) for installation?

    c. A round Jacuzzi with a diameter of 8 ft and a depth of 4 ft has what volume?

    d. You're considering renting a storage unit and you want to get the one that has the largest volume. The rooms are all rectangular. One unit is 5 feet by 5 feet by 6 feet; another is 7 feet by 5 feet by 4 feet. Which unit would you get?

    e. Rewrite 2^3 using multiplication signs. What does it equal?

3. **Challenge Yourself** (🖩) — The prism shown has a base that is a *regular* polygon. While you've not ever had to find the surface area/volume of this shape before, you know all thats needed to solve it.

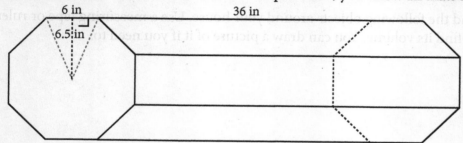

   a. What type of prism is this? *Hint*: What is the name of the base?

   b. What is the area of the base of the prism? *Hint*: Remember how we can view shapes like this as multiple triangles.

   c. What is the area of each rectangular section along its depth?

   d. What is its total surface area? *Hint*: Remember to add up the areas you can't see as well as those you can.

   e. What is its volume? *Hint*: Remember that volume is found by multiplying the area of the base times the depth.

4. **Flashcards** — Review all your flashcards; set aside those that might need more practice.

CHAPTER 18. Solid Objects and Volume
LESSON 3. Cubic Unit Conversion
Worksheet 18.3

You may use a calculator on this worksheet whenever you see this symbol (🖩).

1. **Conversions** (🖩)

    a. Convert 18,000 cubic inches to cubic yards.

    b. Convert 5 cubic meters to cubic centimeters.

    c. How many cubic feet is a shed that's 270,912 in^3?

    d. How many cubic feet is a shed that's 500,000 in^3?

2. **Miscellaneous Problem** (🖩)

    a. How much belting would it require to go around these two pulleys? Assume the belting is going around half of each of the circles.

    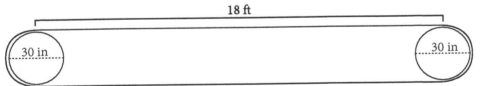

b. What is the volume of the prism shown? The prism's base is a right triangle with a height of 8 ft and a base of 6 ft.

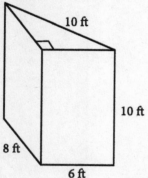

c. What is the volume of the triangular prism above in cubic yards?

d. What is the total surface area of the triangular prism above? *Hint*: Remember the sides you can't see as well as those you can!

3. **Skill Sharpening**

   a. Rewrite $(-6)^3$ using multiplication signs. What does it equal?

   b. Find $\sqrt{36}$

4. **Flashcards** — Review any flashcards that may be rusty.

PRINCIPLES OF MATHEMATICS

CHAPTER 18. Solid Objects and Volume
LESSON 4. Measuring Capacity in the U.S. Customary System

Worksheet 18.4

You may use a calculator on this worksheet whenever you see this symbol (🖩).

Be careful when solving that you don't get dry and liquid units mixed up—remember that even though some of the names are the same, they mean a different quantity.

1. **Conversions**

   a. Convert 3 gallons to quarts.

   b. How many tablespoons would it take to make $\frac{1}{4}$ cup? *Note*: You might need to know this some day when you are in the kitchen—I've had to use a tablespoon to measure $\frac{1}{4}$ cup before because I couldn't find our $\frac{1}{4}$ cup measurer!

   c. You need 8 gallons of water; how many fluid ounces is that?

   d. The instructions on a medicine bottle say to take with 8 fluid ounces of water. How many cups is that?

   e. How many pints are in 2.5 gallons?

2. **Translate the Recipe** — Use this recipe for corn cakes from an 1800s cookbook to answer the questions.

   Two or three eggs; three tablespoons of sugar beaten together; one pint of milk; one pint flour; one pint Indian meal; two teaspoons of yeast powder, or instead one even teaspoonful of soda, and two rising teaspoonfuls cream of tartar; one even teaspoon of salt; butter size of an egg, melted. To economize, use lard instead of butter; it will be nearly as good. Bake in little tins or cake pans. Sufficient for six persons.1

   a. How much milk does the recipe call for?

   b. Convert the milk amount to cups.

---

1   New England Mother, *Aunt Mary's New England Cook Book: A Collection of Useful and Economical Cooking Receipts, All of Which Have Been Practically Tested by a New England Mother* (Boston: Lockwood, Brooks, & Co., 1881), p. 26. Found on Google Books (accessed 1/15/13).

3. **More Conversions**
   a. At 8 cents a pint, what will 4 quarts of milk cost? *Note*: Don't you wish we still had these prices?

   b. How many pint bottles would it take to hold 5 gallons of gasoline?

   c. You're trying to half a recipe that calls for $\frac{1}{4}$ cup flour, so you need to measure $\frac{1}{8}$ cup. You don't have the ability to measure $\frac{1}{8}$ cup. How else could you measure the flour, and how much would you measure? *Hint*: You could use a tablespoon.

   d. Assuming you need a bushel of apples and the apples are sold for $9.99 a bushel or for $2.50 a peck, is it cheaper to buy the apples by the bushel or by the peck?

4. **Practicing the Skill** (▦) — Find the area of 4a–4c, and the volume of 4d–4f. Figure 4c is a parallelogram, 4d is a rectangular prism, and the base of 4e is a right triangle.

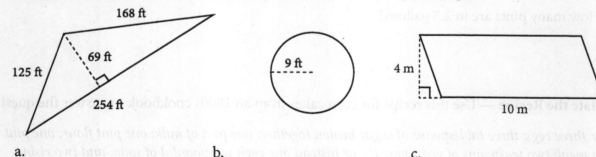

a.   b.   c.

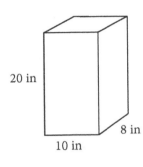
d.

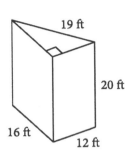
e.

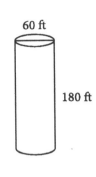
f.

5. **Surface Area** (🖩)

   a. What is the surface area of the rectangular prism in problem 4?

   b. What is the surface area of the triangular prism in problem 4?

6. **Fluid Intake** (🖩) — After realizing you need to drink more water, you decide to start measuring how much you drink over the course of three days. Here are your results:

   Day 1: 8 fl oz + 9 fl oz + 16 fl oz + 8 fl oz

   Day 2: 4 fl oz + 16 fl oz + 8 fl oz + 15 fl oz + 9 fl oz

   Day 3: 8 fl oz + 9 fl oz + 4 fl oz + 19 fl oz + 22 fl oz

   a. Over these three days, how many fluid ounces of water did you drink on average in a day? Round to the nearest ounce.

   b. If you need to drink 60 fluid ounces of water a day, how much do you need to increase your average by?

   c. How many actual gallons of water did you drink over the course of all three days combined?

7. **Flashcards** — Add flashcards for any of the bolded conversion ratios in Lesson 18.4 you don't know; continue reviewing other flashcards as needed.

8. **Questions**

    a. What was the gallon we use in the U.S. today historically called?

    b. Are all of the British capacity measurements the same as those used in the U.S.?

CHAPTER 18. Solid Objects and Volume
LESSON 5. Conversion to and from Cubic Units

Worksheet 18.5

You may use a calculator on this worksheet whenever you see this symbol (🖩).

1. **Learning the Skill** (🖩)

    a. A container 789 in³ can hold how many gallons of water?

    b. 5 liquid quarts is how many cubic inches?

    c. 10 gallons is how many cubic inches?

    d. 20 liquid pints is how many cubic inches?

2. **Volume Fun** (🖩)

    a. You want a container to hold 2 gallons of liquid. How many cubic inches does it need to be?

    b. Suppose you have a fish tank that's 30 inches long, 36 inches wide, and 1.5 feet tall. How many quarts of water can it hold?

    c. Will a cube-shaped shipping box that measures 13 in on all sides hold 1 bushel of wheat?

PAGE 277

3. **Miscellaneous Problems** (🖩)

   a. A rectangular prism 4 ft long, 3 ft wide, and 5 ft high holds how many gallons?

   b. A cistern in the form of a hexagonal prism is 7 ft. deep. Both its bases are the regular hexagon shown. How many gallons can it hold?

   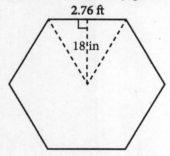
   2.76 ft
   18 in

   c. How many tons of coal will a bin 20 ft by 16 ft by 8 ft hold, if you can fit 1 ton in 80 cubic feet?

4. **Skill Sharpening**

   a. If the parallelogram shown were the base of a prism 8.5 m deep, what would the prism's volume be?

   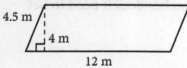
   4.5 m
   4 m
   12 m

   b. Rewrite $(-3)^4$ using multiplication signs. What does it equal?

   c. Find $\sqrt{121}$

5. **Flashcards** — Add flashcards for any of the conversion ratios in Lesson 18.5 you don't know; continue reviewing other flashcards as needed.

CHAPTER 18. Solid Objects and Volume
LESSON 6. Measuring Capacity in the Metric System
Worksheet 18.6

You may use a calculator on this worksheet whenever you see this symbol (📱).

1. **Conversions** (📱)

    a. Convert 586 milliliters to liters.

    b. Convert $4\frac{1}{3}$ liters to milliliters. Give your answer as a decimal.

    c. A 5-gallon tank holds how many liters?

    a. 3 teaspoons is how many milliliters?

2. **Gassing the Tank** (📱) — If you fill up for gas in a foreign country at the equivalent price of $1.19 per liter, how much are you paying per gallon (i.e., 3.78541 liters)?

3. **A Kite** (📱) — A kite is made of the shape shown in the diagram, the semicircle having a radius of 9 inches, and the kite a total height of 34 inches (25 in for the triangular portion, plus the 9 in radius). Find the area of the kite.

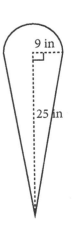

4. **Miscellaneous Problems** (🖩)

   a. The excavation for a house is to be a rectangular prism 40 ft long, 28 ft wide, and 6 ft deep. How many cubic yards of earth must be removed?

   b. What is the cost for the removal in problem 4a if it costs $5 a cubic yard of earth removed?

   c. A stone-mason builds 4 circular pillars, each with a radius of 3 ft and a height of $7\frac{1}{2}$ ft. He also builds a front wall (rectangular prism) 36 feet long, 3 feet high, and 1.5 feet thick. He charges $150 per 128 cubic foot for the work and materials, tax included. What is the total cost?

5. **Exponents and Roots**

   a. Rewrite $(\frac{62}{108})^2$ using multiplication signs. What does it equal?

   b. Find $\sqrt{324}$

6. **Flashcards** — Add flashcards for any of the bolded units in Lesson 18.6 you don't know; continue reviewing other flashcards as needed.

CHAPTER 18. Solid Objects and Volume
LESSON 7. Weight and Mass

Worksheet 18.7

You may use a calculator on this worksheet whenever you see this symbol (🖩).

1. **Weight and Mass** (🖩)

   a. If your car weighs 3,579 pounds, how many tons does it weigh?

   b. If you need 54 ounces of cheese for a recipe, how many 1-pound packages do you need to buy?

   c. If cheese is sold in grams instead (perhaps you're in another country!), how many grams do you need to purchase to get 7 lb of cheese?

   d. If you weigh 128 pounds, what is your mass in kilograms?

2. **Recipe Time**[1] (🖩) — Convert these measurements into cups. Give your answer as a whole number or a fraction.

   a. 120 ml of water

   b. 256 grams of flour if 128 grams of flour equals 1 cup

   c. 55 g of packed brown sugar if 220 grams of packed brown sugar equals 1 cup

---

[1] The conversion ratios in 2b and 2c are based on allrecipes.com, "Cup to Gram Conversions" http://allrecipes.com/howto/cup-to-gram-conversions/ (accessed 12/16/14).

3. **Watering Crops** () — If you've ever looked outside a window from a plane, you may have noticed sections of land that look like circles. Farmers sometimes use center pivot irrigation systems to water their crops. These systems are basically large sprinklers that form the radius of a circle and move around in a circular motion, watering the crops. Find the area sprinklers that these dimensions would cover.

   a. 7 ft

   b. 130 ft

   c. 900 ft

4. **More Conversion** ()

   a. Although we take a lot of measurements in feet, land is sold in acres. There are 43,560 ft² in 1 acre. Knowing this and what you've learned about unit conversion, how many acres do all of the sprinklers in problem 3 *combined* cover?

   b. How many fluid ounces can a 2-liquid pint container hold? (1 pt = 16 fl oz)

   c. How many gallons can 800 ft³ contain?

   d. How many cups are in a gallon?

5. **Formulas** — List the formula for finding the following.

   a. The area of a square

   b. The area of a parallelogram

   c. The volume of a prism

   d. The area of a triangle

6. **Average Weight** (🖩) — An online store is trying to choose which postal carrier they should use to ship their packages. They've concluded that Service A is cheaper if they ship packages 10 pounds or more in weight, and Service B is cheaper if they ship packages less than 10 pounds in weight. In the last week, they shipped packages weighing 6.3 lb, 12.8 lb, 12.5 lb, 24.9 lb, 6.4 lb, 12.6 lb, 12.3 lb, 12.5 lb, 24.6 lb, and 8.5 lb. Round your answers to the tenths place, as all of the poundage was given accurate to that place.

   a. What was the average package weight?

   b. What was the median?

   c. What was the mode?

   d. If this was a typical week, which service do you think would save them money?

7. **Out of the Box** — Find one food item that is described in fluid ounces; one in grams; one in pounds or ounces; and one in gallons, quarts, pints, or cups. (The same item may be described multiple ways.) Write down the measurements from the packages.

8. **Flashcards** — Add flashcards for any of the bolded units in Lesson 18.7 you don't know. Get someone to quiz you on the flashcards you made this chapter.

CHAPTER 19. Angles
LESSON 1. Measuring Angles
Worksheet 19.1

You may use a calculator on this worksheet whenever you see this symbol (📱).

1. **Measuring Angles** — Measure the marked angles in these pictures. Round your answer to the nearest multiple of 5. *Example*: 43° rounds to 45°.

a.

b.

c.

d.

2. **Skill Sharpening** (📱) — *Hint*: Simplify fractions before you multiply

   a. Find the volume of a triangular prism whose base is the triangle shown and whose height is 10 ft.

   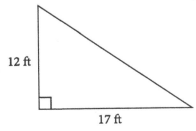

   b. How could you represent $\frac{5}{75}(\frac{5}{75})(\frac{5}{75})$ using an exponent? What does it equal?

   c. Find $\sqrt{225}$.

   d. Find the volume of a 36-in tall cylinder whose base has a radius of 24 in.

   e. How many gallons of water will the cylinder in the previous problem hold?

PAGE 285

f. Convert 60 fluid ounces to pints. (1 pt = 16 fl oz)

g. Convert 10 liters to gallons.

3. **Soccer Field Math** (🖩)

    a. One time, I decided to get some exercise by walking around the soccer fields while my brother practiced. As we road home, I began wondering how far I had walked. How could I possibly figure this out? Suddenly I remembered…math! I could use the Internet to look up a typical size for a soccer field and use those dimensions to figure out how far I had walked.

    If a soccer field is a rectangle 68 m by 105 m and you circled it three times, how far will you have walked?[1]

    [rectangle: 68 m by 105 m]

    b. What about if you circled it five times?

    c. If you wanted to go on a 2 mile walk each day, how many times would you have to walk around the field?

4. **Comprehension Check**

    a. If you make 180° turn, where do you end up facing?

    b. What's another unit that can be used to divide a circle into smaller sections?

---

[1] Dimensions based on those given by the Fédération Internationale de Football Association in "Regulations FIFA U-20 World Cup New Zealand 2015," 28. Found on FIFI.com, http://resources.fifa.com/mm/document/tournament/competition/02/45/23/69/regulationsfu20wc2015newzealand_e_011014_neutral.pdf (accessed 12/16/14).

PRINCIPLES OF MATHEMATICS

CHAPTER 19. Angles
LESSON 2. More with Angles
Worksheet 19.2

You may use a calculator on this worksheet whenever you see this symbol (🖩).

1. **Drawing Angles**

    a. Draw a 120° angle.

    b. Draw a 90° angle.

    c. Draw a 70° angle.

    d. Draw a 30° angle.

    e. Draw a 180° angle.

    f. Draw a 225° angle.

    g. What is the sum of the angles you drew in 1a and 1b?

    h. How many degrees would you have to add to a 40° angle to form a 90°?

    i. How many degrees would you have to add to a 40° angle to form a 180°?

2. **Acute, Right, and Obtuse**

    a. Which angle(s) in problem 1a–1f were acute?

    b. Which angle(s) in problem 1a–1f were right?

    c. Which angle(s) in problem 1a–1f were obtuse?

    d. Which angle(s) in problem 1a–1f were straight?

3. **Angles in Leaves**

    a. Use your protractor to measure each marked angle in this leaf. Round to the nearest multiple of 5.

    b. What is the sum of the angles?

PAGE 287

4. **Scale Drawings** — Give a scale drawing a try by drawing a scale drawing of a rectangular window that's 4.5 ft by 3 ft. This time, don't use graph paper—instead, use your ruler and compass to get straight lines and right angles. *Hint*: Use an inch to represent a foot to keep it easy. Start by drawing a line 4.5 inches long. Then draw a 90° angle, making the line forming the angle 3 inches long. Draw another 90° angle, making that line extend 4.5 inches long. Then draw another 90° angle, making that line extend 3 inches long. You'll end with a rectangle, with four 90° angles and 2 sets of parallel sides!

5. **Skill Sharpening (🖩)**
    a. What is the volume in cubic yards of a rectangular prism whose dimensions are 4 ft by 5 ft by 7 ft?

b. What is the surface area of this rectangular prism?

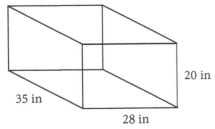

c. Find $\sqrt{441}$.

d. How many cups are in 7 gallons?

e. How many cubic inches does it take to hold 7 gallons?

f. What is the volume of a cylinder whose base has a diameter of 8 ft and whose height is 20 ft?

6. **Skill Sharpening 2** (🖩)

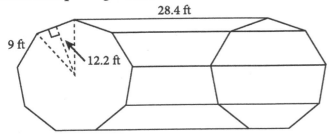

a. Find the surface area of the prism; its bases are regular nonagons.

b. Find the volume of the prism; its bases are regular nonagons.

c. Find √441.

6. **Skill Sharpening 2**

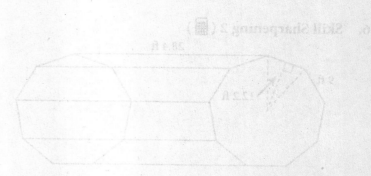

PRINCIPLES OF MATHEMATICS

CHAPTER 19. Angles
LESSON 3. Angles in Pie Graphs
Worksheet 19.3

You may use a calculator on this worksheet whenever you see this symbol (🧮).

1. Angles and Pie Graphs (🧮)

Pre-election Poll	
Candidate	Percent of Voters
Candidate A	40%
Candidate B	35%
Undecided	25%
Total	100%

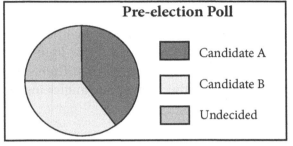

a. Without using a protractor, find the angle of the section of the pie graph representing the percent of voters in favor of Candidate A.

b. Without using a protractor, find the angle of the section of the pie graph representing the percent for Candidate B.

c. Without using a protractor, find the angle of the section of the pie graph representing the percent of undecided voters.

d. Use a protractor to find the different angles in the pie graph. If you did your math correctly above, your angles should be the same as you found previously.

Angle of Candidate A Section: _____

Angle of Candidate B Section: _____

Angle of Undecided Section: _____

e. Specify whether each angle in the graph is right, obtuse, or acute.

Angle of Candidate A Section: _____

Angle of Candidate B Section: _____

Angle of Undecided Section: _____

f. What is the sum of all the angles in the pie graph?

2. **Draw Your Own** (🖩) — Suppose there is rainy weather an average of 60% of the time in a city, sunny weather an average of 30% of the time, and other weather (snow, sleet, etc.) an average of 10% of the time. Draw a pie graph showing this. Use colored pencils or crayons to shade the sections of the circle. Be sure to include a legend that shows what color represents what. *Hint*: Figure out what percent of 360 each percent represents. Then use that information to draw the appropriate angles inside the circle!

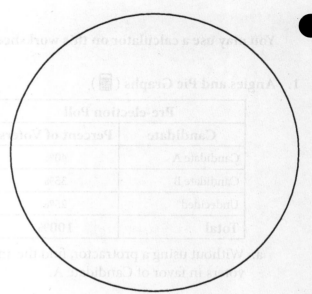

3. **Thinking It Through** (🖩)

    a. Find the area of the pie graph in problem 2 if it has a 3-inch diameter.

    b. What is the area of the portion of the graph representing the percentage of rain? *Hint*: You know what portion of the whole area that section represents!

    c. Find the circumference of the pie graph in problem 2.

    d. What portion of the circumference is part of the section representing rain?

4. **Draw Your Own 2** (🖩) — Suppose your spending for the month is as follows: $160 on household, $100 on groceries, $80 in gifts, and $60 in savings.

    a. Finish filling out the chart.

Category	Dollar Amount Spent	Ratio of Spending in Catagory to Total Spending	Percent of Total Spent in Category (Convert Ratio to a Percent)
Household	$160	$\frac{\$160}{\$400}$	40%
Gifts	$80	———	
Groceries	$100		
Savings	$60	———	
Total Spending	$400		

    b. On a seperate sheet of paper, draw a pie graph representing the spending in 4a. This time, use a compass to draw your own circle! *Hint*: Start by figuring out the angles to use in the graph.

CHAPTER 19. Angles
LESSON 4. Expanding Beyond
Worksheet 19.4

PRINCIPLES OF MATHEMATICS

You may use a calculator on this worksheet whenever you see this symbol (🖩).

1. **Angles Alive**

    a. If light bounces onto a reflective surface at the marked angle shown, draw the angle at which it will bounce off the surface.

    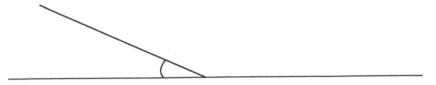

    b. If light bounces onto a reflective surface at the marked angle shown, draw the angle at which it will bounce off the surface.

    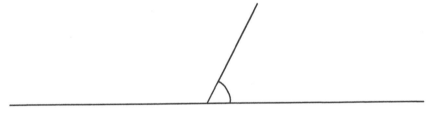

2. **Out of the Box** — Take a flashlight into a dark room with a mirror. Shine it at the mirror and look for the reflection on one of the other walls. You should be able to see the reflection of the light off the mirror at the same angle as you shown the light onto the mirror.

3. **Compass Time**

    a. To what degree is the marked hand of this compass pointing?

    b. If you turn 10° further towards the south, what reading would you be at? That is, what would the compass read if you moved the dial 10 degrees toward the south.

PAGE 293

4. **Drawing Shapes with Correct Angles** — Use what you learned about angles and shapes to draw the following without graph paper. Use your protractor to measure each angle and each line.

   a. Rectangle *Hint*: In a rectangle, each angle must be a right angle, or 90°; sides can be any length so long as both pairs of opposite ones are the same.

   b. Regular Pentagon *Hint*: In a pentagon, each angle is 108°; sides can be any length, so long as they are all the same length.

   c. Regular Hexagon *Hint*: In a regular hexagon, each angle is 120°; sides can be any length so long as they are all the same length.

   d. 100° angle

5. **Exploring Angles** — Refer back to problem 4 as needed.

   a. What is the sum of all the angles in a regular pentagon?

   b. Is each angle in a regular pentagon acute, obtuse, or right?

   c. What is the sum of all the angles in a regular hexagon if each one is 120°?

   d. Is each angle in a regular hexagon acute, obtuse, or right?

6. **Skill Sharpening** (🧮)

   a. Find the volume of this cylinder.

   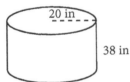

   b. How many gallons of water could the cylinder hold?

   c. How many liters of water could it hold?

   d. Rewrite $(-2.45)^5$ using multiplication. What does it equal?

   e. If you swim 50 meters with the following times, what is your average speed? Round your answer to the nearest whole second.
   45 sec   42 sec   44 sec   46 sec   45 sec

CHAPTER 19. Angles
LESSON 5. Chapter Synopsis and Faulty Assumptions
Worksheet 19.5

You may use a calculator on this worksheet whenever you see this symbol (🖩) .

1. **Recording Real-life Shapes** — Find and measure five different angles around your house (edges of tables, picture frames, lines in pictures, etc.). Specify if each angle is acute, obtuse, right, or straight.

2. **Drawing** — Draw a triangle with all 60° angles. What is the sum of all the angles?

3. **Convert a Metric Recipe** — Express your answers as a whole number, fraction, or mixed number.
   a. How many teaspoons is 80 mL of water?

   b. How many cups is 180 mL of milk?

   c. How many cups is 840 mL of water?

4. **Installing Floor Trim** (🖩) — You're installing floor trim in a rectangular front entryway. You need the trim to go around the edges of the 5 ft by 8 ft entryway that has a 4 ft opening for stairs, and two 3.5 ft openings (one for a door and one into another room). You do not need trim for the openings.

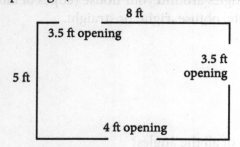

   a. How many feet of trim do you need to go around the perimeter of the entryway? Remember, you do not need trim for the openings.

   b. The trim costs $5.99 a foot installed, plus 5% tax. How much will it end up costing?

   c. Another company sells the trim for $0.50 a foot, plus 5% tax, but they also charge $75 (not taxable) for installations under 100 ft. How much will you save using them?

5. **Skill Sharpening** — Solve these problems mentally.

   a. 17 + 15 + 22

   b. 101 − 88

   c. Total = $32.52   Give = $33   Change =   Bills and Coins =
   d. Total = $32.52   Give = $33.02   Change =   Bills and Coins =

PAGE 298

CHAPTER 20. Congruent and Similar
LESSON 1. Congruency
Worksheet 20.1

You may use a calculator on this worksheet whenever you see this symbol (🖩).

1. **Congruency Marks** — Use the tick marks showing which sides are congruent to find the length of the sides requested.

    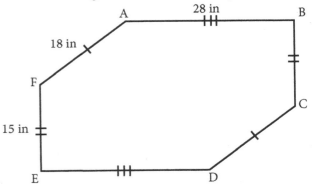

    a. What is the length of $\overline{ED}$?

    b. What is the length of $\overline{DC}$?

    c. What is the length of $\overline{BC}$?

    d. Is this an irregular hexagon, octagon, or pentagon?

2. **Miscellaneous**

    a. Using your protractor, what is the measure of ∠BCA and ∠CAB?

    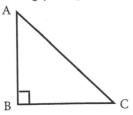

    b. Draw arc lines to show that ∠BCA and ∠CAB are congruent.

    c. Rewrite $(-7)^3$ using multiplication. What does it equal?

PAGE 299

3. **Blueprints** (🖩) — Use the following blueprint to answer the questions.

   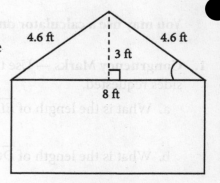

   a. Use your protractor to measure the slope of the shed's roof; the angle you should measure is marked.

   b. What is the perimeter of the visible section of the shed's roof (i.e., the triangle shown)?

   c. What is the area of the visible section of the shed's roof?

   d. If roofing costs $4.50 a square foot, plus 8.5% tax and $250 labor (not taxed), how much will it cost to roof just the visible section of the roof if you also have a coupon for $15 off the labor fee?

4. **A Mailbox**

   a. Find the area of the outlined section of this mailbox. The top part is a semicircle, and the bottom is a rectangle.

   b. Find the perimeter of the front of this mailbox.

5. **Skill Sharpening** — Solve these problems mentally.

   a. 15(80) *Hint*: Think of this as 5 times 80 and 10 times 80…and remember that 5 is half of 10.

   b. Total = $22.88   Give = $23   Change =   Bills and Coins =
   c. Total = $22.88   Give = $23.03   Change =   Bills and Coins =

PRINCIPLES OF MATHEMATICS

CHAPTER 20. Congruent and Similar
LESSON 2. Corresponding Parts – Applying Congruency

Worksheet 20.2

You may use a calculator on this worksheet whenever you see this symbol (🖩).

1. **Understanding Congruency**

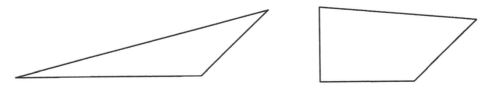

   a. Using your protractor, find the measure of all the angles and write the angle measure inside the shape.

   b. Circle the two angles that are congruent.

2. **Using Congruency** — Find the distance across this valley (the side of the top triangle with the two tick marks) using the information shown.

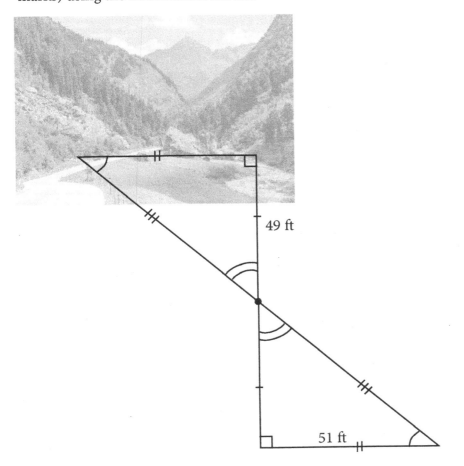

49 ft

51 ft

PAGE 301

3. **Time for Time** — Solve using proportions.

   a. If you can crochet 2 rows in 8 minutes, how long will it take you to crochet 20 rows at the same pace?

   b. If you can run 6 laps around the basketball court in 2 minutes, how long will it take you to run 24 laps at the same pace?

4. **Skill Sharpening**

   a. $(-2)^3$

   b. 4^1

   c. $(\frac{7}{49})^3$

   d. Find $\sqrt{81}$

   e. If a square has an area of 256 ft², what is the length of each of its sides? *Hint*: Find $\sqrt{256 \text{ ft}^2}$

   f. Convert 300 km to ft. (🖩)

   g. Convert 29 km² to m².

5. **Questions.**

   a. Fill in the Blank: Corresponding Parts of Congruent Shapes Are _____.

   b. Why do we look so indepth at shapes, angles, etc?

CHAPTER 20. Congruent and Similar
LESSON 3. Exploring Similar Shapes
Worksheet 20.3

1. **Proportion Fun** — Find the missing numbers in these proportions.

   a. $\dfrac{4 \text{ in}}{2 \text{ in}} = \dfrac{8 \text{ in}}{?}$

   b. $\dfrac{15}{31} = \dfrac{45}{?}$

   c. $\dfrac{20 \text{ min}}{2 \text{ mi}} = \dfrac{? \text{ min}}{10 \text{ mi}}$

   d. $\dfrac{50 \text{ mi}}{1 \text{ hr}} = \dfrac{200 \text{ mi}}{? \text{ hr}}$

2. **Fun with Proportions and Shapes**

   a. If you can jump rope 30 times a minute, how many times can you jump in 20 minutes assuming you can keep the same pace?

   b. These rectangles are similar. Knowing that, find the length of the side marked with a question mark.

   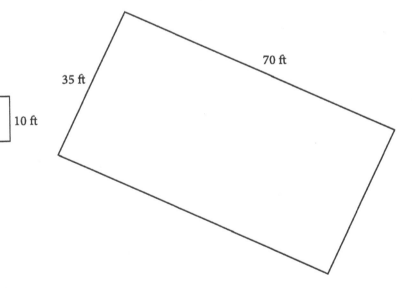

   c. How many times larger is the one rectangle than the other?

d. These triangles are similar. Knowing that, find the length of the side marked with a question mark.

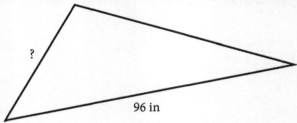

e. How many times larger is the one triangle than the other?

f. Find the length of the unmarked side of the larger triangle.

g. If you know an angle in the large triangle is 50°, what must the corresponding angle in the small triangle be?

h. Draw arcs on the triangles in 2d to show which angles are congruent to each other.

3. **Skill Sharpening** — Solve these problems mentally.

   a. 20(75)   *Hint*: 20 is twice 10…so find 10 times, and then double that.

   b. 20(62)

   c. 20(55)

   d. −7(9)

   e. −3(−22)

   f. Total = $25.79   Give = $26   Change =   Bills and Coins =
   g. Total = $25.79   Give = $26.04   Change =   Bills and Coins =

CHAPTER 20. Congruent and Similar
LESSON 4. Angles in Triangles & AA Similarity

Worksheet 20.4

You may use a calculator on this worksheet whenever you see this symbol (🖩).

1. **Angles in Triangles**

    a. What is the sum of all the angles in this triangle?

    b. What is the measure of the unmarked angle? Do not use a protractor!

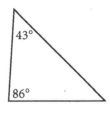

2. **More with Angles**

    a. If a triangle has a 55 and an 80 degree angle, what must the third angle be?

    a. If a triangle has a 70 and a 30 degree angle, what must the third angle be?

3. **Angles in Triangles 2**

    a. What is the sum of all the angles in this triangle?

    b. What is the measure of both the unmarked angles? Do not use a protractor. *Hint*: Look at what angles are congruent.

    c. Is this triangle a right triangle? How do you know?

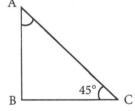

4. **AA Similarity**

    a. What is the AA Similarity Theorem?

    b. Circle all the triangles shown that you know are similar with the information given. Do not use a protractor or ruler—just use what you've learned about triangles to tell.

  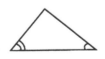 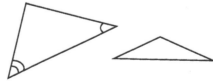

5. **Installing a Window** (🖩) — You decide to get a new window. How much will it cost considering the following: The window is a rectangle that is 7 ft tall by 5 ft wide, with a 2.5-ft-tall semicircle window on top. The window costs $5 a square foot, plus 11% tax and a flat $100 installation (the installation portion is not taxed). What will be your total cost? *Hint*: Break this problem down into steps.

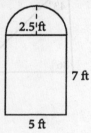

6. **Skill Sharpening** (🖩)

   a. If 1 dollar = 6.23570 Chinese yuan, how many yuans will you get for $50 dollars?

   b. How many dollars will you have spent if you buy something 56 yuans, assuming the conversion ratio in 6a?

   c. Convert 5 m to yd.

   d. Convert 50 m² to cm².

   e. Find the area of a circle whose radius is 4 m.

   f. What would the volume of a cylinder whose base is the circle in 6e, with a height of 5 m?

   g. What would the length of each side of a square with an area of 121 in² be?

PRINCIPLES OF MATHEMATICS

CHAPTER 20. Congruent and Similar
LESSON 5. Similar Triangles in Action – Finding the Height of a Tree

Worksheet 20.5

You may use a calculator on this worksheet whenever you see this symbol (🖩).

1. **Finding the Height of a Tree** (🖩) — Using the two similar triangles in the picture below, set up a proportion to find the height of the tree.

2. **Term Time** — How did the AA Similarity Theorem help us find the height of the tree in problem 1?

3. **Mastering the Skill** — Do *not* use a calculator or protractor. Instead, use what you know about congruency to answer the questions.

   a. What is the measure of $\overline{BC}$?

   b. What is the measure of $\angle ABC$?

   c. What is the measure of $\overline{EF}$?

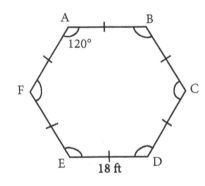

4. **How Long**

   a. If you drive an average of 40 miles an hour, how long will it take you to go 200 miles?

   b. How much will it cost you in gas to go the 200 miles if your car can make it 20 miles on 1 gallon of gas and gas costs $3.30 a gallon?

PAGE 307

5. **Conversion Time**
    a. How many liters are in 2 gallons?

    b. How many liters are in 10 gallons?

6. **Mental Conversions** — Find the answers mentally.
    a. How many quarts are in 4 gallons?

    b. How many quarts are in 3 gallons?

    c. How many tablespoons would it take to make $\frac{1}{2}$ cup?

    d. How many tablespoons would it take to make $\frac{1}{4}$ cup?

CHAPTER 20. Congruent and Similar
LESSON 6. Chpater Synopsis and a Peek Ahead
Worksheet 20.6

You may use a calculator on this worksheet whenever you see this symbol (🖩).

1. **Drawing** — Draw a rectangle that's 4 inches by 2 inches. Do not use graph paper; use your protractor to get the angles 90°.

2. **Recognizing Congruency and Similarity** — Circle all the congruent shapes; draw a rectangle around all the similar ones.

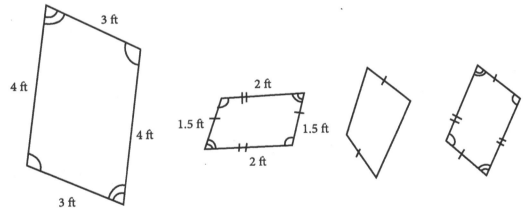

3. **Skill Sharpening**

    a. Find the volume of a cylinder with a radius of 7 in that has a height of 12 in. (🖩)

    b. How many gallons of water could the cylinder from problem 3a hold? (🖩)

c. Find the surface area of the rectangular prism shown.

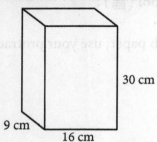

d. If a square has an area of 64 ft², what are the lengths of each of its sides?

e. Rewrite 23² using a multiplication sign, and solve.

4. Triangle Time

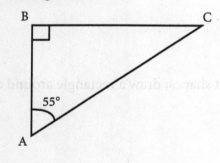
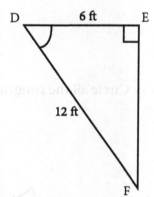

a. What is the measure of ∠ACB?

b. From the information given, can you tell if these triangles are similar? If so, how?

c. From the information given, can you tell if these triangles are congruent? If so, how?

PRINCIPLES OF MATHEMATICS

Review of Chapters 17–20 — Worksheet 20.7

Use these problems to help you review. If you have questions, review the concept in the *Student Textbook*. The lesson numbers in which the various concepts were taught are in parentheses.

You may use a calculator on this worksheet whenever you see this symbol (🖩).

1. **Solids** (🖩) — Use the right triangular prism shown to answer the questions.

    a. What is the surface area? (Lesson 18.1)

    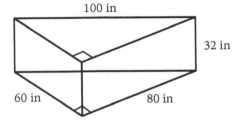

    b. What is the volume? (Lesson 18.20)

    c. How many gallons could it hold? (Lesson 18.5)

    d. What is its volume in cubic feet? (Lesson 18.3)

2. **Kitchen Fun** (🖩)

    a. What shape would best describe the containers shown? (Lesson 18.2)

    b. If each container has a diameter of 2.5 in, what is the area of its base? (Lesson 17.3)

    c. If the containers are 4 in high, what is their volume? (Lesson 18.2)

    d. How many containers would it take to hold 1 gallon of liquid? (Lesson 18.5)

3. **Capacity and Weight** (🖩)

   a. If gas costs $1.60 a liter, and you put in 20 liters, how many gallons did you put in? (Lesson 18.6)

   b. How much did it cost you per gallon?

   c. If loose-leaf tea is sold for $2.48 an ounce, and you buy 1 pound, how much will you pay? (Lesson 18.7)

   d. If you want to make 4 gallons of lemonade, and the instructions say you need 2 lemons per quart, how many lemons do you need? (Lesson 18.4)

   e. If you want to make 4 pints of tea and you need one tea bag per 8-oz cup of tea, how many tea bags do you need? (Lesson 18.4)

4. **Measuring Angles**

   a. Use a protractor to find the measure of the marked angle in this design. Round to the nearest multiple of 5. (Lesson 19.1)

   b. Is the angle acute, obtuse, or right? (Lesson 19.2)

   c. What size angle would need added to the angle in problem 4a to form a 90° angle? (Lesson 19.2)

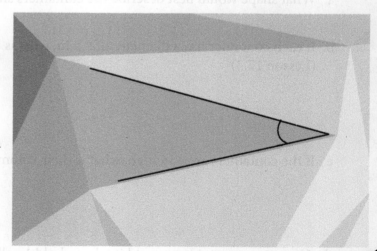

PAGE 312

d. Use a protractor to find the measure of the marked angle in this design. (Lesson 19.1)

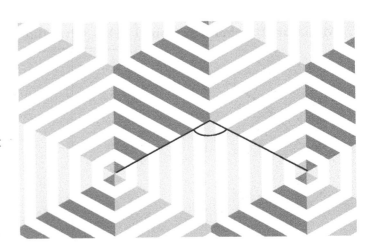

e. If a triangle has an 80° and a 15° angle, what must the third angle equal? (Lesson 20.4)

5. **Congruent** — △ABC and △DEF are congruent

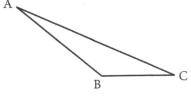 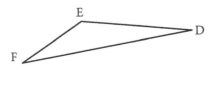

   a. Which angle is congruent to ∠ABC? (Lesson 20.2)

   b. Which line is congruent to $\overline{AB}$? (Lesson 20.2)

   c. Use tick marks to mark all the congruent sides. (Lesson 20.1)

6. **Similar Shapes** (🖩)

   a. If you know the marked sides of these two boxes are similar rectangles, find the length of the side marked with ? in. Round your answer to the nearest tenth. (Lesson 20.3)

   b. Are the angles in a rectangle obtuse, acute, or right? (Lesson 19.2)

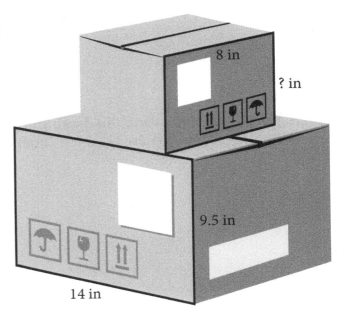

7. **Miniature Baseball Diamond** (🖩) — Give your answers in centimeters.

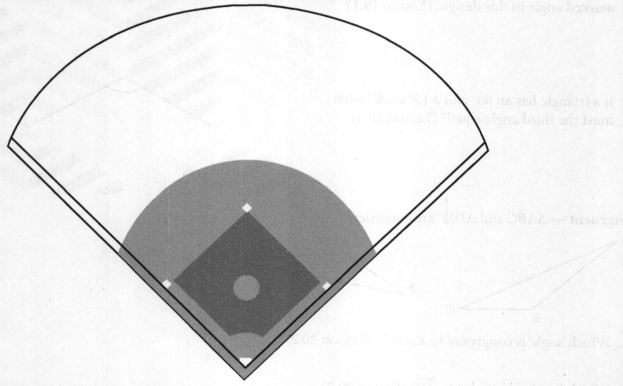

a. What is the perimeter of this scale drawing if the outer edge of the arc is 42% of a circle with a radius of 3.5 cm? Use a ruler to find the other dimensions. (Lessons 17.2 and 17.3)

b. If you decided to enlarge this drawing by a scale of 3 to 1, how long would you make the distance from home plate to third base? It is currently 3 cm.

PRINCIPLES OF MATHEMATICS

CHAPTER 21. Review
LESSON 1. Arithmetic Synopsis

Worksheet 21.1

Use these problems to help you review. If you have questions, review the concept in the *Student Textbook*. The lesson numbers in which the various concepts were taught are in parentheses.

You may use a calculator on this worksheet whenever you see this symbol (🖩).

1. **Place Value (Lessons 1.5 and 1.6)**

    a. In the decimal system, what does the 3 in 8,350 represent?

    b. What does it mean that our number system is a base-10 system?

    c. What would it mean if we used a base-2 system?

2. **Problem Solving and Operations (Lessons 2.6 and 4.5)**

    a. If in Thomaston, 3,980 clocks were made during the first week in September, 3,986 the second week, 4,016 the third week, and 4,221 the fourth week, how many clocks were made there during those four weeks?

    b. If the sales price of each clock was $129.99 and the cost of each clock was $25, about how much did the company make during those four weeks? Round $129.99 to the nearest dollar before you solve.

    c. If it costs $8.95 to download 20 songs, what is the price per song?

3. **More Problem Solving and Operations** (🖩) — Suppose you decide to start knitting socks for sale. Suppose it costs $3.99 plus 5% tax for a ball of yarn, it takes $\frac{1}{3}$ a ball to make 1 pair of socks, and it costs $1.50 to list each pair of socks in an online store. If you sell the socks for $9.99, how much money would you make on each pair?

   a. Define.

   b. Plan.

   c. Execute.

   d. Check.

4. **Even More Problem Solving and Operations**

   a. Let's continue to explore the scenario from problem 3. Suppose you have the opportunity to sell your socks on your own website instead of on an external online store. Ignoring the time taken in maintaining your own website, if it costs you $9.50 a month to have your own website, plus $6 a year in miscellaneous fees, would it be more economical to get your own website over paying the $1.50 per pair to list them on another online store, assuming you sell 200 pairs of socks each year?

   b. You want a picture centered across the width of an $8\frac{1}{3}$ foot long wall. Where should the center of the picture (and hence the nail) go? (Lesson 6.1)

5. **Ratios and Proportions** — Set up a proportion to find the missing information. (Lesson 8.2)

   a. If there's 1 congressman per 30,000 people, how many congressmen should there be for 330,000 people?

   b. If you travel 50 miles per hour, how far will you go in 10 hours?

6. **Pumpkin Division** — Suppose you decide to make your own pumpkin puree from scratch. All your favorite recipes call for 1 can of pumpkin. You need to figure out how many cups that is so that you know how to portion out your own pumpkin puree for use later. The back of a can says the serving size is $\frac{1}{2}$ cup, and that there are $3\frac{1}{2}$ servings in the can.

   a. How many cups are in 1 can?

   b. What is the cost per cup if the can costs $1.99 plus 4.5% tax?

   c. If a whole fresh pumpkin costs $4.99 plus 4.5% tax and makes 20 cups of pumpkin puree, how much does it cost per cup to make the pumpkin puree fresh?

   d. In this scenario, is it cheaper to make or buy pumpkin puree?

7. **Carbonate of Potash** (🖩) — "If the weight of oak ashes is 0.03 of the weight of the wood burned, and the weight of carbonate of potash contained in the ashes is 0.065 of the weight of the ashes, how many pounds of carbonate of potash are there in the ashes of 1 cord of oak, if stacked oak wood weighs 2 tons per cord?"

8. **Mental Math** — Solve these problems mentally. (Lessons 3.1, 3.2, 8.5, and 9.5)

   a. 87 − 28

   b. 62 + 89

   c. 5(78)

   d. What is 4 • $7.99?

   e. Total = $38.92   Give = $39   Change =   Bills and Coins =
   f. Total = $38.92   Give = $39.02   Change =   Bills and Coins =

   g. What is 20% of $45.89? Round to the nearest dollar before finding.

   h. If you have a meeting at 2 p.m. EST and it's 9 a.m. PST, how long do you have before the meeting begins? (Lesson 2.2)

CHAPTER 21. Review
LESSON 2. Geometry Synopsis
Worksheet 21.2

Use these problems to help you review. If you have questions, review the concept in the *Student Textbook*. The lesson numbers in which the various concepts were taught are in parentheses.

You may use a calculator on this worksheet whenever you see this symbol (🖩).

1. **Name Time** — What two-dimensional shape best describes the outlined section of each image? Be as specific as possible. Name triangles both by angles and by sides. (Lessons 13.3 and 13.4)

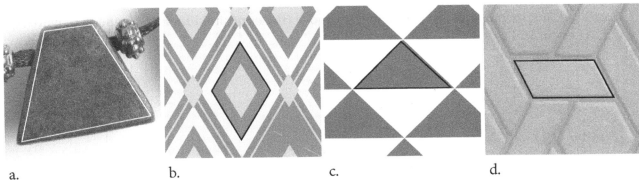

a. b. c. d.

2. **Pool Time (Lessons 17.3, 18.2, and 18.5)** (🖩)

   a. Suppose a park has a circular pool with a radius of 50 feet and a height of 4 feet. How many gallons of water does it take to fill up the pool to the very top if 1 gallon equals 0.13 ft^3? Round your answer to the nearest whole number.

   b. What is the circumference of the pool above?

   c. Suppose the park wanted to cover the top of the pool. How many square yards would they need to cover?

3. **Miscellaneous Problems**

   a. Using the passage below, find the height, length, and width of Solomon's temple in feet, assuming 1 cubit equals 1.5 feet. *Note*: A "score" means 20, so "threescore" means 3 • 20, or 60. (Lesson 14.2)

   "And the house which king Solomon built for the LORD, the length thereof was threescore cubits, and the breadth thereof twenty cubits, and the height thereof thirty cubits." 1 Kings 6:2

b. Find the area of this parallelogram. (Lesson 15.4)

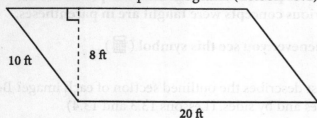

c. Find the perimeter of the parallelogram above (the opposite sides in a parallelogram are equal). (Lesson 15.1)

4. Similar (🖩)

   a. If the shapes shown are squares, what is the length of $\overline{DC}$? (Lesson 20.3)

   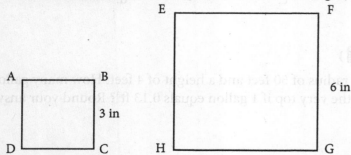

   b. If the triangles below are similar, what are the lengths of the sides marked with question marks? (Lesson 20.3)

   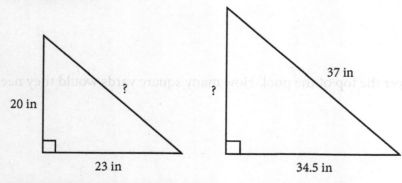

   c. What is the area of the smaller triangle? (Lesson 17.1)

d. What would the volume of a triangular prism built off the smaller triangle be if the prism was 45 in high? (Lesson 18.2)

5. **Skill Sharpening** (🖩)

    a. Convert 20 square kilometers into square miles. (Lesson 16.3)

    b. Convert 1,330 inches cubed into feet cubed. (Lesson 18.3)

6. **Land Challenge** (🖩) — A rectangular piece of land is 220 yd wide by 800 yd long. If in a scale drawing of it, the width is made $1\frac{3}{4}$ in, what will the length be?

CHAPTER 21. Review
LESSON 3. Life of Johannes Kepler
Worksheet 21.3

1. **Johannes Kepler**

    a. In your own words, share one lesson we can learn from the life of Kepler. (There are many—pick any one.)

    b. What major accomplishment did Kepler achieve as a result of perseverance under difficulties?

2. **Report Assignment** — Pick a career that you'd like to learn more about, and then research how math is used in it. For example, a veterinarian might have to measure out medicine for a dog…or figure out what dose of medicine a small puppy verses a large lab would need.

    Ask your parent/teacher for ideas. You might try looking online, asking a librarian, asking someone at church or that you know who works in the field you've chosen, or even visiting or calling someone you don't know from that field and asking them how they use math (just explain that you're doing a school project, and they will likely be happy to answer a couple of questions). Every career uses math to some extent or another, so keep asking until you find some examples.

    Today, pick your career and do some initial research. Your report will be due on the day you take your final exam.

3. **Study** — Study for the final exam. See the study day suggestions on page 4 of the *Student Workbook*.

PRINCIPLES OF MATHEMATICS

CHAPTER 21. Review
LESSON 4. Course Review

Worksheet 21.4A

You may use a calculator on this worksheet whenever you see this symbol (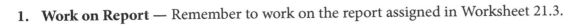).

1. **Work on Report** — Remember to work on the report assigned in Worksheet 21.3.

2. **Notations**
   a. Express 5% as a decimal.

   b. Express $-\frac{2}{3}$ as a decimal.

   c. Express $\frac{4}{18}$ as a percent.

   d. Express 17 divided by 25 as a fraction.

   e. Express $8\frac{1}{3}$ as an improper fraction.

   f. Express 25% as a fraction in lowest terms.

   g. Express $-5.7 \cdot -5.7 \cdot -5.7$ using an exponent.

   h. Express 789,560,000 in scientific notation.

   i. Express 1.25 as a percent.

   j. Express 0.035 as a percent.

3. **Symbols** — Write the correct symbol (>, <, =) in between each pair of numbers to compare them.

   a. $\frac{1}{3}$ 0.25

   b. 4^2 16

   c. 0.8 $\frac{3}{4}$

   d. Length of $\overline{AB}$ Length of $\overline{CD}$

   A————————————B   C————————————————D

PAGE 325

e.   −7.8   5.6

f.   Length of $\overline{AB}$ Length of $\overline{CD}$

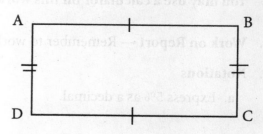

4. **Symbols 2**

   a. Show three different ways to represent 4 times 5.

   b. Show three different ways to represent 4 divided by 5.

5. **Terms**

   a. What is the identity property of multiplication?

   b. What do we call the bottom number in a fraction?

   c. What do we call a four-sided, closed, two-dimensional shape with all right angles and two pairs of straight, parallel sides?

6. **Conventions**

   a. 7(9)

   b. 6 + −2(−3)

   c. $8(3 + 4)^2$

   d. $25^2 \cdot -3$

   e. Rewrite ft^3 as a multiplication.

7. **Operations and Algorithms** (🖩) — Solve the following. If the question contains fractions, leave your answer in fractions, but simplify the fractions as much as possible. Round decimal answers to the nearest hundredth.

   a. $\dfrac{4}{5} \cdot \dfrac{6}{8}$

   b. $\dfrac{9}{7}$ divided by $\dfrac{2}{3}$

   c. $\dfrac{40}{50} + \dfrac{87}{200} + \dfrac{7}{100}$

   d. $\dfrac{7}{12} - \dfrac{3}{20}$

   e. $608.5 \div 51.1$

   f. $20 \div 25$

   g. $8 \cdot -4$

   h. 4.25% of $89.67

   i. Convert 50 inches into yards.

   j. Convert 8,000 feet to meters.

   k. Convert 5 gallons to cups.

   l. True or false: There's only one way to multiply on paper.

8. **Question** — What are the two overarching principles we've looked at this year.

PRINCIPLES OF MATHEMATICS

CHAPTER 21. Review
LESSON 4. Course Review
Worksheet 21.4B

You may use a calculator on this worksheet whenever you see this symbol (🖩).

1. **Miscellaneous**

   a. Round 0.875 to the nearest hundredth.

   b. Can you divide by 0?

   c. Solve via the distributive property (show your work): 4(2+3)

   d. Find the least common denominator of $\frac{7}{44}$ and $\frac{8}{16}$.

   e. Add $\frac{7}{44}$ and $\frac{8}{16}$.

   f. Find the greatest common factor of 18 and 48.

   g. Simplify $\frac{18}{48}$.

   h. 4 + – –8

   i. $4\frac{1}{5} \cdot \frac{16}{8}$

   j. Name one thing –9 could represent.

   k. –3 • –2

   l. –3 • 2

2. **Ratios and Proportions** (🖩)

   a. If you want to make a scale model of a building that's 400 ft tall, and you want 1 in of the model to represent 15 ft of the building (i.e., a scale ratio of 1 in:15 ft), how many inches tall should your model be?

   b. Use graph paper to draw a scale drawing of a table that's 7.5 feet long and 3.5 feet wide.

3. **Mental Math** — Solve these problems mentally.

   a. 48 + 26

   b. 32 − 18

   c. $4.60 • 3

   d. $2.99 • 8

4. **Sets and Sequences**

   a. Draw a Venn Diagram to show how the set of rectangular prisms, the set of prisms, and the set of triangular prisms relate to each other.

   b. What are the next two numbers in this sequence, assuming the pattern continues? {4, 8, 12, 16…}

   c. What is the common difference in the sequence from 4b?

5. **Graphs** (🖩)

   a. According to the graph, what percent of the income came from lawn mowing?

   b. If the total income was $12,300, how many dollars came from lawn mowing?

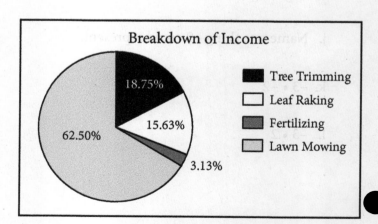

Breakdown of Income

- Tree Trimming: 18.75%
- Leaf Raking: 15.63%
- Fertilizing: 3.13%
- Lawn Mowing: 62.50%

6. **Accurate Stats** — If a survey of a sample of medicine users concludes that a medicine is 45% effective, and the margin of error is 10%, what expected range of effectiveness does the survey show?

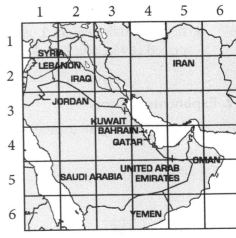

7. **Labeling Coordinates** — In what coordinate does the word "Iran" fall?

8. **Averages (🖩)** — If someone makes $567.62 one month, $678.25 the next month, and $456.48 the next month from a small hobby business, what is their average monthly income from the hobby business for those three months?

9. **Percents**

   a. If you buy a $14.99 shirt on a 40% off sale, how much will you end up paying if a 4.5% tax is added after the discount is deducted?

   b. How much *total* (tip included) will you pay if you leave an 18% tip on a $34.95 bill? Round your bill to the nearest dollar before finding the tip.

10. **Grocery Multiplication**

    a. If a pound of apples costs $1.99, how much would $\frac{1}{2}$ pound cost?

    b. If a pound of apples cost 500 Thai bahts, and the exchange rate is 1 Thai baht per 0.031 U.S. dollar, how much would $\frac{1}{4}$ pound cost in U.S. dollars?

11. **Conversions**

    a. How many hours are there in 10 days?

    b. If 1 U.S. dollar equals 1.09 Canadian dollars, how many U.S. dollars would it cost to buy an item priced at 45 Canadian dollars?

12. **Exponents & Roots**

    a. Rewrite using an exponent and solve: 10 x 10 x 10

    b. Find $\sqrt{400}$.

13. **Shape Time** (🖩)

    a. Find the perimeter and area of this irregular polygon, which consists of a rectangle and a triangle.

    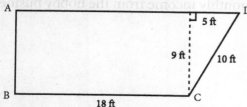

    b. What is the measure of ∠ADC? Round to the nearest multiple of 5.

    c. What would you have to add to that angle to reach 180°?

    d. Draw an angle the size you found in 13c.

    e. If the measure of two angles in a triangle are 25° and 43°, what must the third angle measure?

14. **Engine Pumps** — If an engine pumps $2\frac{1}{4}$ gallons of water per stroke and makes 46 strokes per minute, how many gallons will it deliver in an hour?

Quizzes and Tests

PRINCIPLES OF MATHEMATICS

QUIZ 1 — Chapters 1 and 2 | Q1

1. **Comparing Numbers** — Use the symbols <, >, or = to show how these quantities compare.
    a. 56 + 8   2 + 60
    b. 88 − 4   49 + 17
    c. VII   IX

2. **Place Value**
    a. Describe how a place-value system works.

    b. What does it mean if a number is written in a base-12 place-value system?

3. **Time for Time**
    a. If a luncheon starts at 10:30 a.m. and lasts 2 hours, when will it end?

    b. If a TV show is airing at 7 p.m. PST and you're in EST, at what time is it airing in your time zone?

    c. If it is 1600 military time, what time is it in 12-hour clock?

4. **Keeping a Checkbooks** — Find the ending balance of this checkbook register.

Check Number	Date	Memo	Payment Amount	Deposit Amount	$ Balance
	2/1	Opening Balance			5,612
120	2/5	Music Lessons	57		
	2/10	Birthday Check		75	
121	2/15	Groceries	104		
	2/15	Interest		1	
	2/15	Paycheck		508	

5. **Bonus Question** — Why does math work?

QUIZ 2
Chapter 3 | Q2

1. **Representation** — Solve the following problems.

    a. 2(2 + 2) + 8

    b. 2($4 + $8) + $7

    c. 4(10 − 3 x 3)

2. **Mental Math** — Solve the following problems mentally.

    a. 24 + 8 + 6

    b. 85 + 20

    c. 85 − 12

    d. Round 754 to the nearest ten.

    e. Round 754 to the nearest hundred.

3. **Making Jewelry** — Suppose you decided to start making and selling jewelry. To make one particular necklace, you need beads that cost $6, $8, and $10, plus wiring that costs $4. Using those beads, you can make 7 necklaces.

    a. How much will all the supplies for those necklaces cost you?

    b. How much does each of the 7 necklaces cost to make (not counting your time)?

    c. How much will you have to sell the necklaces for if you want to make $5 a necklace?

PAGE 337

4. **Design Your Own** — Write your own multiplication word problem and solve it.

5. **How Many Years?** — Christopher Columbus discovered America in 1492.
   a. How many years had it been since that discovery in 2012?

   b. In what year was the quadricentennial (400th anniversary)?

Bonus Question — Define properties.

PRINCIPLES OF MATHEMATICS

QUIZ 3
Chapter 4
Q3

1. **Construction** — If 12 books occupy on average 1 foot of shelf room, about how many shelves 4 feet long will it require to hold 288 books?

2. **Wedding Cakes**
    a. If a wedding cake that feeds 171 guests costs $855, how much does the cake cost per guest?

    b. Assuming you could reduce the size of the cake so that the cost per guest remained the same, how much would it cost if you had 135 guests instead?

3. **Land** — If you buy 160 acres of land for $127,980 and then buy another 60 acres of land for $50,000, what is the price you paid altogether per acre?
    a. Define:

    b. Plan:

    c. Execute:

    d. Check:

PAGE 339

4. **Language Check**
   a. Solve using the order of operations: 8(4 + 3 + 6)

   b. Solve using the distributive property, showing your work: 8(4 + 3 + 6)

5. **Skill Check** — Solve.
   a. 5)896

   b. 897 x 56

   c. 5 x 0

   d. 85 + 0

QUIZ 4
Chapter 5 — Q4

1. **Expressing Fractions** — Describe the following as a fraction. Simplify as much as possible.

    a. 26 divided by 8 *Example Meaning:* 26 muffins divided by 8 people

    b. 5 divided by 10 *Example Meaning:* 5 chapters divided by 10 weeks

2. **Division**

    a. Rewrite $8\overline{)55}$ as an improper fraction.

    b. Rewrite $8\overline{)55}$ as a mixed number.

3. **Factoring and Simplifying**

    a. In $\frac{44}{50}$, what is the greatest common factor of the numerator and the denominator?

    b. Simplify $\frac{44}{50}$

    c. In $\frac{16}{24}$, what is the greatest common factor of the numerator and the denominator?

    d. Simplify $\frac{16}{24}$

4. **Adding Fractions**

    $\dfrac{10}{84} + \dfrac{22}{42}$

    a. Rewrite so each fraction has the same denominator.

    b. Add. Simplify the answer if needed.

5. **Subtracting Fractions**

    $\dfrac{15}{30} - \dfrac{5}{60}$

    a. Rewrite so each fraction has the same denominator.

    b. Subtract. Simplify the answer if needed.

6. **Exploring** $\dfrac{1}{24} + \dfrac{7}{20}$

    a. Rewrite each denominator as a product of its prime factors.

    b. What is the least common denominator?

    c. Rewrite the fractions so they have the same denominator.

    d. Add. Simplify your answer if needed.

7. **Payments** — A man pays $30 a month for phone and $18 a month for trash. How much does he pay for phone and trash in 10 years?

8. **Payments 2** — If the man in the previous problem changes to a telephone plan that only costs $28 a month, how much will he save in 5 years? In 10 years?

9. **Fabrics** — If you buy $\frac{3}{4}$ of a yard of fabric, and only end up using $\frac{2}{3}$ of a yard, how much fabric do you have left?

Bonus Question — Share 1 thing you learned about the history of fractions.

QUIZ 5
Chapter 7 Q5

You may use a calculator on this worksheet whenever you see this symbol (🖩).

1. **Skill Check**

    a. $0.56 + 1.78$

    b. $0.23 \cdot 0.65$

    c. $78.53 \div 2.4$

    d. $1.78 - 0.56$

2. **Conversions** 🖩 — Express the following measurements as decimals.

    a. $\frac{7}{8}$ mile

    b. $\frac{2}{3}$ cup

    c. $1\frac{1}{4}$ cup

PAGE 345

3. **Applying the Skills**
   a. If you head into a convention with $45.67 in your wallet and leave with $5.08, how much did you spend?

   b. Suppose you decide to go into the lawn mowing business. You charge $25 a lawn. You have to spend $3.25 a gallon on gas, but one gallon lasts you 2 lawns. Assuming gas is your only cost, how much will you make mowing 10 lawns?

PRINCIPLES OF MATHEMATICS

QUIZ 6 — Chapter 8

1. **Computing Totals and Giving Change** — Solve these problems mentally.

    a. A customer purchases 3 pounds of strawberries at $2.50 a pound. How much do they owe you?

    b. The customer in problem 1a gives you a $20 bill. How much change should you give them?

2. **Ratios**

    a. Express the following as a ratio: $4 per 12 plates

    b. Convert the ratio to a decimal number.

3. **Ratios and Proportions** — Express the following as proportions and solve.

    a. If a 4.6 foot sidewalk costs $35, what is the cost of a 9.2 foot sidewalk if the price per foot stays the same?

    b. If it costs $56 for 2 tickets to a show, how much will it cost for 10 tickets, assuming there are no quantity discounts?

    c. If a scale drawing of a bookshelf using a 1 centimeter to 5 inch scale shows a screw being inserted 2 centimeters from the top, how many inches from the top should you insert the screw?

4. **Planting Crops**

   a. If it takes 15,000 seeds to cover $\frac{1}{4}$ of an acre, how many seeds do you need to cover $3\frac{1}{2}$ acres?

   b. If a bag of seeds costs $30 and contains 30,000 seeds, how much will it cost you to buy the number of bags you need to plant the section discussed problem a?

   c. Another company sells bags of seeds for $32. Their bags contain 40,000 seeds. Which way is the cheaper way to buy the seeds?

Bonus Question — Describe one way proportions help describe God's creation.

| PRINCIPLES OF MATHEMATICS

QUIZ 7
Chapter 9

1. **Expression** — Express each ratio as a fraction (in simplest terms), as a decimal, and as a percent.
    a. 5 out of 10 people

    b. $250 out of $1,000

2. **Skill Check** — Solve the following problem mentally. Round to the nearest whole dollar first. What is 20% of $112.45?

3. **More Skill Check** — Find the requested percentages.
    a. What is 25% off $89.50?

    b. What is 15% of $56.15?

    c. If you buy something for $14.50 and a 5% tax is added to it, what will your total be?

4. **Conversion** — Express the following as percents.
    a. $\frac{4}{10}$

    b. 1.89

PAGE 349

5. **Investing** — Let's say you and a friend decide to go into business together. Between the two of you, you can rummage up 52% of the $400 you need to get started. How much more money do you need to raise?

6. **Management** — You are in charge of managing a division of a lawn company. You've been told to cut costs by 20%. If your current costs are $865, by how many dollars do you need to cut your expenses?

QUIZ 8 Chapter 10 — Q8

1. **How Could I Represent...**

    a. If I represent a profit as +3, how could I represent a loss of 3?

    b. If I represent 6 months in the future as +6, how could I represent 6 months in the past to differentiate it?

2. **How Do We Use...** — Name three things you could represent with negative numbers.

3. **Dots** — Draw a dot representing –2.5, 3, and –4 on the number line.

    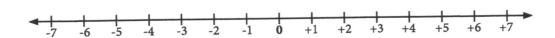

4. **Learning the Skill** — Find the answers to these expressions.

    a. $(-5) + (-3)$

    b. $-10 + 5$

    c. $-2 \cdot 3$

    d. $-2 \cdot -3$

    e. $45 \div -5$

f. $--5$

g. Rewrite as a decimal: 3.25%

h. $\frac{-27}{9}$

5. **Absolute Value** — Find the absolute value of these numbers.
    a. $|-3|$

    b. $|-5.78|$

    c. $|5.78|$

6. **More Application**
    a. Suppose you sell a picture on a stock photography site for $5.50. If the site takes 30% of the sales price, how much will you make?

    b. If you could sell the same picture on a different site for $4.25 where the site takes a flat $1.75 fee, which way will get you more money, and by how much?

Bonus Question — What does each negative sign mean?

QUIZ 9
Chapter 11 — Q9

PRINCIPLES OF MATHEMATICS

1. **Understanding Sets** — Suppose you're trying to organize pictures on a computer using a program in which the same picture can be put into multiple folders. List each picture letter under all appropriate folders. *Note*: The folders are really sets—collections of photos.

   Folders:

   Plants
   {contains any picture that has a plant in it}

   Trees
   {contains any picture that has a tree in it}

   Flowers
   {contains any picture that has blooming flowers in it}

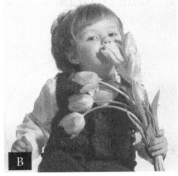

2. **Understanding Subsets** — In problem 1, did the set "Plants" have any subsets?

3. **Putting Numbers into Sets** — Circle all the sets listed that can be used to describe the given number.

   a. 27

   Whole Numbers   Prime Numbers   Odd Integers

   b. $89\frac{1}{2}$

   Whole Numbers   Prime Numbers   Positive Numbers

PAGE 353

4. **Venn Diagram** — Based on this Venn Diagram, is the set named A a subset of the set named B? How do you know?

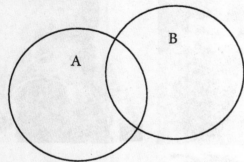

5. **More Sets** — Would the number 90 be part of this set? {<65}

6. **Sequences** — What are the next three numbers in the sequences below assuming the pattern continues?

   a. {2, 2, 5, 8, 8, 11, 14, 14, 17, ____, ____, ____}

   b. {1, 8, 15, 22, ____, ____, ____}

7. **Term Time**

   a. Is there a common difference in either sequence in problem 6? If so, what is it?

   b. What is a sequence?

Bonus Question — Name one real-life example of sequences.

PAGE 354

PRINCIPLES OF MATHEMATICS — QUIZ 10, Chapter 12 — Q10

Today's quiz is going to be a little different than normal. Rather than give you questions, it's time to apply what you learned on your own!

Here's your assignment:

1. **Find at least one real-life graph.** — To find one, pull out a newspaper or business magazine. You may also want to look for statistics online. For example, www.fedstats.gov offers links to many of the different statistics our government tracks.

2. **Write a three-paragraph analysis of the graph.** — Possible questions to address in your analysis:

    - What does the graph show?

    - What type of graph was used? Was it a good choice?

    - What scale did the graph use?

    - Was a conclusion drawn off the graph? If so, was the conclusion drawn accurate? Why or why not? Is there something that should be researched further before you would know if it was accurate?

    - If there was no conclusion drawn, what kind of conclusions (if any) could be drawn…or what would need further explored in order to draw a conclusion?

QUIZ 11
Chapter 13 — Q11

1. **Name That Shape**

    a. Name three words that could be used to describe the computer monitor (some will be more specific than others).

    b. Describe the shape in the pattern shown based on its angles.

    c. Describe the shape in the pattern shown based on its sides.

2. **Term Time** — How many sides do the following figures have?

    a. decagon

    b. quadrilateral

    c. nonagon

    d. Are the wings of a butterfly an example of reflection, rotation, translation, or dilation?

3. **Labeling**

    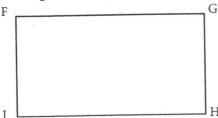

    a. Put a star on $\overline{IH}$.

    b. Circle $\angle FGH$.

PAGE 357

4. **Term Time** — Which is true of these terms in Euclidean geometry?
   a. A line
      i. A line doesn't have a starting or an ending point (it goes on in both directions indefinitely).
      ii. A line is a great circle with a definite starting and ending point.

   b. A point
      i. A point has width.
      ii. A point has no width or length.

5. **Bonus Question** — Give one example of a real-life object and the shape that describes it. You cannot use any from this quiz, but examples you remember from the text or worksheets are fine.

QUIZ 12
Chapter 14 — Q12

You may use a calculator on this worksheet whenever you see this symbol (🖩).

You may use the "Units of Measure" reference sheet (pages 447–448) to help answer the questions.

1. **Mastering the Skill** (🖩)

    a. How many millimeters is 4 meters?

    b. How many feet is 0.5 mile?

    c. How many centimeters is 24 inches?

2. **Visiting Russia** (🖩)

    a. According to Google Maps, the driving distance from Moscow to Yakutsk is 8,290 km; how far is that in miles?

    b. While in Russia, you go to the store. You purchase items for 50 rubles and 1,000 rubles. They charge you 18% more in tax. How much will your total be in Russian rubles?

    c. How much in U.S. dollars did it cost you for the purchase (including tax) in the previous problem? Use a conversion rate of 1 Russian ruble = 0.03 U.S. dollars.

    d. If you're traveling at 200 kilometers per hour, how far will you travel in 25 minutes?

3. At the Fabric Store (🖩)

   a. You are making a quilt and need $\frac{1}{4}$ yard of fabric. If the fabric costs $3.99 a yard, how much will $\frac{1}{4}$ yard cost?

   b. You need 5 inches of fabric. If the fabric costs $3.99 a yard, how much will 5 inches cost you? *Hint*: Watch your units!

4. How Big Was the Ark of the Covenant (🖩)

   "And they shall make an ark of shittim wood: two cubits and a half shall be the length thereof, and a cubit and a half the breadth thereof, and a cubit and a half the height thereof." Exodus 25:10

   Convert the dimensions of the Ark of the Covenant into feet, assuming 1 cubit equals 18 inches.

QUIZ 13
Chapter 15 — Q13

You may use a calculator on this worksheet whenever you see this symbol (🖩).

You may use the "Units of Measure" reference sheet (pages 447–448) to help answer the questions.

1. **Mastering the Skills**

   a. What is the area of a square with a 7-ft side?

   b. What is the perimeter of a square with a 7-ft side?

   c. What is the area of the parallelogram pictured?

   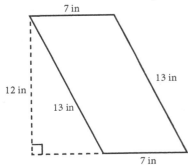

   d. What is the perimeter of the parallelogram pictured?

2. **Turf on a Football Field** (🖩)

   a. A community center is trying to re-seed their football field. If the field is 360 ft by 160 ft, how many bags of seeds will they need if each bag can cover about 1,000 square feet?

   b. How much will the seeds cost them if each bag costs $70?

PAGE 361

c. If they go with a different company, they can get the same bag for $30. How much will they save?

d. If the coach has players run around the perimeter of the field, how far will they have run?

3. **Formulas**
   a. Express how to find the perimeter of a rectangle as a formula (i.e., using only symbols).

   b. Express that the pizza orders equal the sub orders plus 2 as a formula (i.e., using only symbols).

Bonus Question — Why are we able to use formulas in real life?

PRINCIPLES OF MATHEMATICS

QUIZ 14
Chapter 16 | Q14

You may use a calculator on this worksheet whenever you see this symbol (🖩).

You may use the "Units of Measure" reference sheet (pages 447–448) to help answer the questions.

1. **Exponents** — Rewrite these exponents as repeated multiplications and solve.

   a. 7^3

   b. 88^2

   c. 10^1

2. **Square Roots** — Find $\sqrt{81}$

3. **Area of a Square and Its Sides** — How long are the sides of a square if its area is 144 ft²?

4. **Scientific Notation**

   a. Convert to decimal notation: 1.56×10^8

   b. Convert to scientific notation: 564,000,000

c. Which is greater, 1.78×10^7 or 4.6×10^5?

5. **Converting Square Units** (🖩)

   a. A garden is 120 yd². You need to purchase fertilizer that is sold in ft². How many square feet do you need?

   b. A room is 132 ft². You need to buy carpet that is typically sold in square yards. How many square yards do you need? Assume you cannot buy a portion of a square yard.

Bonus Question — What's one aspect of God's creation scientific notation helps us describe numerically?

QUIZ 15
Chapter 17 — Q15

You may use a calculator on this worksheet whenever you see this symbol (🖩).

You may use the "Units of Measure" reference sheet (pages 447–448) to help answer the questions.

1. **Mastering the Skills: Perimeter and Area** (🖩)

   a. Find the perimeter of the triangle shown.

   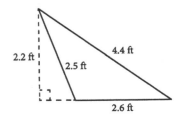

   b. Find the area of the triangle shown.

   c. Find the area of the triangle shown in square inches.

2. **Mastering the Skills: Even More Perimeter and Area** (🖩) — What additional information would you need to know in order to find the area of this triangle?

   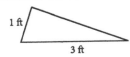

3. **Maps** (🖩) — The lines in the map below mark the streets at a busy intersection. The gray area marks the section upon which there are plans to build a store. Find the area of the gray section. *Hint:* Think of the gray area as 2 triangles and 1 rectangle.

   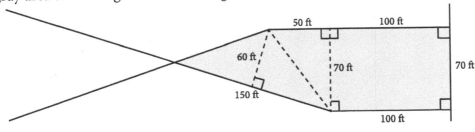

PAGE 365

4. **Car Tires** (🖩) — Tires can be thought of as circles. The picture shows the radius of both the outer tire and the inner rim.

   a. What is the circumference of the entire tire?

   b. What is the area of the entire tire?

   c. What is the circumference of just the inner rim?

5. **Irrational Numbers** (🖩) — True or false: $\frac{1}{12}$ is an irrational number.

Bonus Question — Give one fact about π's history.

QUIZ 16
Chapter 18 — Q16

You may use a calculator on this worksheet whenever you see this symbol (📱).

You may use the "Units of Measure" reference sheet (pages 447–448) to help answer the questions.

1. **Conversion Time** (📱)

    a. If a recipe calls for 2 quarts, and you're making 8 times the recipe, how many *gallons* will you need altogether?

    b. If you buy 10 bushels of apples, how many dry quarts of apples have you bought?

    c. A 10-ounce block of cheese is how many grams?

    d. If your car can hold 10 gallons of gas, how many liters can it hold?

2. **Surface Area and Volume** (📱)

    a. Find the surface area in square inches of the block shown if its base is a square with 1.5 in sides.

    b. Find the volume of the block in cubic inches.

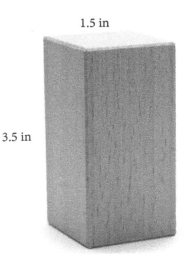

1.5 in

3.5 in

PAGE 367

3. **More Surface Area and Volume** (🖩)

   a. Find the surface area of the triangular prism. The dashed line marking the height of the triangular base is 0.9 in.

   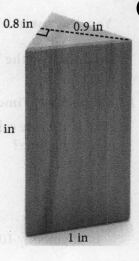

   b. Find the volume of the triangular prism. The dashed line marking the height of the triangular base is 0.9 in.

4. **Capacity** (🖩)

   a. How many gallons would it take to fill a pool that's 20 ft by 10 ft by 5 ft?

   b. If you have to add 1 cup of a cleaning solution per 1,000 gallons, how many cups will you need to add to the pool in problem 4a?

Bonus Question — How does surface area help us see God's design in birds?

QUIZ 17
Chapter 19 — Q17

You may use the "Units of Measure" reference sheet (pages 447–448) to help answer the questions.

1. **Measure These Angles** — Measure the marked angles in these pictures. Round measurements to nearest multiple of 5.

a.

b.

2. **Naming Angles**

    a. Was the marked angle in the bookshelf in 1a right, obtuse, or acute?

    b. Was the marked angle in the drawing in 1b right, obtuse, or acute?

    c. How many degrees is a right angle?

3. **Adding Angles** — What would the two angles you measured in problem 1 added together equal?

PAGE 369

4. **Drawing Angles**

   a. Draw a 120° angle.

   b. Draw a 45° angle.

5. **Pie Graph Fun** — Suppose at a restaurant one evening 45% of customers ordered blueberry ice cream, 30% chocolate ice cream, and 25% mint ice cream.

   a. What would the angle be on a pie graph for the section representing the 45% who ordered blueberry ice cream?

   b. What would the angle be on a pie graph for the section representing the 30% who asked for chocolate ice cream?

   c. What would the angle be on a pie graph for the section representing the 25% who asked for mint ice cream?

Bonus Question — How did a wrong view of math stifle science in the Middle Ages?

QUIZ 18
Chapter 20 — Q18

You may use the "Units of Measure" reference sheet (pages 447–448) to help answer the questions.

1. **Finding Missing Sides** — Use the knowledge that these triangles are congruent to answer the questions. Do *not* use a calculator or protractor.

    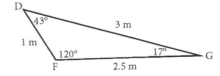

   a. What is the length of $\overline{AC}$?

   b. What is the measure of ∠ACB?

2. **Finding Missing Sides 2** — Do *not* use a calculator or protractor.

   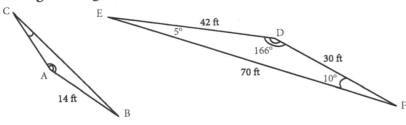

   a. From the information given, can you tell if these triangles are similar? If so, how?

   b. What is the length of $\overline{BC}$?

   c. What is the measure of ∠CBA?

   d. What is the perimeter of △ACB? *Hint*: First find the measure of $\overline{AC}$.

3. **Finding the Missing Side** — If these are both similar right triangles, find the height of the tree.

   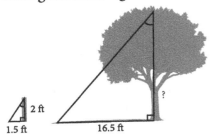

4. **Showing Congruency** — List all the sides that are congruent in these shapes, both within a shape and between the shapes.

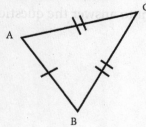
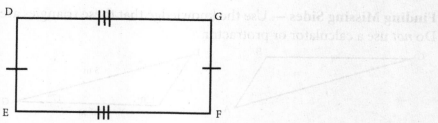

5. **Triangles**

   a. What is the sum of all the angles in a triangle?

   b. What is the measurement of the missing angle in a triangle with two 45° angles?

6. **Mastering the Skill** — Do *not* use a calculator or protractor. Instead, look at the tick marks and arcs.

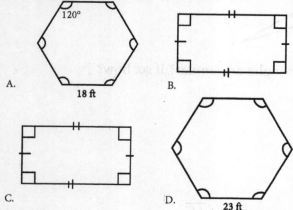

   a. Which shapes are similar?

   b. Which shapes are congruent?

Bonus Question — What do reasoning and proofs start with?

PRINCIPLES OF MATHEMATICS

TEST 1
Chapters 1 through 6 — T1

1. **Fractions in Action**

   a. You have divided up your land into sections and have evaluated the lighting and soil conditions. You have $1\frac{1}{2}$ acres on one side of the farm and $\frac{1}{3}$ of an acre on the other that you've determined are ideal for planting strawberries. How many acres of strawberries will you have altogether if you plant both sections?

   b. If you're cooking and want to triple a recipe that calls for $1\frac{2}{3}$ cup flour, how much flour should you use?

   c. If you need $\frac{1}{2}$ a yard of trim for one part of a dress and another $\frac{2}{3}$ a yard for another part, how many yards altogether should you buy?

   d. If you bought $12\frac{1}{2}$ inches of wood and used $5\frac{3}{4}$ inches, how much do you have left?

2. **Pricing Items for Sale** — You are trying to price tomatoes you're growing to sell at a farmer's market. You spent $16 on seeds, $60 on starter containers, $13 on fertilizer, and $11 on potting soil. You have 10 plants, which according to the package should yield about 20 pounds of tomatoes each. How much should you charge per pound to make 8 times your expenses? (You need to charge more than your expenses to cover your actual cost…including overhead costs such as your time in planting and selling, the water you used to water the tomatoes, etc.…plus make money!)

3. **Keeping Track of the Checkbook** — Input these transactions into the checkbook register, updating the balance column as you go.
   07/01 Opening Balance: $24,587
   07/02 Deposit Sales for Week: $1,568
   07/02 Pay Farmer Supply Company $120 with check 292
   07/03 Pay Tractor Repair Company $134 with check 293

Check Number	Date	Memo	Payment Amount	Deposit Amount	$ Balance

4. **Computing a Total Mentally** — Solve these problems mentally.
   a. 89 cents – 62 cents
   b. 78 cents – 25 cents
   c. 32 cents + 65 cents
   d. Round 56 to the nearest ten.
   e. Round 35 to the nearest ten.

5. **Checking the Skills** — Remember, all fractional answers should be simplified.
   a. $\frac{3}{4} \times \frac{7}{9}$
   b. $\frac{8}{9} \div \frac{4}{5}$
   c. $\frac{2}{5} + \frac{7}{21}$
   d. $4(25 - 6 \times 2)$
   e. $(2 + 7)8$
   f. What is the greatest common factor of 88 and 66?

   g. What is the least common multiple of 88 and 66?

Bonus Question — Name a biblical truth that helps shape our view of math.

PRINCIPLES OF MATHEMATICS

TEST 2
Chapters 7 through 11

T2

You may use a calculator on this worksheet whenever you see this symbol (🖩).

1. **Craft Time** (🖩)

    a. If you typically fit 15 pictures on 3 scrapbook pages, how many pages will you need for 75 pictures?

    b. If you can buy scrapbook paper for $3.50 per 5 pages, how much will it cost you to get all the pages you'll need?

    c. If instead you can buy scrapbook paper for $0.99 per 2 pages, how much will you save?

    d. You've been given $75 for your birthday and would like to spend half of it on ribbon for your hat-making business and half of it on fun activities. If you can get ribbon for $1.99 per spool, how many spools can you get?

2. **Computing Totals and Giving Change** — Solve these problems mentally.

    a. You buy 3 pounds of meat at $2.50 a pound. How much do you owe?

    b. If you give a $10 bill for the meat purchased in 2a, how much should you get back?

3. **Fertilizing Costs**

    a. The fertilizer package says to use 1.5 bags per acre of land. If you need to fertilize 48 acres, how many bags do you need?

    b. If each bag of fertilizer costs $3, plus a flat fee of $5.99 shipping for the entire order, how much will it cost you to fertilize in the situation described in 3a?

    c. If there's an additional 5.5% tax added to everything in the previous problem except the shipping, what will be the pre-shipping total?

4. **Finding the Total** — If you have a coupon for 25% off your total, how much will a $5.67, $7.89, and $10.20 purchase cost?

5. **Sets** — According to the diagram, what is a subset of apples?

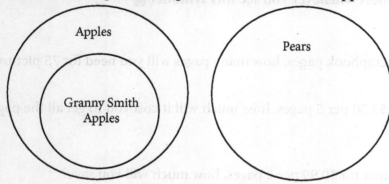

6. **Temperature** — If it warms up from −5° to 35°, by how many degrees did it change?

7. **Miscellaneous**

   a. If you use positive numbers to refer to locations above a picture, what type of numbers would you use to refer to locations beneath it?

   b. How would you express $7.50 that someone owed someone?

   c. Find the next two numbers in this sequence, assuming the pattern continues: {24, 27, 30, 33…}

   d. What is the common difference in the sequence from problem 7c?

   e. Is 4.5 an integer?

   f. $\dfrac{-10}{2}$

   g. $-4 \cdot -2$

   h. $-9 + 7$

PRINCIPLES OF MATHEMATICS

TEST 3 — Chapters 12 through 16 — T3

You may use a calculator on this worksheet whenever you see this symbol (🖩).

You may use the "Units of Measure" reference sheet (pages 447–448) to help answer the questions.

1. **Frequency Table (🖩)** — Suppose a zoo's gift shop kept track of the continent their first 15,000 visitors were from. Find the relative frequency for each continent.

Continent	Frequency	Relative Frequency
Europe	750	
Asia	900	
Africa	750	
North America	11,400	
Australia	450	
South America	750	
Antartica	0	

2. **Interpreting the Graph** — From the following graph, how many sunny days were there in July?

   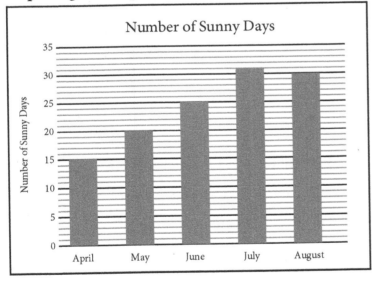

3. **Averages (🖩)** — The growth of 10 plants that formed a part of an experiment is shown.

   5 in, 4 in, 4 in, 5 in, 8 in, 3 in, 5 in, 7 in, 4 in, 4 in

   a. What was the average growth (i.e., the mean)?

   b. What was the mode?

PAGE 377

c. What was the median? *Hint*: When two numbers fall equally in the middle, take the average of those two numbers.

4. **Getting a Cab** (📱)
   a. Suppose a cab charges 4.10 Brazillian reals a kilometer, and that this amount is the equivalent of $1.40 a kilometer. If you have to go 35 kilometers, what will it cost in U.S. dollars, assuming you are required to leave a 5% tip on top of the charge?

   b. How many miles is 35 kilometers?

   c. If you travel an average of 43.5 miles in an hour, how long will it have taken you in minutes to travel the distance you found in 4b?

5. **At the Store** (📱) — Use the exchange rate of 1 Thai baht per 0.031 U.S. dollar to answer the questions.
   a. If strawberries are 200 bahts a pound, how many bahts would $\frac{1}{3}$ pound cost?

   b. How many dollars would the $\frac{1}{3}$ pound of strawberries cost?

   c. If an item costs 250 bahts plus 8% tax, how many dollars is that?

6. **Conversion** (🖩) — Convert the following units.

   a. 300 miles to kilometers

   b. 50 ft² to yd²

   c. 405 in² to ft²

7. **Coordinates** — How would you describe the location shown using coordinates?

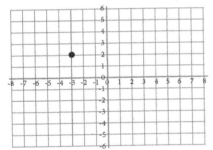

8. **Shape Time**

   a. Find the area of the rectangle in square feet.

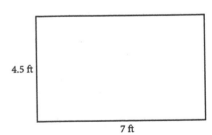

   b. What is the perimeter of the rectangle?

   c. Name at least two additional names that could be used to describe the rectangle.

9. **More Shape Time** — What is the area of this parallelogram?

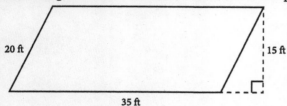

10. **Checking the Language**

    a. $\sqrt{225}$

    b. 8^3

    c. Rewrite in decimal notation: 4.569×10^{16}

11. **Term Time**

    a. If the margin of error is 5% on an opinion survey that lists 33% as holding one opinion, what is the expected range of the entire population that holds that opinion?

    b. How many sides does a quadrilateral have?

    c. Are the blades in a ceiling fan an example of scaling or rotation?

PRINCIPLES OF MATHEMATICS

TEST 4
Chapters 17 through 20 — T4

You may use a calculator on this worksheet whenever you see this symbol (🖩).

You may use the "Units of Measure" reference sheet (pages 447–448) to help answer the questions.

1. **Capacity**

   a. If a jar holds 4 cups, how many pints does it hold?

   b. If you need to drink 60 fl oz of water a day, how many cups is that?

2. **Area Time** (🖩) — Find the area of these figures in square feet.

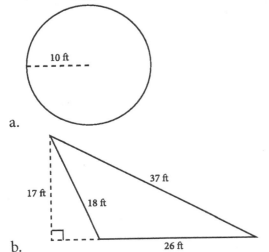

   a.

   b.

3. **Volume** (🖩) — Use the triangle in 2b as the base to answer these questions.

   a. What would the volume of a triangular prism built off the triangle in problem 2b be if its height was 20 ft?

   b. What would the volume be in cubic inches?

   c. How many gallons of water could the triangular prism hold?

PAGE 381

4. **Volume** (🖩) — Use the circle in 2a as the base to answer these questions.

   a. What would the volume of a cylinder built off the circle in 2a be if its height was 15 ft?

   b. What would the volume be in yards?

5. **Angle Time** — Use a protractor to measure the marked angles inside this spider web. Round to the nearest multiple of 5.

6. **More Angles**

   a. Is angle a in the previous problem obtuse, acute, or right?

   b. If a triangle has a 65° and a 35° angle, what must the third angle measure?

7. **Similar Shapes** (🖩) — If you measure the shadow formed by a cell tower and find it to be 30 ft, and then place a yardstick (3 ft) vertically nearby and measure the shadow cast by it as 0.6 ft, how tall is the cell tower?

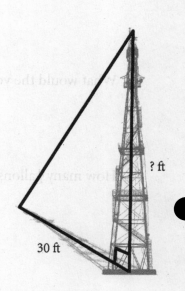

PAGE 382

8. **Similar Shape** (🖩) — Use these similar rectangles to answer the questions.

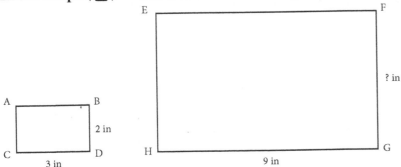

   a. Find the missing length of side $\overline{FG}$. Do not use a ruler.

   b. Find the area of the smaller rectangle.

9. **How Much Spray?** (🖩) — The figure below shows the footprint of a home. Suppose you needed to spray the perimeter for bugs. How many feet would you need to spray? Dimensions are given in feet.

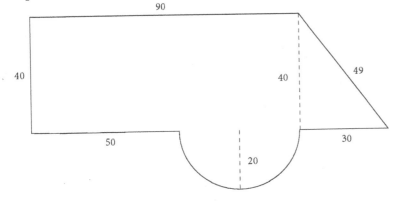

10. **Architecture** (🖩) — Use the picture shown to answer the questions. *Note*: Measurements are hypothetical.

    a. Suppose the building's owner was considering paying a sculptor to add a design to the triangular portion at the top. How much area would the sculptor have to work with?

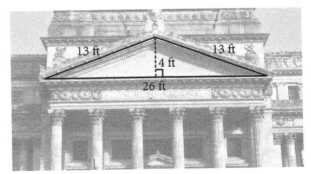

    b. How much molding would he need to do a perimeter just around the triangle?

PAGE 383

c. If the circular columns in the picture are made out of marble, and the radius of each of the circular columns is 1.5 ft and the height of the column 15 ft, how many cubic feet of marble does each column contain?

d. If marble cost $24 a cubic foot, how much would the marble in all 6 columns have cost?

11. Drawing Angles — Draw a 50° angle.

12. Missing Lengths — What is the length of $\overline{EF}$?

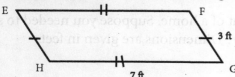

13. Surface Area — What is the surface area of this rectangular prism?

PAGE 384

PRINCIPLES OF MATHEMATICS

TEST 5 Final Exam | T5

You may use a calculator on this worksheet whenever you see this symbol (🖩).

You may use the "Units of Measure" reference sheet (pages 447–448) to help answer the questions.

1. **Household Shapes** (🖩)

    a. The tissue box is a rectangular prism. What is its volume?

    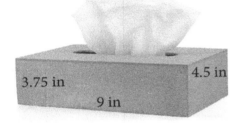

    b. The outlined section of this flag holder is a triangular prism. What is its volume?

    c. If the measure of two of the triangle's angles are 60°, what must the third angle measure?

2. **Painting a Box** (🖩)

    a. Say you needed to paint 30 boxes green as part of a set you were putting together for a play. If each box is 15 inches tall, 12 inches wide, and 18 inches long, how many square inches will you paint if you paint all six sides of all 30 boxes?

    b. Your paint can tells you it can cover 100 square feet. How many cans of paint will it take to paint all the boxes?

PAGE 385

3. **Circles** (🖩)

   a. How much fabric could a cross-stitch hoop with an 8-in diameter hold?

   b. What is the circumference of the hoop?

4. **Fish Pond** (🖩)

   a. Find the volume of a fish pond if it is a cylinder 3 ft high and has a radius of 5 ft.

   b. How many gallons can the pond hold?

5. **Miscellaneous** (🖩)

   a. The Washington Monument is about 555.43 feet (6,665.16 in) tall. If you want to make a scale drawing where every side is 1,000 times less than the original, how many inches tall should your drawing be?

   b. If you buy an item for $534, what will your total with tax be if the tax rate is 12.5%?

   c. Let's say you help out a friend selling products at a trade show, and he offers to pay you 10% of all you sell. If you sell $324.23, how much should your friend pay you?

6. **Average** (🖩) — If you took a 4-night trip and you slept 7 hours one night, $8\frac{1}{2}$ hours the next, 4 hours the next, and 10 hours the last night, how much did you sleep on average per night during that trip? Give your answer as a decimal number rounded to the nearest tenth.

7. **On a Trip** (🖩)
   a. If a sign informs you the border is 25 kilometers away, how many miles is it?

   b. If you're traveling 50 miles per hour, how many minutes will it take you to go the distance in the previous problem?

   c. If you fill up with 6 liters of gas, how many gallons did you buy?

   d. What was the equivalent cost per gallon if you paid 29.50 South African rands and the exchange rate was 1 dollar per 11 rands?

   e. If the exchange rate is 1 dollar per 2,684 Iranian rials, how many total dollars would you pay if you bought items that cost 8,089 rials, 4,567 rials, 1,896 rials, and 23,900 rials?

   f. If you buy 4 pounds of apples sold at 4,000 rials a pound, how many dollars will it cost you given the exchange rate given in 7e?

g. A box in the grocery store weighs 100 grams. How many pounds is that?

8. **Miscellaneous**

   a. Express 0.56 as a fraction in simplest terms.

   b. Express 7% as a decimal.

   c. Express $\frac{7}{8}$ as a decimal.

   d. Round 0.484 to the nearest hundredth.

   e. Find $7 \cdot -3$

   f. Find $\frac{-8}{-2}$

   g. Express 90 miles divided by 3 hours as an improper fraction; you do not need to simplify.

   h. Rewrite $\frac{89}{2}$ as a mixed number and as a decimal.

   i. Find $\frac{7}{8} - \frac{1}{24}$

   j. Find $\sqrt{100}$

   k. Find $8 + 6 \cdot 3^2$

   l. Convert $\frac{1}{2}$ a gallon to cups.

9. **Coordinates** — Label each coordinate using the standard format. Draw lines connecting the points to form a triangle.

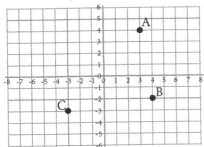

10. **Stats** — Suppose the following graph was presented at a home owner's meeting in support of why a new park should be built. List at least one additional pieces of information you might want to know about the graph (i.e., questions to ask the presenter).

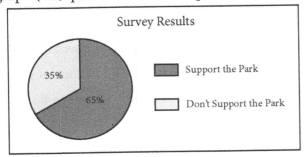

11. **Sets** — Is the following diagram accurate? If not, redraw it so that it is.

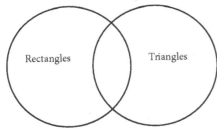

12. **Around Town**

    a. If you have a meeting at 5 p.m. EST and it's 1 p.m. PST, how long do you have before your meeting?

    b. If you bought $\frac{2}{3}$ a bushel of apples, and a family member bought $\frac{3}{4}$ a bushel of apples, how many bushels did you both buy combined? Give your answer as a mixed number.

c. If you want to make a woodworking pattern $1\frac{1}{2}$ times larger, and it calls for a length to be $\frac{9}{12}$ of a foot, what length should you make it? Give your answer as a mixed number.

d. If you make $34,000 a year, how much do you make per month?

Extra Credit: Problems from the Past (🖩) — These problems are from a math book published in the early 1900s. They are included here with very minimal modification. Have fun!

a. If a solution for spraying fruit trees and plants is made up of 4 lb of lime, 4 lb of copper sulphate, $\frac{1}{4}$ lb of Paris green, and enough water to bring it all to 50 gallons, what will 100 gallons of such a mixture cost if lime costs $0.25 a pound, copper sulphate $0.40 a pound, and Paris green $0.75 a pound?

b. What is the cost per tree for the mixture in the previous problem if a sufficient amount to spray one tree is $2\frac{1}{2}$ gallons?

c. A [rectangular] field 180 ft by 200 ft is enclosed by four lines of galvanized iron wire. Eight yards of this wire weigh a pound, and the wire cost 8 cents a pound. What was the total cost of the wire?

d. If each hen laid on an average 100 eggs per year, what was the average monthly return per hen if the eggs were sold for 85 cents a dozen? *Hint*: Think about how many eggs are in a dozen.

Answer Key

General Grading Notes

Please use your own judgment when grading. Below are some general principles to keep in mind.

- **Different Strategies** — There is often more than one legitimate approach to a problem. You want to evaluate if students are learning the concepts and solving the problems carefully, correctly, and logically.

- **Open-Ended Questions** — On open-ended questions, answers may vary significantly from what is listed.

- **Partial Credit** — Feel free to give partial credit if a student set up the problem correctly but made a calculation error.

- **Units of Measurement** — If a unit is given in the problem (dollars, feet, etc.), students need to **include the unit in their answer**. For example, if a student lists "6" instead of "6 in" on a problem where the answer key lists "6 in," their answer is only partially correct. Watching their units carefully will serve them well, both in real life and in upper-level courses.

- **Word Problems** — Mental arithmetic should be encouraged, but when solving word problems, students should still always show their work, writing down the equation(s) they solved so you can see what process they followed. It's a very helpful habit to develop, as it makes it easier to find any errors. However, unless requested in the problem, it's not necessary for them to write down every step that is shown in the answer key—*just enough steps that you can tell how they approached the problem.*

- **Decimals** — From Worksheet 7.4 on, decimal answers should be rounded to the hundredth digit unless otherwise specified. Also, unless otherwise specified, it doesn't matter if students round their answer at the end or after each step.

  Be aware that **answers may be slightly off the answer in the key due to differences in rounding at the end or after each step**. This is not a problem. The important thing is that students followed instructions and solved the problem accurately. Exceptions: When finding a percentage, students should not round the percent amount, and when doing unit conversion, students should not round a conversion ratio until the end.

  For example, if told to find a 7.5% sales tax, students should use 0.075 to calculate the tax, and if told to convert between pints and cubic inches (1 pint = 28.875 in^3), students should not round 28.875 (they can, however, round their answer).

- **Fractions** — From Worksheet 5.3 on, fractional answers should be denoted in simplest terms, unless otherwise specified. This includes writing mixed numbers as improper fractions.

  Not only will this make it easier to grade and avoid confusion, it will also provide the student with practice forming equivalent fractions.

  Even after decimals are covered in Chapter 7, students should continue solving problems given in fractions as fractions, so as to become proficient in working with fractions. If the problem includes both fractions and decimals, however, students may give their answer in either.

Assigning a Grade

The grade column in the Suggested Schedule (page 6–18) is available for you to keep track of a student's grade should you choose to do so. Feel free to use whatever method for grading you've chosen to adopt, or to leave those columns blank if you prefer not to assign grades.

Extra-Credit Assignments

Throughout the course, some of the worksheets include extra-credit assignments. It is up to you to decide how the assignment should affect the student's grade. For example, you could decide that completing an assignment will raise their worksheet grade by a certain number of points, or that it will increase their quarter or final grade by a certain amount.

Additional Resources and Course Notes

Please see http://www.christianperspective.net/math/pom1 for links to helpful online resources (such additional drill worksheets, an online abacus, and an online scientific calculator), along with additional notes and information related to this course. There is also a way to contact the author there.

Chapter 1: Introduction and Place Value

Worksheet 1.1

1. Possibilities include numbers on an alarm clock, dates on a milk carton, grams of sugar on a cereal box, Bible verse numbers, speed-limit signs, prices at a grocery store, zip codes and street numbers on envelopes, page numbers in books, rulers (including rulers in computer programs), and font sizes. Other ideas are measuring ingredients, figuring out how many places to set for company, figuring out how long you have left before an appointment, and keeping track of money.
2. Answer should be a dictionary definition of "worldview."
3. Math is neutral; a biblical math curriculum is the same as any other, with a Bible verse or problem thrown in now and then; and math is a textbook exercise.
3. I can use a calculator any time I see the symbol 🖩.

Worksheet 1.2

1. Numerous possibilities were given within the text. Examples should not be repeated from yesterday's worksheet.
2. Math notebook should be prepped.
3. Math is a way of describing the consistent way this universe operates; it works outside of a textbook because God is faithful to uphold all things.

Worksheet 1.3

1. Within math, there's a battle to remember our dependency on the Lord.
2. Should be a dictionary definition of "naturalism" and "humanism."
3. Abacus needs to be prepped or located.

Worksheet 1.4

1. a. 311,050,977
   b. 13,561,600,000,000
   c. 1,336,718,015
2. a. twenty-seven million, two hundred fifty-three thousand, nine hundred fifty-six
   b. twenty-five million, one hundred forty-five thousand, five hundred sixty-one
   c. nineteen million, three hundred seventy-eight thousand, one hundred two
3. a. <
   b. >
   c. >
4. Check text for possible symbols.

Worksheet 1.5

1. a. 3,827
   b. 6,913
   c. 4,058
   d. 3,645

2. a. [abacus image]
   b. [abacus image]
   c. [abacus image]
   d. [abacus image]

3. a. Sixty-one thousand, two hundred seventy-two
   b. Seventeen thousand, seven hundred twenty-seven
   c. 12,021
   d. 47,821
4. a. >
   b. <
   c. =
5. decimal system (or Hindu-Arabic decimal system)
6. the city with 123,000
7. Answer should communicate that in a place-value system, the place, or location, of a number determines its value.
8. Since students won't actually have to use Egyptian hieroglyphics again and their purpose here is simply to help students understand that there are different ways to express quantities, it does not matter if every detail is the same. Just check to make sure that the 𝕌 symbols are on the left of the ⋂ ones, and that there are the appropriate number of each.

   a. 000
      000   ⋂⋂
   b. 000   ⋂⋂⋂⋂
      00   ⋂⋂⋂
   c. 000   ⋂⋂⋂⋂
      000   ⋂⋂⋂⋂
      000

9. a. [clock showing 3:45 with Roman numerals]
   b. 1998
   c. A chord based off the fourth note of a scale.
   d. 2:00

PAGE 393

10. All the different number systems remind us not to start looking at our current system as math itself, but rather as one way of describing God's creation.

Worksheet 1.6

1. a. 1 set(s) of 8 = 1 x 8 = 8
   1 set(s) of 4 = 1 x 4 = 4
   0 set(s) of 2 = 0 x 2 = 0
   0 set(s) of 1 = 0 x 1 = 0
   1100 in binary is the same 12 in the decimal system.

   b. 1 set(s) of 16 = 1 x 16 = 16
   0 set(s) of 8 = 0 x 8 = 0
   0 set(s) of 4 = 0 x 4 = 0
   0 set(s) of 2 = 0 x 2 = 0
   0 set(s) of 1 = 0 x 1 = 0
   10000 in binary is the same as 16 in the decimal system.

   c. 1 set(s) of 16 = 1 x 16 = 16
   0 set(s) of 8 = 0 x 8 = 0
   1 set(s) of 4 = 1 x 4 = 4
   0 set(s) of 2 = 0 x 2 = 0
   0 set(s) of 1 = 0 x 1 = 0
   10100 in binary is the same as 20 in the decimal system.

2. a. 12 set(s) of 16 = 12 x 16 = 192
   5 set(s) of 1 = 5 x 1 = 5
   C5 in hexadecimal is the same as 197 in the decimal system.

   b. 14 set(s) of 16 = 14 x 16 = 224
   9 set(s) of 1 = 9 x 1 = 9
   E9 in hexadecimal is the same as 233 in the decimal system.

   c. 15 set(s) of 16 = 15 x 16 = 240
   15 set(s) of 1 = 15 x 1 = 15
   C5 in hexadecimal is the same as 255 in the decimal system.

3. a. >
   b. >
   c. <
   d. =

4. a. 3,796,742
   b. 92,960,000
   c. 432,200

5. a. 4,625
   b. 2,080
   c. 7,500
   d. 9,326

6. a. MMXIV
   b. MLXXVI
   c. DXCII

7. a. It would mean each place was worth 5 of the previous place's value.
   b. I would need five digits. *Example*: 0, 1, 2, 3, 4

Chapter 2: Operations, Algorithms, and Problem Solving

Worksheet 2.1

1. a. 4 and 9 are the addends, and 13 is the sum.
   b. 15 is the minuend, 9 is the subtrahend, and 6 is the difference.
   c. 8 and 5 are the addends, and 13 is the sum.
   d. 17 is the minuend, 6 is the subtrahend, and 11 is the difference.

2. a. 11
   b. 7
   c. 4
   d. 10
   e. X
   f. VIII
   g. IX

3. a. = or 8 = 8
   b. > or 9 > 8
   c. > or 5 > 4
   d. < or 8 < 9
   e. < or 11 < 13

4. Hebrews 1:3 and Jeremiah 33:25-26 should have been added to notebook.

Worksheet 2.2

1. a. 8 p.m.
   b. 12 p.m. (noon)
   c. 1:15 p.m.
   d. 4 hours

2. a. God was in the beginning and created day and night.
   b. Yes, time as we know it with day and night will have an end.
   c. No, eternity will not have an end.
   d. We should diligently seek to be found of God in peace, without spot, and blameless.

3. a. 6 a.m.
   b. 6 p.m.
   c. 5 p.m.
   d. 5 p.m.
   e. 2 hours
   f. Student should have added time zones to notebook and made flashcards to learn those within the continental United States.

   Extra Credit — Write out at least one interesting tidbit on the history of time zones.

Worksheet 2.3

1. Students were told to solve these problems on an abacus.
   a. 27
   b. 1,012
   c. 1,257

2. Students were told to solve these problems on an abacus.
   a. 708
   b. 448
   c. 1,101

3. Students were told to solve these problems using the traditional written methods. The first three are shown as examples.
   a. 1
      1,489
      + 2,008
      3,497
   b. 1 1 1 1
      89,871
      + 689
      90,560
   c. 1
      450,123
      + 589,256
      1,039,379
   d. 66,138
   e. 208,079
   f. 683
   g. 368
   h. 80
   i. 433
   j. 85

4. Students were told to solve on an abacus.
   66 + 130 = 196

5. a. 3:30 p.m.
   b. 9:30 p.m.
   c. 5:20 p.m.
   d. 1300
   e. 2100

6. a. > or 854 > 750
   b. < or 1,139 < 1,239

7. They are consistent because Jesus is consistently upholding all things by the Word of His Power. Hebrews 1:3

Worksheet 2.4

Wording in the memo column may vary slightly.

1.

Check Number	Date	Memo	Payment Amount	Deposit Amount	$ Balance
	9/1	Opening Balance			3,456
250	9/10	Donation – My Favorite Charity	150		3,306
	9/10	Deposit – Babysitting		59	3,365
	9/17	Deposit – Raking		9	3,374
	9/25	Deposit – Mowing		45	3,419
251	9/26	Dillon's Department Store – Bike	179		3,240
252	9/30	C's Computer – Laptop	1,549		1,691

2. a.–d. The grayed transactions should be highlighted or checked off with a checkmark, the bolded text should be added, and there should be a mark on the statement and in the check register that the balances reconcile.

Check Number	Date	Memo	Payment Amount	Deposit Amount	$ Balance	
	9/1	Opening Balance			3,456	
250	9/10	Donation – My Favorite Charity	150		3,306	
	9/10	Deposit – Babysitting		59	3,365	
	9/17	Deposit – Raking		9	3,374	
	9/25	Deposit – Mowing		45	3,419	
251	9/26	Dillon's Department Store – Bike	179		3,240	
252	9/30	C's Computer – Laptop	1,549		1,691	
	9/30	Interest		1	1,692	rec

Statement Balance + Unprocessed Deposits – Unprocessed Payments = \$3,241 + 0 – \$1,549 = \$1,692

Note: We know the checkbook reconciles because the statement balance plus unprocessed deposits minus unprocessed payments equals the checkbook register balance.

3. a. 7 p.m.
   b. 5 p.m.
   c. Students should have reviewed time zone flashcard.

Worksheet 2.5

1. a. Ones 0 + 5 = 5 5
      Tens 2 + 3 = 5   + 50
      55

   b. Ones 6 + 5 = 11   11
      Tens 1 + 1 = 3   + 20
      31

2. Students were told to solve these using the traditional written methods.
   a. 745
   b. 1,682
   c. 806
   d. 1,796
   e. 10,928
   f. 374,320

3. a. 3 hours (The meeting is occurring at 5 p.m. EST, which is 3 hours later than 2 p.m. EST.)
   b. 6:30 p.m.
   c. Students should have reviewed time zone flashcard.

4. The method must accurately describe consistencies God created and sustains.

5. Wording in the memo column may vary slightly.

Check Number	Date	Memo	Payment Amount	Deposit Amount	$ Balance
	11/1	Beginning Balance			3,025
150	11/2	Sporting Goods	45		2,980
151	11/6	Grocery Store	27		2,953
	11/8	Auto Deposit		124	3,077
	11/19	Deposit – Rebate		15	3,092
	11/28	Withdrawal	25		3,067
152	12/3	Cell Phone	54		3,013
	12/7	Auto Deposit		124	3,137

6. a.–d. The grayed transactions should be highlighted or checked off with a checkmark, the bolded text should be

added, and there should be a mark on the statement and in the check register that the balances reconcile.

Check Number	Date	Memo	Payment Amount	Deposit Amount	$ Balance	
	11/1	Beginning Balance			3,025	
150	11/2	Sporting Goods	45		2,980	
151	11/6	Grocery Store	27		2,953	
	11/8	Auto Deposit		124	3,077	
	11/19	Deposit – Rebate		15	3,092	
	11/28	Withdrawal	25		3,067	
152	12/3	Cell Phone	54		3,013	
	12/7	Auto Deposit		124	3,137	
	11/30	Interest		1	3,138	rec

Statement Balance + Unprocessed Deposits – Unprocessed Payments = $3,068 + $124 – $54 = $3,138

Note: We know the checkbook reconciles because the statement balance plus unprocessed deposits minus unprocessed payments equals the checkbook register balance.

Worksheet 2.6

See word problem bullet in the "General Grading Notes" on page 391.

1. a. Define:
   chairs for my family = 5
   chairs for visitors = 8
   total chairs = ?
   b. Plan:
   chairs for my family + chairs for visitors = total chairs
   c. Execute: 5 + 8 = 13
   d. Check: 13 is a reasonable number of chairs for two families of this size. 13 – 5 = 8

2. a. Define:
   party favor bags made = 12
   party favor bags used = 9
   party favor bags left over = ?
   b. Plan:
   party favor bags made – party favor bags used = party favor bags left over
   c. Execute: 12 – 9 = 3
   d. Check: 3 is a reasonable number. 9 + 3 = 12

3. a. Define:
   cost of gift 1 = $45
   cost of gift 2 = $37
   difference between gifts = ?
   b. Plan:
   cost of gift 1 – cost of gift 2 = difference between gifts
   c. Execute: $45 – $37 = $8
   d. Check: $8 is a reasonable answer. 37 + 8 = 45

4. a. $80 – $13 = $67
   b. $80 – $13 – $3 = $64

5. a. 13,083 mi – 5,300 mi = 7,783 mi
   b. 13,630 mi – 7,954 mi = 5,676 mi

6. Answer should be an addition and a subtraction problem.

7. Answer should be the solutions to the word problems written in problem 6.

PAGE 396

8. a. $1\frac{1}{2}$ hours
   b. 3 p.m.
   c. Students should have reviewed time zone flashcard.

9. Problem-solving steps should be written in math notebook.

10. a chest of tools

Chapter 3: Mental Math and More Operations

Worksheet 3.1

1. a. 35
   b. 215
   c. 17

2. a. 9
   b. 6
   c. 3
   d. 2
   e. 7
   f. 6
   g. 5
   h. 3
   i. 8

3. Students were told to solve these problems mentally.
   a. 8 *Mental Process*: Start at 19. Add 1 to get to 20 and 7 more to get to 27. 1 + 7 = 8
   b. 8 *Mental Process*: Start at 27. Add 3 to get to 30 and 5 more to get to 35. 3 + 5 = 8
   c. 9 *Mental Process*: Start at 39. Add 1 to get to 40 and 8 more to get to 48. 1 + 8 = 9
   d. 22 *Mental Process*: Start at 28. Add 2 to get to 30 and 20 more to get to 50. 2 + 20 = 22
   e. 15 *Mental Process*: Start at 55. Add 5 to get to 60 and 10 more to get to 70. 5 + 10 = 15

4. a. Define:
   monthly gas = $25
   monthly insurance = $100
   monthly wear and tear = $120
   total monthly cost = ?
   b. Plan:
   monthly gas + monthly insurance + monthly wear and tear = total monthly cost
   c. Execute: $25 + $100 + $120 = $245
   d. Check: $245 looks reasonable;
   $245 – $25 – $100 – $120 = 0

Worksheet 3.2

Students should solve problems 1–6 mentally.

1. a. 570
   b. 990
   c. 250
   d. 800

2. a. 600
   b. 1,000
   c. 300
   d. 800

3. a. 560 *Mental Process*: 500 + 60 = 560
   b. 950 *Mental Process*: 900 + 50 = 950
   c. 640 *Mental Process*: 540 + 100 = 640

d. 370 *Mental Process:* 400 − 30 = 370
e. 40 *Mental Process:* 60 − 20 = 40

4. a. 42 *Mental Process:* Round 19 up 1 and add: 23 + 20 = 43; Subtract 1 we added: 43 − 1 = 42
    b. 62
    c. 31
    d. 132
    e. 97
5. a. 30 *Mental Process:* Add 2 to 18 to get to 20; add 28 more to get to 48. 2 + 28 = 30.
    b. 52
6. 75¢ − 59¢ = 16¢
7. $253 − $89 = $164
8. Unlike God, we can't keep track of every detail.

Worksheet 3.3

1. a. 63
    b. 12
    c. 27
    d. 56
    e. 30
2. 8 x $8 = $64
3. 3 x $9 = $27
4. a. > *or* 25 > 23
    b. > *or* 6 > 5
    c. = *or* 24 = 24
5. a. 3
    b. 5
    c. 8
    d. 9
    e. 4
    f. 6
    g. 7
6. Students were told to solve these problems mentally.
    a. 7¢
    b. 17¢
    c. 11¢
    d. 48¢
7. a. factors
    b. product
8. 350
9. The multiplication table and Napier's rods record what happens when we multiply quantities; the fact that multiplication is consistent enough to be recorded in a table or on rods points to God's faithfulness and power in governing all things.

Worksheet 3.4

1. a. 11
    b. 6
    c. 8
    d. 9
2. 72 ÷ 8 = 9; 9 pages a day
3. 56 ÷ 7 = 8; 8 favors per guest

4. 64 ÷ 8 = 8; 8 cookies per person
5. a. 1
    b. 7
    c. 5
    d. 3
    e. 8
6. Students were told to solve these problems mentally.
    a. 36 *Mental Process:* Start at 1964; add 6 to get to 1970 and 30 to get to 2000. 6 + 30 = 36
    b. 48
    c. 37
    d. 33
    e. 1,900
7. 24 is the dividend, 6 is the divisor, and 4 is the quotient.
8. The Egyptians divided by repeated subtraction.

Worksheet 3.5

1. a. =
    b. commutative property
    c. =
    d. associative property
    e. ≠
    f. =
    g. identity property of multiplication
    h. =
    i. identity property of addition
    j. =
    k. =
2. a. no
    b. no
    c. addition, multiplication
    d. 1
    e. 0
    f. A characteristic that hold true in every situation; a name to describe a consistency God holds together.
3. a. 5
    b. 8
    c. 7
    d. 1
    e. 3
4. Students were told to answer these problems mentally.
    a. $200
    b. $20
    c. $33
    d. $19
    e. 800
    f. 780
5. The properties from Lesson 3.5 should be written in math notebook.

Worksheet 3.6

1. a. 4 • 2(4 + 5 − 3) = 4 • 2 (9 − 3) = 4 • 2 (6) = 8(6) = <u>48</u>
    b. 5 + 8 x 5 ÷ 5 = 5 + 40 ÷ 5 = 5 + 8 = <u>13</u>
    c. 8 • 3 ÷ 6 • 2 = 24 ÷ 6 • 2 = 4 • 2 = <u>8</u>
    d. 50 ÷ (5 + 5) x 2 = 50 ÷ 10 x 2 = 5 x 2 = <u>10</u>
    e. (7 − 2 x 3) + 8 − 2(2) = (7 − 6) + 8 − 2(2) = 1 + 8 − 4 = <u>5</u>

2. No; the order of operations is to do multiplication before subtraction. You would get a different answer otherwise.
3. a. 4(7 + 8 + 3) = 4(18) = <u>72</u>
   b. (4 + 2)6 = (6)6 = <u>36</u>
4. Check to see that the problem was written with parenthesis.
   20($4 + $6) = 20($10) = $200
5. a. 9
   b. 6
   c. 3
6. a. Define:
      total pages in book = 315
      pages read = 47
      pages left = ?
   b. Plan:
      total pages in book − pages read = pages left
   c. Execute: 315 − 47 = 268
   d. Check: The answer seems reasonable. 47 + 268 = 315
7. The order of operations should be written in math notebook.

Chapter 4: Multi-digit Multiplication and Division
Worksheet 4.1

1. a.

   71 x 5 = 355

   b.

   45 x 60 = 2,700

   c.
   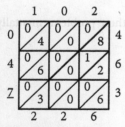
   102 x 463 = 47,226
   Note: We mentally added 10 to the underlined 7 from the previous diagonal.

2. a.
   ```
        2
      305
   x   50
      000
   + 15250
     15,250
   ```

   b.
   ```
        3
      415
   x    6
    2,490
   ```

   c.
   ```
        1
      202
   x    9
    1,818
   ```

   d.
   ```
       1 2
       25
   x   25
      125
   +  500
      625
   ```

3. a. Define:
      24 = number of dictionaries
      $18 = cost of dictionary
      total cost = ?
   b. Plan:
      number of dictionaries x the cost per dictionary = total cost
   c. Execute: 24 x $18 = $432
   d. Check: $432 is a reasonable cost for 24 dictionaries.
4. $875 x 12 = $10,500; his salary will be $10,500.
5. 82 x 196 pounds = 16,072 pounds
   32 x 16,072 pounds = 514,304 pounds; there are 514,304 pounds of flour on the trainload.
6. a. 6
   b. 8
   c. 7
7. a. 48(141) = 6,768
   b. 108(42) = 4,536
8. a. An algorithm is a step-by-step method.
   b. Multiplication algorithms work because they describe the consistent way God causes objects to multiply.

Worksheet 4.2

1. Check to make sure problems were solved using the method shown.
   a. 3(2 + 6) = 3(8) = 24
   b. 3(2 + 6) = 3(2) + 3(6) = 6 + 18 = 24
   c. (17 + 3)5 = (20)5 = 100
   d. (17 + 3)5 = (17)5 + (3)5 = 85 + 15 = 100
2. the commutative property
3. Check to make sure problems were solved using the distributive property.
   a. 7(24) = 7(20 + 4) = 140 + 28 = 168
   b. 3(45) = 3(40 + 5) = 120 + 15 = 135
   c. 4(304) = 4(300 + 4) = 1,200 + 16 = 1,216
4. Students were told to solve these problems mentally.
   a. 425
   b. 248
   c. 168
5. a. Define:
      yearly cost of magazine = $108
      monthly cost = ?

b. Plan:
   yearly cost of magazine ÷ 12 months = *monthly cost*
c. Execute: $108 ÷ 12 = $9.00
d. Check: The answer seems reasonable. 12 x $9 = $108

Worksheet 4.3

1. a.
   ```
        45
   5)225
     - 20
       25
     - 25
        0
   ```
   b.
   ```
        89
   7)623
     - 56
       63
     - 63
        0
   ```
   c.
   ```
        63
   8)504
     - 48
       24
     - 24
        0
   ```

2. a. 80 pages ÷ 4 = 20 pages
   b. 297 miles ÷ 9 = 33 miles

3. a. (39)25 = 975
   b. 450 ÷ 5 = 90
   c. Be sure the distributive property was used. 8(789) = 8(700) + 8(80) + 8(9) = 5,600 + 640 + 72 = 6,312

4. Students were told to solve these problems mentally.
   a. 13 cents *Mental Process:* Add 8 to get to 70 and 5 to get to 75. 8 + 5 = 13
   b. 8 cents *Mental Process:* Add 3 to get to 20 and 5 to get to 25. 3 + 5 = 8
   c. 48 cents *Mental Process:* Seeing 26 and 4 totaled to 30, it's easy to group them together first and then add the 18 to 30.

5. It told us that God gave man work to do.

Worksheet 4.4

1. a.
   ```
           85
   36)$3,060
      - 288
        180
      - 180
          0
   ```
   b.
   ```
         16
   28)448
      - 28
       168
     - 168
         0
   ```
   c.
   ```
        57 r 11
   15)866
     - 75
      116
    - 105
       11
   ```
   d.
   ```
        11 r 20
   80)900
     - 80
      100
    -  80
       20
   ```

2. a. unsolvable, as cannot divide by 0
   b. 0
   c. 0
   d. 0
   e. 852
   f. 987

3. a. Define:
      total dollars = $13,500
      total books = 4,500
      cost per book = ?
   b. Plan:
      cost per books = *total dollars* ÷ *total books*
   c. Execute: $13,500 ÷ 4,500 = $3. Each book cost $3.
   d. Check: $3 is reasonable for a book cost.

4. a. 160 pages ÷ 20 = 8 pages
   b. *selling price* = 9 x *cost per book* = 9 x $3 = $27

5. a. 48(29 + 15 ÷ 3 x 1) = 48(29 + 5 x 1) = 48(29 + 5) = 48(34) = 1,632
   b. 45 ÷ 15 + 8 + 0 = 3 + 8 + 0 = 11
   c. Check to make sure problem was solved using the distributive property.
      12(50) + 12(6) = 600 + 72 = 672

Worksheet 4.5

1. a. Define: *starting distance apart* = 8,888 miles
      boat 1 daily increase = 144 mi
      boat 2 daily increase = 166 mi a day
      total distance apart at end of week = ?
      week = 7 days
   b. Plan:
      boat 1 weekly increase = *boat 1 daily increase* • 7
      boat 2 weekly increase = *boat 2 daily increase* • 7
      boat 1 weekly increase + *boat 2 weekly increase* + *starting distance apart* = *total distance apart at end of week*
   c. Execute:
      boat 1 weekly increase = 144 mi x 7 = 1,008 mi
      boat 2 weekly increase = 166 mi x 7 = 1,162 mi
      1,008 mi + 1,162 mi + 8,888 mi = 11,058 mi
      They will be 11,058 mi apart.
   d. Check: 11,058 mi sounds reasonable given how far apart they started and how much the distance between them increased each day.
      1,008 ÷ 7 = 144; 1,162 ÷ 7 = 166
      11,058 − 8,888 − 1,162 = 1,008

2. a. *cake* = $1,024
      brochette = 3 x $8 = $24
      chips = 3 x $4 = $12
      luncheon meet = 4 x $17 = $68
      cheese = 3 x $4 = $12
      crackers = 5 x $4 = $20
      strawberries = 10 x $4 = $40
      dipping chocolate = 5 x $10 = $50
      total cost = $1,024 + $24 + $12 + $68 + $12 + $20 + $40 + $50 = $1,250
      cost per guest = $1,250 ÷ 250 = $5
   b. *cost per guest* = ($1,250 + $250) ÷ 300 = $5
      Note: The $250 came from first finding how much more we need to add to our total cost for the rolls. If we bought enough for each guest and each pack has 60, we

would need 300 ÷ 60, or 5 packages, which would cost 5 x $50 = $250.
Remember, we can use parentheses to show operations to do first. We could have also written this problem out as two separate problems like this:
new cost = $1,250 + $250 = $1,500
cost per guest = $1,500 ÷ 300 = $5
The parentheses just saved us a step!
c. cost per guest = ($1,250 − $500) ÷ 250 = $3

3. a. *1 month furnished* = $2,000 + $600 = $2,600
1 month unfurnished = $800 + $2,600 = $3,400
$2,600 < $3,400
The furnished apartment is the more economical option.
b. *7 months furnished* = $2,600 x 7 = $18,200
7 months unfurnished = $800 x 7 + $2,600 = $8,200
$18,200 > $8,200
The unfurnished apartment is the more economical option.
$18,200 − $8,200 = $10,000
The unfurnished apartment is cheaper by $10,000.
c. *6 months month-by-month* =
6 x $800 = $4,800
6 months pre-paid = $3,600
savings = $4,800 − $3,600 = $1,200

4. Students were told to solve these problems mentally.
a. 71 *Mental Process*: Add 4 to get to 20, 60 to get to 80 and 7 to get to 87. 4 + 60 + 7 = 71
b. 38
c. 63
d. 17

5. a. 15(860 ÷ 43 + 15) = 15(20 + 15) = 15(35) = 525
b. 89
c. Check that solved using the distributive property.
4(12) + 4(3) = 48 + 12 = 60

Worksheet 4.6

1. *cost of car* = $4,565
estimated life of car = 5 years
cost of car ÷ estimated life of car = cost per year of car
$4,565 ÷ 5 = $913

2. a. *equipment cost* = $255
cost per cone = $1
sales price per cone = $4
profit per cone at $4 = $4 − $1 = $3
initial cost ÷ profit per cone = number of cones to recuperate cost
$255 ÷ $3 =
85 cones to recuperate the cost
b. *alternative sales price per cone* = $5
profit per cone at $5 = $5 − $1 = $4
$255 ÷ $4 =
63 r3, so 64 cones to recuperate the cost
c. *expense per cone* = $1 x 230 = $230
expense to start = $255
total expenses = $230 + $255 = $485
total made = number sold x selling price = 230 x $5 = $1,150

total made − total expenses = profit
$1,150 − $485 = $665

3. On problems 3a–3d, check that the method shown was used.
a. 7(3 + 2) = 7(5) = 35
b. (7 x 3) + (7 x 2) = 21 + 14 = 35
c. 100 + 2 (14) = 100 + 28 + 128
d. 100 + (2 x 9) + (2 x 5) = 100 + 18 + 10 = 128
e. 6 + 12 = 18
f. 0
g. unsolvable, as cannot divide by 0
h. 65 r5
i. 1,275

4. To find the mileage on your family car, you need to know the total distance driven for a certain number of gallons. This is done by recording the mileage when filling up at the gas station two times in a row. The difference between the mileage will be the total distance gone using the gallons filled up with the second time (that's the amount of gas used between the two fill ups).

Example (numbers will vary):
1,234 (mileage at first fill)
1,334 (mileage at second fill)
Mileage gone between fills = 1,334 − 1,234 = 100
Gallons car took at second fill = 10
Miles per gallon = 100 (total miles) ÷ 10 (total gallons) = 10 miles per gallon

Extra Credit — Students can get extra credit by making Napier's rods and using them to solve
34 x 896 (which equals 30,464). See example in Lesson 4.6.

Chapter 5: Fractions and Factoring

Worksheet 5.1

1. a. $\frac{5}{20}$
b. $\frac{$30}{$50}$
c. $\frac{80}{100}$
d. $\frac{76}{1}$
e. $\frac{34}{48}$
f. $\frac{$10,680}{12}$
g. $\frac{130}{1}$
h. $\frac{856}{428}$
i. $\frac{4}{1}$
j. $\frac{4}{4}$

2. a. $\frac{3,072}{256}$
b. $\frac{150}{300}$
c. $\frac{30}{60}$
d. $\frac{1}{3}$

3. a. $10,680 ÷ 12 = 890
b. 856 ÷ 428 = 2
c. 4 ÷ 4 = 1
d. 3,072 ÷ 256 = 12

4. a. The number 5 in 1a and $30 in 1b should be circled.

b. 1d, 1f, 1g, 1h, 1i, and 1j are improper fractions and should have a star next to them.
c. $\frac{80}{100}$ would be read eighty hundredths, eighty over one (or a) hundred, or eighty divided by one (or a) hundred.
d. $\frac{34}{48}$ would be read thirty-four forty-eighths, thirty-four over forty-eight, or thirty-four divided by forty-eight.
e. No. In this course, whole numbers start at 1; however, 0 is sometimes defined as a whole number in other materials.

5. a. A notation is a "series or system of written symbols used to represent numbers, amounts, or elements…" [*New Oxford American Dictionary*, 3rd edition (Oxford University Press, 2012), Version 2.2.1 (156) (Apple, 2011), s.v., "notation."]
   b. Three ways of looking at fractions are as partial quantities, a notation, and division.

6. Check that these problems were solved using the distributive property.
   a. 42(5) + 42(92) = 210 + 3,864 = 4,074
   b. 12(6) + 12(85) = 72 + 1,020 = 1,092

Worksheet 5.2

1. a. Fractions that should be circled: $\frac{25}{7}, \frac{3}{2}, \frac{46}{5}, \frac{5}{5}$
   b. $\frac{25}{7} = 3\frac{4}{7}$; $\frac{3}{2} = 1\frac{1}{2}$; $\frac{46}{5} = 9\frac{1}{5}$; $\frac{5}{5} = 1$
   c. $1\frac{8}{9} = \frac{17}{9}$; $5\frac{2}{3} = \frac{17}{3}$

2. a.
   $$\begin{array}{r} 97 \\ 7\overline{)684} \\ -63 \\ \hline 54 \\ -49 \\ \hline 5 \end{array} = 97\frac{5}{7}$$

   b.
   $$\begin{array}{r} 35 \\ 25\overline{)879} \\ -75 \\ \hline 129 \\ -125 \\ \hline 4 \end{array} = 35\frac{4}{25}$$

3. a. 11
   b. 1
   c. 9

4. 88, 9, and 54

5. a. $12\overline{)\$700}$
   b. $\frac{\$700}{12}$
   c. $\$58\frac{4}{12}$

6. a. Define:
   2,250 = *total books*
   120 = *books per box*
   \$2 = *cost per box*; ? = *total cost of boxes*
   b. Plan:
   total cost of boxes = cost per box • number of boxes
   number of boxes = total books ÷ books per box
   c. Execute:
   number of boxes = 2,250 ÷ 120 = 18 r90 (19 *boxes*)
   total cost of boxes = \$2 • 19 = \$38
   d. Check: \$38 is reasonable given the data.
   38 ÷ 2 = 19

7. We need so many mathematical "tools" because God created a complex universe, and it takes a lot of different tools to describe it.

Worksheet 5.3

From now on, unless otherwise stated, all fractional answers should be simplified to their lowest terms, and improper fractions should be converted to whole or mixed numbers.

1. circled fractions: $\frac{18}{27}$; $\frac{8}{12}$

2. Answers should not be simplified.
   a. $\frac{28}{40} \div \frac{4}{4} = \frac{7}{10}$
   b. $\frac{3}{4} \times \frac{12}{12} = \frac{36}{48}$

3. The identity property of multiplication says that multiplying any number by 1 doesn't change its value.

4. Because multiplying by 1 (or a fraction worth 1) doesn't change the value of the number.

5. a. $\frac{5}{20} \div \frac{5}{5} = \frac{1}{4}$
   b. $\frac{\$30}{\$50} \div \frac{10}{10} = \frac{\$3}{\$5}$
   c. $\frac{80}{100} \div \frac{20}{20} = \frac{4}{5}$
   d. $\frac{34}{48} \div \frac{2}{2} = \frac{17}{24}$

6. a. $\frac{17}{2} = 8\frac{1}{2}$
   b. $\frac{8}{4} = 2$
   c. $\frac{20}{11} = 1\frac{9}{11}$

7. a. $\frac{75}{100} = \frac{75}{100} \div \frac{25}{25} = \frac{3}{4}$
   b. $\frac{50}{100} = \frac{50}{100} \div \frac{25}{25} = \frac{1}{2}$
   c. $\frac{25}{100} = \frac{25}{100} \div \frac{25}{25} = \frac{1}{4}$

8. It makes sense because a quarter is worth a quarter ($\frac{1}{4}$) of a dollar.

9. *total cost* = 5 x \$12 + 7 x \$16 + \$5 = \$60 + \$112 + \$5 = \$177

Worksheet 5.4

1. Factor trees may vary slightly, but the circled numbers should be the same.
   a. 55
      ⑤ x ⑪
   b. 78
      6 x ⑬
      ②x③
   c. 112
      28 x 4
      4 x⑦ ②x②
      ②x②
   d. 85
      ⑤ x ⑰
   e. 29 is a prime number.

PAGE 401

f. 1,000
   100 ∧ 10
   10 ∧ 10 ⑤∧②
   ⑤∧②⑤∧②

2. The prime numbers under 100 should be written in math notebook.

3. A prime number is what we call a whole number greater than 1 that cannot be evenly divided by any whole number except by itself and 1. A prime factor is a factor that is a prime number.

4. a. $9\overline{)7}$
   b. $7\overline{)56}$
   c. $1\frac{1}{2} = \frac{3}{2} = 2\overline{)3}$
   d. $4\frac{8}{9} = \frac{44}{9} = 9\overline{)44}$
   e. $26\overline{)468}$
   f. $25\overline{)2,175}$

5. because God made us in His image

6. a. 18
   b. 87
   c. Should be solved using the distributive property.
      $36(4) + 36(9) = 144 + 324 = 468$

7. a. Define:
      $30 = monthly storage
      $40 = monthly post office box
      $10 = monthly website
      12 months = 1 year, so 3 years = 36 months
      cost in 3 years = ?
   b. Plan:
      total number of months in three years • (monthly storage + monthly post office box + monthly website) = cost in 3 years
   c. Execute: $36($30 + $40 + $10) = 36($80) = $2,880$
   d. Check: $2,880 is a reasonable answer. $2,880 \div 80 = 36$

Worksheet 5.5

1. a. 55, 11, or 5 (Only one of these needs to be listed.)
   b. 32, 16, 8, 4, or 2 (Only one of these needs to be listed.)

2. a. $\frac{3 \times 13}{2 \times 2 \times 13}$
   b. 13 should be circled in both the numerator and denominator.
   c. Greatest Common Factor: 13
   d. $\frac{3}{4}$

3. a. $\frac{2 \times 2 \times 2 \times 7}{2 \times 2 \times 3 \times 7}$
   b. 2, 2, and 7 should be circled in both the numerator and denominator.
   c. Greatest Common Factor: 28
   d. $\frac{56}{84} \div \frac{28}{28} = \frac{2}{3}$

4. a. $\frac{2 \times 37}{2 \times 2 \times 2 \times 3 \times 7}$
   b. 2 should be circled in both the numerator and denominator.
   c. Greatest Common Factor: 2
   d. $\frac{74}{168} \div \frac{2}{2} = \frac{37}{84}$

5. factors

Worksheet 5.6

1. a. $\frac{11}{12} - \frac{1}{2} = \frac{11}{12} - \frac{6}{12} = \frac{5}{12}$
   b. $\frac{1}{2} + \frac{1}{4} = \frac{2}{4} + \frac{1}{4} = \frac{3}{4}$
   c. $\frac{2}{3} + \frac{4}{5} = \frac{10}{15} + \frac{12}{15} = \frac{22}{15} = 1\frac{7}{15}$
   d. $\frac{9}{10} + \frac{5}{16} + \frac{46}{80} = \frac{72}{80} + \frac{25}{80} + \frac{46}{80} = \frac{143}{80} = 1\frac{63}{80}$

2. $\frac{1}{2}$ acre = $\frac{4}{8}$ acre
   $\frac{4}{8}$ acre + $\frac{1}{8}$ acre = $\frac{5}{8}$ acre

3. $\frac{1}{2} + \frac{1}{4} + \frac{2}{3} = \frac{6}{12} + \frac{3}{12} + \frac{8}{12} = \frac{17}{12} = 1\frac{5}{12}$ ton

4. a. $60 = 2 \times 2 \times 3 \times 5$
      $90 = 2 \times 3 \times 3 \times 5$
      GCF $= 2 \times 3 \times 5 = 30$

5. $25

6. a. $1 + \frac{6}{9} + \frac{5}{9} = \frac{9}{9} + \frac{6}{9} + \frac{5}{9} = \frac{20}{9} = 2\frac{2}{9}$
   b. $15 - \frac{8}{9} = \frac{135}{9} - \frac{8}{9} = \frac{127}{9} = 14\frac{1}{9}$

Worksheet 5.7

1. a. $21 = 3 \times 7$
      $84 = 2 \times 2 \times 3 \times 7$
      LCM $= 2 \times 2 \times 3 \times 7 = 84$
   b. $15 = 3 \times 5$
      $27 = 3 \times 3 \times 3$
      LCM $= 3 \times 3 \times 3 \times 5 = 135$

2. a. $12 = 2 \times 2 \times 3$
      $16 = 2 \times 2 \times 2 \times 2$
   b. LCD $= 2 \times 2 \times 2 \times 2 \times 3 = 48$
   c. $\frac{20}{48} - \frac{9}{48}$
   d. $\frac{11}{48}$
   e. GCF $= 2 \times 2 = 4$

3. a. $68 = 2 \times 2 \times 17$
      $22 = 2 \times 11$
   b. LCD $= 2 \times 2 \times 11 \times 17 = 748$
   c. $\frac{55}{748} + \frac{238}{748}$
   d. $\frac{293}{748}$
   e. GCF $= 2$

4. a. $176 = 2 \times 2 \times 2 \times 2 \times 11$
      $312 = 2 \times 2 \times 2 \times 3 \times 13$
   b. LCD $= 2 \times 2 \times 2 \times 2 \times 3 \times 11 \times 13 = 6,864$
   c. $\frac{195}{6,864} - \frac{66}{6,864}$
   d. $\frac{129}{6,864}$, which simplifies to $\frac{43}{2,288}$
   e. GCF $= 2 \times 2 \times 2 = 8$

5. a. $\frac{56}{120} = \frac{56}{120} \div \frac{8}{8} = \frac{7}{15}$
   b. $\frac{350}{2,000} = \frac{350}{2,000} \div \frac{50}{50} = \frac{7}{40}$
   c. $\frac{48}{12} = 48 \div 12 = 4$
   d. $\frac{120}{8} = 120 \div 8 = 15$
   e. $\frac{36}{25} = 1\frac{11}{25}$

6. a. $\frac{5}{6}$
   b. $\frac{5}{21}$

7. a. Greatest Common Factor (GCF) — The greatest factor two numbers share. We find it by multiplying the prime factors common to both numbers.
   b. Least Common Multiple (LCM) — The least number of which all the numbers we're dealing with is a multiple of; found by multiplying each prime factor the same number of times it is included in any one of the numbers.
   c. Least Common Denominator — The least common denominator means the same thing as the least common multiple, only it lets us know we're specifically dealing with denominators.
8. $3 \times \$30 \times 12 = \$1,080$

Chapter 6: More with Fractions

Worksheet 6.1

1. a. $\frac{2}{3} \times \frac{4}{5} = \frac{8}{15}$
   b. $\frac{9}{1} \times \frac{1}{2} = \frac{9}{2} = 4\frac{1}{2}$
   c. $\frac{2}{3} \times \frac{5}{9} = \frac{10}{27}$

2. a. Take $\frac{1}{3}$ two times; $\frac{2}{3}$
   b. Take 2, $\frac{1}{3}$ times (i.e., find $\frac{1}{3}$ of 2); $\frac{2}{3}$

3. a. $\frac{1}{2} \times 2 = 1$, so a half note would be worth 1 beat.
   b. $\frac{1}{4} \times 2 = \frac{1}{2}$, so a quarter note would be worth a half a beat.
   c. $\frac{1}{8} \times 2 = \frac{1}{4}$, so an eighth note would be worth a quarter of a beat.

4. a. $\frac{3}{1} \times \frac{2}{3}$ cup $= \frac{6}{3}$ cups $= 2$ cups
   b. $\frac{1}{2} \times \frac{2}{3}$ cup $= \frac{2}{6}$ cup $= \frac{1}{3}$ cup

5. a. Define:
     shirt 1 = $12
     shirt 2 = $\frac{1}{2}$ x $12
     both shirts = ?
   b. Plan:
     both shirts = shirt 1 + shirt 2
   c. Execute: $12 + \$12 \times \frac{1}{2} = \$12 + \frac{\$12}{2} = \$18$
   d. Check: Yes, this seems reasonable.

6. a. $15 = 3 \times 5$
      $7 = 7$
      $12 = 2 \times 2 \times 3$
      LCD $= 2 \times 2 \times 3 \times 5 \times 7 = 420$
   b. $\frac{4}{15} + \frac{6}{7} + \frac{5}{12} = \frac{112}{420} + \frac{360}{420} + \frac{175}{420} = \frac{647}{420} = 1\frac{227}{420}$
   c. There are no common factors.

7. a. $\frac{5}{6}(\frac{1}{8} + \frac{2}{7}) = \frac{5}{6}(\frac{7}{56} + \frac{16}{56}) = \frac{5}{6}(\frac{23}{56}) = \frac{115}{336}$
   b. Students were told to solve this problem mentally.
      $50(3) + 2(3) = 150 + 6 = 156$

Worksheet 6.2

1. a. $\frac{8}{3} + \frac{15}{8} = \frac{64}{24} + \frac{45}{24} = \frac{109}{24} = 4\frac{13}{24}$
   b. $\frac{28}{3} - \frac{20}{7} = \frac{196}{21} - \frac{60}{21} = \frac{136}{21} = 6\frac{10}{21}$
   c. $\frac{20}{3} \times \frac{21}{4} = \frac{420}{12} = 35$

2. a. $3 \times 1\frac{1}{2}$ c $= 3 \times \frac{3}{2}$ c $= \frac{9}{2}$ c $= 4\frac{1}{2}$ c
   b. $1\frac{2}{3}$ c $+ \frac{1}{2}$ c $= \frac{5}{3}$ c $+ \frac{1}{2}$ c $= \frac{10}{6}$ c $+ \frac{3}{6}$ c $= \frac{13}{6}$ c $= 2\frac{1}{6}$ c

3. a. $10\frac{1}{4}$ ac $- 2\frac{1}{2}$ ac $= \frac{41}{4}$ ac $- \frac{5}{2}$ ac $= \frac{41}{4}$ ac $- \frac{10}{4}$ ac $= \frac{31}{4}$ ac $= 7\frac{3}{4}$ ac
   b. We could use the $\frac{1}{4}$ c measurer 6 times ($6 \times \frac{1}{4}$ c $= \frac{6}{4}$ c $= 1\frac{1}{2}$ c) or the $\frac{1}{2}$ c measurer 3 times ($3 \times \frac{1}{2}$ c $= \frac{3}{2}$ c $= 1\frac{1}{2}$ c)
   c. width $= 4\frac{1}{3}$ in $- \frac{1}{8}$ in $- \frac{1}{8}$ in $= 4\frac{8}{24}$ in $- \frac{3}{24}$ in $- \frac{3}{24}$ in $= 4\frac{2}{24}$ in $= 4\frac{1}{12}$ in
      length $= 7\frac{1}{2}$ in $- \frac{1}{8}$ in $- \frac{1}{8}$ in $= 7\frac{4}{8}$ in $- \frac{1}{8}$ in $- \frac{1}{8}$ in $= 7\frac{2}{8}$ in $= 7\frac{1}{4}$ in

4. a. $26 = 2 \times 13$; $52 = 2 \times 2 \times 13$
   b. LCM $= 2 \times 2 \times 13 = 52$
   c. $\frac{10}{52} - \frac{5}{52}$
   d. $\frac{5}{52}$
   e. GCF $= 2 \times 13 = 26$

Worksheet 6.3

1. Fractions should have been simplified while multiplied.
   a. $\frac{\cancel{4}^1}{12} \times \frac{\cancel{6}^1}{\cancel{24}_2} = \frac{1}{12}$
   b. $\frac{\cancel{3}^1}{4} \times \frac{\cancel{12}^1}{\cancel{36}_{\cancel{12}_4}} = \frac{1}{4}$
   c. $\frac{\cancel{4}^1}{\cancel{3}_1} \times \frac{\cancel{27}^9}{\cancel{12}_3} = \frac{9}{3} = 3$

2. Answer should be a word problem and solution with two fractions being multiplied.

3. a. $\frac{1}{10} \times \$40 = \$4$
   b. $\frac{1}{10} \times \$50\frac{1}{2} = \frac{1}{10} \times \frac{\$101}{2} = \frac{\$101}{20} = \$5\frac{1}{20}$
   c. $\$100 \times 10 = \$1,000$

4. $4\frac{12}{16}$ in $- 3\frac{1}{4}$ in $= 4\frac{3}{4}$ in $- 3\frac{1}{4}$ in $= 1\frac{2}{4}$ in $= 1\frac{1}{2}$ in

5. a. $\frac{1}{15} + \frac{1}{2} = \frac{2}{30} + \frac{15}{30} = \frac{17}{30}$
   b. $2 \times 2 \times 5 \times 5$

Worksheet 6.4

1. a. $\frac{9}{8}$
   b. $\frac{1}{4}$
   c. $\frac{7}{27}$

2. a. $\frac{\frac{2}{3}}{\frac{8}{9}}$
   b. $\frac{\frac{5}{6}}{\frac{45}{88}}$

3. width $= \frac{1}{2} \times 2\frac{4}{5}$ in $= \frac{1}{2} \times \frac{\cancel{14}^7}{5}$ in $= \frac{7}{5} = 1\frac{2}{5}$ in
   height $= \frac{1}{2} \times 7\frac{1}{8}$ in $= \frac{1}{2} \times \frac{57}{8}$ in $= \frac{57}{16}$ in $= 3\frac{9}{16}$ in

4. Add $\frac{1}{4}$ inch to both sides:
   $4\frac{2}{4}$ in $+ \frac{1}{4}$ in $+ \frac{1}{4}$ in $= 4\frac{4}{4}$ in $= 5$ in
   Note: We rewrote $4\frac{1}{2}$ as $4\frac{2}{4}$ so we could add.

5. Check to make sure students simplified while multiplying.

a. $\frac{\cancel{2}^1}{\cancel{16}_2} \times \frac{\cancel{24}^3}{\cancel{56}_8} = \frac{3}{16}$

b. $\frac{\cancel{2}^1}{\cancel{14}_1} \times \frac{\cancel{28}^{\cancel{31}}}{\cancel{30}_{\cancel{10}\,1}} = \frac{1}{5}$

Worksheet 6.5

1. a. $\frac{3}{4} \div \frac{2}{3} = \frac{3}{4} \times \frac{3}{2} = \frac{9}{8} = 1\frac{1}{8}$

   b. $\frac{1}{2} \div \frac{4}{7} = \frac{1}{2} \times \frac{7}{4} = \frac{7}{8}$

   c. $\frac{2}{3} \div \frac{5}{6} = \frac{\cancel{2}}{\cancel{3}} \times \frac{\cancel{6}^2}{5} = \frac{4}{5}$

   d. $\frac{1}{2} \div 4 = \frac{1}{2} \times \frac{1}{4} = \frac{1}{8}$

   e. $\frac{9}{8} \div 2 = \frac{9}{8} \times \frac{1}{2} = \frac{9}{16}$

2. $5\frac{1}{2}$ inches by $4\frac{1}{4}$ inches

   Explanation: Size after we fold the longer (11 inch) side in half: 11 in ÷ 2 = $\frac{11}{2}$ in = $5\frac{1}{2}$ in by $8\frac{1}{2}$ in (the other side didn't change)

   The longer side is now the $8\frac{1}{2}$-inch side. Folded in half, we'd get $8\frac{1}{2}$ in ÷ 2 = $\frac{17}{2}$ in × $\frac{1}{2}$ = $\frac{17}{4}$ in = $4\frac{1}{4}$ in. Our card is now $5\frac{1}{2}$ in (that side didn't change when we did our second fold) by $4\frac{1}{4}$ in.

3. $56 \div 3\frac{3}{4} = 56 \div \frac{15}{4} = 56 \times \frac{4}{15} = \frac{224}{15} = 14\frac{14}{15}$

   14 garments, with some leftover

4. a. Define:
   number of pieces = 4
   length of each piece = 8 ft
   wood for each birdhouse = $2\frac{1}{2}$ ft
   total birdhouses = ?

   b. Plan:
   total wood ÷ wood for each birdhouse = total birdhouses
   total wood = number of pieces • length of each piece

   c. Execute:
   total wood = 4 × 8 ft = 32 ft
   total birdhouses = 32 ft ÷ $2\frac{1}{2}$ ft = 32 ft ÷ $\frac{5}{2}$ ft =
   32 × $\frac{2}{5}$ = $\frac{64}{5}$ = $12\frac{4}{5}$, so 12 birdhouses, with some leftover.

   d. Check: Yes, 12 birdhouses is a reasonable answer.
   $\frac{64}{5} \div \frac{2}{5} = \frac{64}{5} \times \frac{5}{2} = 32$

5. $\frac{1}{\cancel{4}_2} \cdot \frac{\cancel{2}^1}{3} c = \frac{1}{6} c$

6. a. $650 = 5 \times 10 \times 13$
   $130 = 2 \times 5 \times 13$

   b. LCM = 2 × 5 × 10 × 13 = 1,300

   c. $\frac{360}{1,300} - \frac{130}{1,300}$

   d. $\frac{230}{1,300} = \frac{23}{130}$

   e. GCF = 5 × 13 = 65

7. Division can be seen as splitting up a quantity or as seeing how many times "one number is contained in another." [*Merriam-Webster Online Dictionary*, http://www.merriam-webster.com/dictionary/division, (accessed 11-20-14), s.v., "division."]

Worksheet 6.6

1. a. $\frac{1}{3} \div \frac{1}{4} = \frac{1}{3} \times \frac{4}{1} = \frac{4}{3} = 1\frac{1}{3}$

   b. $\frac{8}{9} + 2\frac{3}{4} = \frac{8}{9} + \frac{11}{4} = \frac{32}{36} + \frac{99}{36} = \frac{131}{36} = 3\frac{23}{36}$

   c. $1\frac{1}{7} \times \frac{4}{5} = \frac{8}{7} \times \frac{4}{5} = \frac{32}{35}$

   d. $\frac{2}{3} - \frac{1}{5} = \frac{10}{15} - \frac{3}{15} = \frac{7}{15}$

2. a. 24 = 2 × 2 × 2 × 3
   b. 89 is a prime number.
   c. They do not have any common factors.
   d. LCM = 2 × 2 × 2 × 3 × 89 = 2,136

3. Answers should show that students simplified while multiplying. Simplification may vary from what is shown here, but the answer should be the same.

   a. $\frac{\cancel{24}^3}{\cancel{16}_{\cancel{8}\,2}} \times \frac{\cancel{46}^{23}}{\cancel{16}} = \frac{69}{16} = 4\frac{5}{16}$ ounces

   b. $\frac{\cancel{4}}{\cancel{16}_1} \times \frac{\cancel{16}^{\cancel{16}}}{1} = 4$

   c. $\frac{20}{\cancel{16}_1} \times \frac{\cancel{16}}{1} = 20$

4. $\frac{\$650}{6 \text{ months}}$

5. $3\frac{1}{6} \times 3 = \frac{19}{\cancel{6}_2} \times \$\cancel{3}^1 = \$\frac{19}{2} = \$9\frac{1}{2}$

6. a. Define:
   sale price = $4 a pound;
   normal price = $6 a pound;
   pounds purchased = $\frac{1}{2}$ pound
   savings = ?

   b. Plan:
   savings = (pounds purchased • normal price) – (pounds purchased • sale price)

   c. Execute: savings = ($\frac{1}{2}$ • $6) – ($\frac{1}{2}$ • $4) = $3 – $2 = $1

   d. Check: Yes, the answer seems reasonable.

7. $1\frac{1}{2} \times \frac{3}{4}$ lb = $\frac{3}{2} \times \frac{3}{4}$ lb = $\frac{9}{8}$ lb = $1\frac{1}{8}$ lb

   b. $\frac{1}{2}$ lb + $\frac{1}{2}$ lb = 1 lb, and we need $1\frac{1}{8}$ lb, so we need 3 blocks of cheese (unless we decide to put in a little less cheese than the recipe calls for in order to get by with 2 blocks).

8. A recipe should have been doubled or halved.

Worksheet 6.7

1. a. That each place represented 2 sets of the previous place.
   b. You need 2 digits, including 0. (0,1)

2. a. False
   b. The purpose of "carrying" digits is to keep track of place value.

3.

Check Number	Date	Memo	Payment Amount	Deposit Amount	$ Balance
	10/1	Opening Balance			181
159	10/5	Shoe Store	26		155
160	10/6	Grocery Store	33		122
	10/11	Deposit - Paycheck		89	211
	10/20	Deposit - Birthday Money		64	275
	11/5	Withdrawal	35		240

4. a. Multiplication is a faster way of looking at repeated additions.

b. Division is a shortcut for writing repeated subtraction.
c. Dividend refers to the starting number in a division problem—the initial quantity we need to divide.
d. Numerator is a name for the top number of a fraction (i.e., the number of parts we're considering/our dividend).
e. Multiplicative inverse is a name for the number that, when multiplied by another number, equals 1. Some people call it the reciprocal of a number.
f. The identity property of multiplication describes the fact that multiplying by 1 doesn't change the value.

5. No, subtraction is not commutative.
6. Students were told to solve these problems mentally.
   a. 120
   b. 115
7. a. $\frac{4 + 5(9)}{10} = \frac{4 + 45}{10} = \frac{49}{10} = 4\frac{9}{10}$
   b. 0
   c. unsolvable, as cannot divide by 0
8. Check work for use of the distributive property.
   $5(\frac{1}{15}) + 5(\frac{2}{30}) = \frac{5}{15} + \frac{10}{30} = \frac{1}{3} + \frac{1}{3} = \frac{2}{3}$
9. $\frac{5}{25} = \frac{1}{5}$
10. a. $49 = 7 \times 7$
    $70 = 2 \times 5 \times 7$
    b. LCM $= 2 \times 5 \times 7 \times 7 = 490$
    c. GCF $= 7$
11. a. $45 = 3 \times 3 \times 5$
    $20 = 2 \times 2 \times 5$
    b. LCM $= 2 \times 2 \times 3 \times 3 \times 5 = 180$
    c. GCF $= 5$
12. a. $1\frac{3}{12} + 2\frac{4}{12} = 3\frac{7}{12}$
    b. $\frac{17}{2} - \frac{21}{8} = \frac{68}{8} - \frac{21}{8} = \frac{47}{8} = 5\frac{7}{8}$
    c. $\frac{1}{\cancel{3}} \times \frac{\cancel{9}^3}{8} = \frac{3}{8}$
    d. $\frac{2}{5} \div \frac{7}{5} = \frac{2}{\cancel{5}} \times \frac{\cancel{5}^1}{7} = \frac{2}{7}$
13. a. $\frac{2}{3}c + 3\frac{1}{4}c = \frac{8}{12}c + \frac{39}{12}c = \frac{47}{12}c = 3\frac{11}{12}c$
    b. $8 \times 1\frac{3}{4}$ yd $= \cancel{8}^2 \times \frac{7}{\cancel{4}_1}$ yd $= 14$ yd
    c. $\$85 - \$10 - \$15 - \$12 - \$13 - \$2 = \$33$

Chapter 7: Decimals

Worksheet 7.1

1. a. $\frac{56}{100}$
   b. $\frac{3}{10}$
   c. $\frac{30}{100}$
   d. $\frac{486}{1,000}$
   e. $\frac{874}{1,000}$
2. a. $\frac{14}{25}$
   b. $\frac{3}{10}$ (already simplified)
   c. $\frac{3}{10}$
   d. $\frac{243}{500}$
   e. $\frac{437}{500}$
3. a. 3.14
   b. Seven and eight hundred, ninety-one thousandths
4. a. 0.5
   b. 0.86
   c. $\frac{1}{2} = \frac{50}{100} = 0.5$
   d. $\frac{12}{25} = \frac{48}{100} = 0.48$
5. a. $\frac{5}{100}$, 0.05
   b. $\frac{65}{10,000}$, 0.0065
6. [Check image showing: Pay to the order of: blank, $1,023.68, One thousand twenty-three and 68/100 dollars]
7. We can explore God's creation with math because God gave us this ability.

Worksheet 7.2

1. a. 879.79
   b. 188.04
   c. 3,084.42
   d. $60 + 78.96 = 138.96$
2. a. $\frac{24}{25}$
   b. 0.96
   c. $0.56 = \frac{56}{100} = \frac{14}{25}$; $0.4 = \frac{4}{10} = \frac{2}{5}$; they are the same value as the fractions in 2a.
   d. It was easier to add decimals, as you don't have to find the lowest denominator.
3. because decimals use the same place-value system to represent partial quantities
4. Problems should have been solved on an abacus.
   a. 14.828
   b. 0.45
   c. 1.89
5. $\$120 + \$4.99 + \$5.60 + \$32.22 = \$162.81$
6.

Check Number	Date	Memo	Payment Amount		Deposit Amount		$ Balance	
	10/1	Opening Balance					1,200	45
300	10/3	Telephone Company	39	99			1,160	46
	10/6	Deposit - Rebate			29	12	1,189	58

Worksheet 7.3

1. a. 6.3
   b. 4.8
   c. 17.1414
   d. $2.5(38.73) = 96.825$
   e. 0.468
2. a. $\frac{7}{4} \cdot 98 = \frac{686}{4} = 171\frac{1}{2}$
   b. $\frac{12}{15} = \frac{4}{5}$

c. $\frac{20}{35} + \frac{21}{35} = \frac{41}{35} = 1\frac{6}{35}$

d. $\frac{45}{18} - \frac{6}{18} = \frac{39}{18} = 2\frac{3}{18} = 2\frac{1}{6}$

3. a. Define:
   number of bags = 50
   sales price per bag = $4.55
   gross sales = ?
   b. Plan:
   gross sales = number of bags • sales price per bag
   c. Execute: 50 • $4.55 = $227.50
   d. Check: Yes, $227.50 is reasonable. $227.50 ÷ 50 = $4.55

4. 12(2) • $59.99 = 24 • $59.99 = $1,439.76

5. 6 months • 80 gallons a month = 480 gallons per year
   480 • $3.56 = $1,708.80 per year

Worksheet 7.4

Check to make sure decimal answers are rounded to the nearest hundredth from now on, unless otherwise specified.

1. a. 45.90
   b. 21.75
   c. 7,892.21

2. a. 6.25
   b. 12.71
   c. 33.67

3. a. $\frac{\cancel{8}^4}{\cancel{9}_3} \cdot \frac{\cancel{3}^1}{\cancel{2}_1} = \frac{4}{3} = 1\frac{1}{3}$

   b. $\frac{7}{\cancel{5}_1} \cdot \frac{\cancel{5}^1}{4} = \frac{7}{4} = 1\frac{3}{4}$

   c. $\frac{18}{13} - \frac{6}{13} = \frac{12}{13}$

4. monthly: $50,000 ÷ 12 = $4,166.67
   weekly: $50,000 ÷ 52 = $961.54
   hourly: $961.54 ÷ 40 = $24.04

5. Students were told to solve these problems mentally.
   a. $8
   b. $2, as $10 – $8 = $2
   c. $7

6. 19 ÷ 2 = 9.5 Note: We can make 9 projects. In this case we rounded down, as we can't make a partial project.

7. a. cost per day = (2 • $3.19) + $69.99 = $76.37
   amount charged per day = $129.99 • 3 = $389.97
   amount made per day =
   $389.97 – $76.37 = $313.60 per day
   b. amount charged per day at special rate =
   $79.99 • 3 = $239.97
   amount made per day at special rate =
   $239.97 – $76.37 = $163.60

8. 60,000 ÷ 24,900 = 2.41

9. Answer should be the estimated total of items in an actual shopping cart.

Worksheet 7.5

1. a. 0.33
   b. 0.92
   c. 0.83

2. a. 3.14
   b. width = 8.5 in – 0.75 in – 0.75 in = 7 in
   length = 11 in – 0.75 in – 0.75 in = 9.5 in

3. a. $\frac{2}{11} \cdot \frac{8.97}{1} = \frac{17.94}{11} = 1.63$
   b. $\frac{1}{4} \cdot \frac{5.2}{1} = \frac{5.2}{4} = 1.3$
   c. 10.78 – 0.47 = 10.31
   d. $\frac{2}{3} \cdot \$2.99 = \frac{\$5.98}{3} = \$1.99$
   e. cost ÷ number of ounces = cost per ounce
   $2.19 ÷ 8.5 = 0.26 cents per ounce
   $2.99 ÷ 13 = 0.23 cents per ounce
   The 13 ounce package is cheaper by 3 cents per ounce.

4. a. $636.56 ÷ $8.99 = 70.81, so they would need to sell 71 shirts to recover their cost.
   b. $636.56 ÷ $6.99 = $91.07, so they would need to sell 92 shirts to recover their cost.

5. 101.2° F – 98.6° F = 2.6° F warmer

6. total price ÷ the number of acres = the price per acre
   $24,790 ÷ 3 = $8,263.33

7. To divide a decimal number by a decimal number, count the number of digits to the right of the decimal point in the divisor and move the decimal point that number of digits to the right in the dividend. Then divide as usual. This is essentially multiplying the dividend and the divisor by a fraction worth one.

Worksheet 7.6

1. a. $5.55 • 4 = $ of hats purchased = $22.20
   $3.25 • 4 = $ of shirts purchased = $13
   $7.99 = the total spent on sunglasses
   hats + shirts + sunglasses = total purchase
   $22.20 + $13 + $7.99 = $43.19
   Note: We multiplied $5.55 and $3.25 by 4 rather than 3 to include the hat and shirt John bought for himself.
   b. If using a coupon: $43.19 – $5.50 = $37.69

2. Answer should be a word problem and solution involving a decimal.

3. a. $1\frac{1}{2} c ÷ 2 = \frac{3}{2} c \cdot \frac{1}{2} = \frac{3}{4} c$
   b. $\frac{2}{3} c ÷ 2 = \frac{2}{3} c \cdot \frac{1}{2} = \frac{2}{6} c = \frac{1}{3} c$

4. cost of supplies = $10.50 + $2.99 + $7.68 + $4.99 = $26.16
   total sales = (23 • $1.50) + (1 • $\frac{1}{2}$ • $1.50) = $34.50 + $0.75 = $35.25
   total sales – cost of supplies = profit
   $35.25 – $26.16 = $9.09
   profit divided amongst three people =
   $9.09 ÷ 3 = $3.03

5. a. 7.06 inches
   b. 0.67 of a cup
   c. 0.25 of an inch

6. a. $30 ÷ 1,000 = $0.03
   b. Each paintball costs 3 cents, so the cost of 30 paintballs would be 30 • $0.03 = $0.90.
   cost of rental + cost of paintballs used = total cost of playing = $0.90 + $29.99 = $30.89

7. $14.5 \left(\frac{2}{3} + \frac{6}{7} - \frac{2}{9}\right) = 14.5(0.67 + 0.86 - 0.22) = 14.5(1.31) = 19$

8. Check project done on the computer.

Chapter 8: Ratios and Proportions

Worksheet 8.1

1. a. $\frac{7 \text{ apples}}{1 \text{ rhubarb}}$ and 7:1
   b. $\frac{14 \text{ sunny days}}{17 \text{ non-sunny days}}$ and 14:17
   c. $\frac{55 \text{ mi}}{1 \text{ hr}}$ and 55:1

2. seven to eight and seven per eight

3. a. Sunflower A: $\frac{13}{8}$
      Sunflower B: $\frac{34}{21}$
      Sunflower C: $\frac{55}{34}$
      Sunflower D: $\frac{89}{55}$
      Sunflower E: $\frac{144}{89}$
   b. Sunflower A: $8\overline{)13.000}^{1.625} \approx 1.63$
      Sunflower B: $21\overline{)34.000}^{1.619} \approx 1.62$
      Sunflower C: $34\overline{)55.000}^{1.617} \approx 1.62$
      Sunflower D: $55\overline{)89.000}^{1.618} \approx 1.62$
      Sunflower E: $89\overline{)144.000}^{1.617} \approx 1.62$

4. a. House: $\frac{5.02}{3.1}$
      Face: $\frac{1.46}{0.9}$
   b. House: $3.1\overline{)5.020}^{1.619} \approx 1.62$
      Face: $0.9\overline{)1.460}^{1.622} \approx 1.62$

Worksheet 8.2

1. a. 45 *Note*: Since the numerator was multiplied by 9, we multiplied the denominator by 9 too to form an equivalent ratio.
   b. 3 *Note*: Since the numerator was divided by 3, we divided the numerator by 3 too to form an equivalent ratio.
   c. 29 *Note*: Since the numerator was divided by 2, we divided the denominator by 2 too to form an equivalent ratio.
   d. 20 *Note*: Since the numerator was divided by 2, we divided the denominator by 2 too to form an equivalent ratio.

2. a. $\frac{38 \text{ passengers}}{1 \text{ roller coaster}} = \frac{380 \text{ passengers}}{10 \text{ roller coasters}}$
   b. $\frac{400 \text{ bushels}}{1 \text{ acre}} = \frac{4,000 \text{ bushels}}{10 \text{ acres}}$

3. a. $\frac{20 \text{ men}}{800 \text{ rounds}} = \frac{250 \text{ men}}{? \text{ rounds}}$
   ? rounds = 10,000 rounds
   Note: We multiplied $\frac{20}{800}$ by $\frac{12.5}{12.5}$ to form the equivalent ratio. We knew to use $\frac{12.5}{12.5}$ because 250 ÷ 20 = 12.5, so if the numerator was multiplied by 12.5, the denominator needed to be as well in order for the ratio to be equivalent.

   b. $\frac{20 \text{ men}}{800 \text{ rounds}} = \frac{? \text{ men}}{15,000 \text{ rounds}}$
   ? men = 375 men
   Note: We multiplied $\frac{20}{800}$ by $\frac{18.75}{18.75}$ to form the equivalent ratio. We knew to use $\frac{18.75}{18.75}$ because 15,000 ÷ 800 = 18.75, so if the denominator was multiplied by 18.75, the numerator needed to be as well in order for the ratio to be equivalent.

   c. $\frac{30 \text{ cars}}{1,100 \text{ tons}} = \frac{? \text{ cars}}{3,300 \text{ tons}}$
   ? cars = 90 cars
   90 − 30 = 60 additional cars
   Note: Since the denominator was multiplied by 3, the numerator needed to be as well to form an equivalent ratio.

   d. $\frac{25 \text{ miles}}{1 \text{ gal}} = \frac{1,000 \text{ miles}}{? \text{ gal}}$
   ? gal = 40 gal
   Note: Since the numerator was multiplied by 40 (which we could find by dividing 1,000 by 25), we needed to multiply the denominator by 40 too in order to form an equivalent ratio.

4. a. ≠
   b. =
   c. ≠

5. a. $2\frac{20}{25} + 7\frac{22}{25} = 9\frac{42}{25} = 10\frac{17}{25}$
   b. $\frac{56}{64} - \frac{9}{64} = \frac{47}{64}$
   c. $28 = 2 \cdot 2 \cdot 7$
      $44 = 2 \cdot 2 \cdot 11$
      GCF = $2 \cdot 2 = 4$
   d. LCM = $2 \cdot 2 \cdot 7 \cdot 11 = 308$
   e. $7.8(2.77) = 21.61$ *Note*: Answer should be given as a decimal number.

Worksheet 8.3

1. a. $8\frac{1}{9} \div 7\frac{2}{3} = \frac{73}{9} \div \frac{23}{3} = \frac{73}{\cancel{9}_3} \times \frac{\cancel{3}^1}{23} = \frac{73}{69} = 1\frac{4}{69}$
   b. $4.6 \div 0.25 = 18.4$

2. a. 13.5 *Note*: We multiplied the denominator by 3, so had to multiply 4.5 by 3 as well to find the missing number.
   b. $\frac{4}{8}$ *Note*: We divided $\frac{6}{12}$ by $\frac{2}{2}$ to get $\frac{3}{6}$, so we needed to divide $\frac{8}{16}$ by $\frac{2}{2}$ to find the missing number.

3. a. =
   b. ≠

4. $\frac{8.5 \text{ square feet}}{\$850} = \frac{17 \text{ square feet}}{\$?}$; $1,700
   Note: Since the numerator was multiplied by 2 (17 ÷ 8.5 = 2), we multiplied the denominator by 2 also to form the equivalent ratio.

5. $\frac{\$3.25}{1 \text{ day}} = \frac{\$100}{? \text{ days}}$; 30.77 days
   Note: Since the numerator was multiplied by 30.77 (which we found by dividing $100 ÷ $3.25 = 30.77), we multiplied our denominator of 1 by 30.77 as well.

6. $\frac{104 \text{ pounds}}{2 \text{ acres}} = \frac{? \text{ pounds}}{15.5 \text{ acres}}$; 806 pounds
   Note: Since the denominator was multiplied by 7.75, we multiplied the numerator by 7.75 too to form an equivalent ratio.

7. a. $45 = 3 \cdot 3 \cdot 5; 78 = 2 \cdot 3 \cdot 13$;
   LCM $= 2 \cdot 3 \cdot 3 \cdot 5 \cdot 13 = 1{,}170$
   b. $15 = 3 \cdot 5; 105 = 3 \cdot 5 \cdot 7$; GCF $= 3 \cdot 5 = 15$
   c. Check to see that the distributive property was used.
   $4(4.78) + 4(5.67) = 19.12 + 22.68 = 41.8$
   d. $0.25 \cdot 1.87 = 0.47$

Worksheet 8.4

1. a. $\frac{1 \text{ inch}}{3 \text{ feet}} = \frac{5 \text{ inch}}{? \text{ feet}}$; 15 feet
   b. $\frac{1 \text{ square}}{\$30} = \frac{5 \text{ squares}}{? \text{ dollars}}$; $150
   c. $\frac{34.3 \text{ m}}{33.6 \text{ m}} \div \frac{300}{300} = \frac{0.114 \text{ m}}{0.112 \text{ m}}$; width = 0.114 m; length = 0.112 m
   Note: We rounded to the nearest thousandth.

2. No, the drawing is not proportional. It is not proportional because the ratios do not equal. $\frac{24}{65} \neq \frac{4}{10}$

3. a. Check that simplified as multiplied.
   $\frac{\cancel{4}^1}{7} \times \frac{5}{\cancel{16}_4} = \frac{5}{28}$
   b. $\frac{8}{25} - \frac{5}{25} = \frac{3}{25}$
   c. $\frac{\cancel{4}^1}{7} \times \frac{34}{\cancel{20}_5} = \frac{34}{35}$

4. a. Scale drawing will vary based on bookcase.
   b. Students should have viewed a house blueprint.

5. a. Ratio of teen's head to rest of his body: $\frac{2 \text{ cm}}{12.5 \text{ cm}}$
   Ratio of baby's head to rest of his body: $\frac{1 \text{ cm}}{3 \text{ cm}}$
   b. Ratio of teen's head to rest of his body as a decimal number: 0.16
   Ratio of baby's head to rest of his body as a decimal number: 0.33
   c. No. *Note:* A proportion is two equivalent ratios. $\frac{2 \text{ cm}}{12.5 \text{ cm}} \neq \frac{1 \text{ cm}}{3 \text{ cm}}$

Worksheet 8.5

1. a. 24 cents; 2 dimes, 4 pennies
   b. 25 cents; 1 quarter
   c. 16 cents; 1 dime, 1 nickel, 1 penny
   d. 1 cent; 1 penny
   e. $3.22; 3 dollars, 2 dimes, 2 pennies
   f. $3.25; 3 dollars, 1 quarter
   g. 77 cents; 3 quarters, 2 pennies
   h. 80 cents; 3 quarters, 1 nickel
   i. 14 cents; 1 dime, 4 pennies
   j. 20 cents; 2 dimes

2. Students were told to solve these problems mentally.
   a. Find a rounded answer: $4 \cdot 4 \approx 16$
      Adjust to find actual: $16 - 0.04 = \$15.96$
   b. 10×10 cents = 100 cents or $1.00
   c. 5×5 cents = 25 cents or $0.25
   d. $8 \div 4$ quarters in a dollar = $2.00
   e. 10×4 quarters in a dollar = 40 quarters
   f. $100 \div 4$ quarters in a dollar = $25

3. $\frac{10 \text{ apples}}{5 \text{ people}} = \frac{? \text{ apples}}{36 \text{ people}}$; 72 apples

4. $\frac{\frac{3}{4} \text{ c}}{1 \text{ gal}} = \frac{? \text{ c}}{78 \text{ gal}}$
   $\frac{3}{\cancel{4}_2} \text{ c} \times \cancel{78}^{39} = \frac{117}{2} = 58\frac{1}{2}$ c

5. a. $\frac{1}{2} + 4 = 4\frac{1}{2}$
   Note: The 4 is the result of the division: $\frac{\cancel{8}^2}{\cancel{25}_1} \times \frac{\cancel{50}^{2}}{\cancel{4}_1} = \frac{4}{1} = 4$
   b. $34 = 2 \cdot 17$
      $62 = 2 \cdot 31$
      GFC = 2
   c. LCM $= 2 \cdot 17 \cdot 31 = 1{,}054$
   d. $0.78 \cdot 5.5 = 4.29$; the nearest whole number is 4.

Worksheet 8.6

1. Students were told to solve these problems mentally.
   a. 13 cents; 1 dime, 3 pennies
   b. 15 cents; 1 dime, 1 nickel
   c. $2.17; 2 dollars, 1 dime, 1 nickel, 2 pennies
   d. $2.25; 2 dollars, 1 quarter
   e. 50 cents; 2 quarters

2. a. 1:8 or $\frac{1 \text{ inch}}{8 \text{ squares}}$
   b. 0.125

3. a. $\frac{400 \text{ bushels}}{1 \text{ acre}} = \frac{? \text{ bushels}}{2.5 \text{ acres}}$; 1,000 bushels
   b. $\frac{400.23 \text{ bushels}}{1 \text{ acre}} = \frac{? \text{ bushels}}{0.5 \text{ acres}}$; 200.12 bushels
   c. $\frac{\$4.95}{250 \text{ yards}} = \frac{\$?}{750 \text{ yards}}$; $14.85

4. a. $\frac{3.5 \text{ yards}}{1 \text{ wreath}} = \frac{? \text{ yards}}{26 \text{ wreaths}}$; 91 yards
   b. $91 \cdot \$0.99 = \$90.09$
      $\$90.09 + \$7.50 = \$97.59$
   c. $5\frac{5}{6}$ acres is the total land, and $\frac{1}{2}$ the land is going to be used for apples. So you have
   $\frac{1}{2} \cdot 5\frac{5}{6}$ ac $= \frac{1}{2} \cdot \frac{35}{6}$ ac $= \frac{35}{12}$ ac $= 2\frac{11}{12}$ ac
   If you plant 40 trees per acre, you can fit
   $2\frac{11}{12} \cdot 40 = \frac{35}{\cancel{12}_3} \times \cancel{40}^{10} = \frac{350}{3} = 116\frac{2}{3}$
   You can plant 116 trees. (You can't plant a portion of a tree.)
   d. $116 \cdot \$2.50 = \$290$
   e. $\frac{1 \text{ inch}}{6 \text{ feet}} = \frac{1.5 \text{ inch}}{? \text{ feet}}$; 9 feet

5. n-o; $\frac{2 \text{ cm}}{3 \text{ cm}} \neq \frac{20 \text{ cm}}{35 \text{ cm}}$

Chapter 9: Percents

Worksheet 9.1

1. 100

2. a. $\frac{4}{20}$, 0.20, 20%
   b. $\frac{2}{10}$, 0.20, 20%
   c. $\frac{8}{80}$, 0.10, 10%
   d. $\frac{70}{250}$, 0.28, 28%
   e. $\frac{200}{100}$, 2, 200%

3. a. 16%
   b. 89%
   c. 2%
   d. 125%
   e. 42%
   f. 96%

4. a. $\frac{25}{100} = \frac{1}{4}$
   b. $\frac{69}{100}$
   c. $\frac{5}{100} = \frac{1}{20}$

d. $\frac{72}{100} = \frac{18}{25}$
e. $\frac{378}{100} = 3\frac{78}{100} = 3\frac{39}{50}$
f. $\frac{6}{100} = \frac{3}{50}$

5. a. 0.25
   b. 0.69
   c. 0.05
   d. 0.72
   e. 3.78
   f. 0.06

6. True

7. If I had a 20%-off-one-item coupon, 20% of the item's price would be deducted from my cost.

8. $\frac{84 \text{ lb}}{14 \text{ castings}} = \frac{? \text{ lb}}{9 \text{ castings}}$; 53.85 which rounds to 54 pounds

9. Students were told to solve these problems mentally.
   a. 14 cents; 1 dime, 4 pennies
   b. 5 cents; 1 nickel
   c. 15 cents; 1 dime, 1 nickel

Worksheet 9.2

1. a. $\frac{50}{100} = \frac{?}{250}$; 125
   b. $\frac{25}{100} = \frac{?}{75}$; 18.75
   c. $\frac{20}{100} = \frac{?}{120}$; 24
   d. $\frac{10}{100} = \frac{?}{420}$; 42
   e. $\frac{3}{100} = \frac{?}{60}$; 1.8
   f. $\frac{1}{100} = \frac{?}{25}$; 0.25

2. $\frac{10}{20}$, 0.50, 50%

Students were told to use a proportion to solve problem 3 and 4.

3. $\frac{20}{100} = \frac{? \text{ tip}}{\$28.45}$; $5.69 or $5.70, depending on when rounded in solving

4. $\frac{30}{100} = \frac{?}{40}$; $12

5. a. The base was $40.
   b. The rate was 30% (or $\frac{30}{100}$).

6. a. 32%
   b. 80%

7. $\frac{2}{3}$ c − $\frac{1}{4}$ c = $\frac{8}{12}$ c − $\frac{3}{12}$ c = $\frac{5}{12}$ c
   You reduced it by $\frac{5}{12}$ of a cup.

8. Students were told to solve these problems mentally.
   a. 48 cents; 1 quarter, 2 dimes; 3 pennies
   b. 50 cents; 2 quarters
   c. 10 cents; 1 dime

Worksheet 9.3

1. Check that method solved matches that shown.
   a. 0.25 • 340 = 85
   b. 0.12 • $500 = $60
   c. 0.15 • 70 = 10.5

2. a. My base was $500.
   b. My rate was 12.5%.

3. Check that method solved matches that shown.
   a. 0.1 • 12 = 1.2
   b. 0.1 • 100 = 10
   c. 0.1 • 220 = 22
   d. 0.1 • 8 = 0.8
   e. 0.1 • 78 = 7.8
   f. When multiplying by 10%, move the decimal one place to the left.

4. a. 145%
   b. 275%

5. 0.05 • $12,000 = $600

6. $\frac{1}{3}$ • 2,500 = 833.33 calories *Note*: Alternately, students could have converted $\frac{1}{3}$ to 33% and then found that percent of 2,500 (0.33 • 2,500 = 825).

7. Students were told to solve these problems mentally.
   a. 23 cents; 2 dimes, 3 pennies
   b. 3 cents; 3 pennies
   c. 5 cents; 1 nickel

8. $\frac{4}{5} = \frac{?}{569}$; 455 *Note*: Answer should be rounded to the nearest whole number.

9. *Rate • Base = Percentage* should be copied into math notebook.

Worksheet 9.4

1. a. 84%
   b. 9%
   c. 87%

2. a. 0.10 • $43.89 = $4.39 Note: we rounded $4.389 to $4.39.
      $43.89 − $4.39 = $39.50
   b. 0.20 • $55.67 = $11.13
      $11.13 + $55.67 = $66.80
   c. 0.07 • $52.67 = $3.69
      $3.69 + $52.67 = $56.36

3. a. 100% − 20% = 80%
      0.80 • $34.80 = $27.84
   b. 100% − 25% = 75%
      0.75 • $14.99 = $11.24
   c. 100% + 10% = 110%
      1.10 • $42 = $46.20
   d. 100% + 6% = 106%
      1.06 • $44.56 = $47.23

4. 100% − 67% = 33%

5. a. 0.008 • 2,000 = 16
   b. 0.005 • $325,000 = $1,625

Worksheet 9.5

1. Students were told to solve these problems mentally.
   a. $3.80
   b. $0.50
   c. $9.20
   d. $13.50

2. a. 15%
   b. 5%
   c. 125%

3. Answer should be a percent found in a newspaper.

4. Students were told to solve these problems mentally.
   a. 89 cents; 3 quarters, 1 dime, 4 pennies
   b. $1; 1 dollar

c. 90 cents; 3 quarters, 1 dime, 1 nickel

5. $\frac{2\frac{1}{4} c}{1 \text{ batch}} = \frac{? c}{3\frac{1}{2} \text{ batches}}$; $2\frac{1}{4}$ c • $3\frac{1}{2} = \frac{9}{4}$ c • $\frac{7}{2} = \frac{63}{8}$ c = $7\frac{7}{8}$ cups.

Chapter 10: Negative Numbers

Worksheet 10.1

1.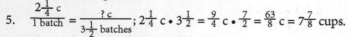

2.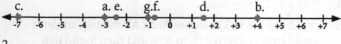

3. a. −$10.50
   b. $5
   c. −40
   d. negative
   e. −10,000

4. 0.20 • 5 pounds = 1 pound

5. 0.05 • 2,400 mg = 120 mg

6. Students were told to solve this problem mentally. $15.80

Worksheet 10.2A

1.

2. a. 4
   b. 8
   c. −11
   d. −2
   e. −5
   f. −4
   g. −3
   h. −7

3. a. −7
   b. 6
   c. 0.5
   d. $-\frac{3}{4}$

4. 7 − 10 = −3. You owe $3.

Worksheet 10.2B

1. a. −5
   b. −7 mi
   c. −3 in
   d. −6
   e. −8
   f. −3
   g. $-\frac{1}{5}$
   h. $-\frac{2}{8}$
   i. $-\frac{2}{4}$

2. 21 + −51 = −30 miles; 30 miles west

3. Students were told not to round or simplify.

a. 0.452, $\frac{452}{1,000}$
b. 0.3125, $\frac{3,125}{10,000}$
c. 1.68, $\frac{168}{100}$
d. 1.6842; $\frac{16,842}{10,000}$
e. 0.5625 • 75 = 42.1875

Worksheet 10.3

1. a. negative; −6.25
   b. negative, positive, negative; −6.25
   c. negative, positive, negative, positive; 6.25

2. a. 8 + 6 = 14
   b. 5 + 2 = 7
   c. 5 + −2 = 3
   d. 6 + $\frac{1}{2}$ = $6\frac{1}{2}$

3. 45.5(1 + 3.2) = 45.5(4.2) = 191.1

4. $3.50 + $4.50 + $7.75 = $15.75;
   1.06 • $15.75 = $16.70

5. Students were told to solve these problems mentally.
   a. 20% of $26 = $2.60 + $2.60 = $5.20
   b. 0.40 • 80 = 32. *Mental Process*: 10% of 80 would be 8, so 40% would be 4 times that, or 32.

6. the opposite of

Worksheet 10.4

1. a. 14° F; −10° C
   b. 59° F; 15° C
   c. −10° F ; −23° C

2. a. 165° F − 60° F = 105° F
   b. 32° F − 40° F = −8° F
   c. 0° C − 20° C = −20° C

3. *Note*: Students were told not to round or simplify on 3e, 3g, and 3i.
   a. 4 + 10 ÷ 2 = 4 + 5 = 9
   b. 46%
   c. 46.7%
   d. 132.5%
   e. 0.325, $\frac{325}{1,000}$
   f. 0.325 • 60 = 19.5
   g. 0.2512, $\frac{2,512}{10,000}$
   h. 0.2512 • 5 inches = 1.26 inches shorter; 5 inches − 1.26 inches = 3.74 inches
   i. 0.055, $\frac{55}{1,000}$
   j. 1.055 • $10 = $10.55
   k. Students were told to solve this problem mentally. $2.45
   l. $-\frac{3}{6} = -\frac{1}{2}$
   m. $-\frac{2}{12} = -\frac{1}{6}$
   n. 5

4. Answer should be something about Galileo—see Lesson 10.4 for ideas.

Worksheet 10.5

1. a. 3
   b. 14.5
   c. 20
2. a. 15 mi − −10 mi = 25 mi; |25 mi| = 25 mi; they are 25 miles apart.
   b. −5° − 25° = −30°; |−30°| = 30°; it increased by 30° F.
   c. −15° − 75° = −90°; |−90°| = 90°; it increased by 90° F.
   d. 10 mi + −20 mi = −10 mi; 10 miles in the negative direction
3. a. Students were told to solve this problem mentally.
      $0.30 \cdot 125 = 37.5$
   b. $0.032 \cdot 25.5 = 0.82$
   c. $45(8 + 2) + 20 \div 2 = 45(10) + 10 = 450 + 10 = 460$
   d. $-\frac{3}{3} = -1$
   e. $-\frac{2}{20} = -\frac{1}{10}$
   f. 7

Worksheet 10.6

1. a. −100
   b. 100
   c. $-1\frac{1}{2}$
   d. $1\frac{1}{2}$
   e. −6
   f. 6
2. a. −2
   b. 2
   c. −4
3. a. −5
   b. −5
   c. −24
   d. −8
4. a. −3
   b. −4
   c. −25
   d. $-\frac{5}{15} = -\frac{1}{3}$
   e. $\frac{1}{3}$
5. the opposite of

Worksheet 10.7

1. a. $(2\frac{3}{4}) + (-5\frac{1}{4}) = (\frac{11}{4}) + (-\frac{21}{4}) = -\frac{10}{4} = -2\frac{2}{4} = -2\frac{1}{2}$
   b. $5\frac{2}{5} - 7\frac{1}{5} = \frac{27}{5} - \frac{36}{5} = -\frac{9}{5} = -1\frac{4}{5}$
2. a. $2\frac{1}{3}$ lb + −4 lb = $\frac{7}{3} + -\frac{12}{3} = -\frac{5}{3} = -1\frac{2}{3}$
   b. $4 + -3\frac{1}{3} = \frac{12}{3} + -\frac{10}{3} = \frac{2}{3}$; there's positive $\frac{2}{3}$ ounces of force on the balloon.
3. a. −10
   b. 35
   c. 27
   d. −27
   e. −4
   f. 4
   g. 5
4. $1.045 \cdot \$45 = \$47.03$

Worksheet 10.8

1. a. $\frac{-50 \text{ mi}}{1 \text{ hr}}$
   b. $\frac{-200 \text{ mi}}{4 \text{ hr}}$
   c. $\frac{1 \text{ day}}{30 \text{ days}}$
   d. $\frac{\$780}{4 \text{ weeks}}$
2. a. −2
   b. $70 + -3 = 70 + 3 = 73$
   c. $60 + -- -3 = 60 + 3 = 63$
   d. $40 + -2 = 38$
3. a. 32
   b. −10
   c. −3
   d. $-\frac{2}{4} = -\frac{1}{2}$
   e. −10
   f. 10
   g. $\frac{5}{6}$
4.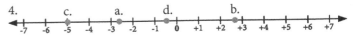
5. Students were told to solve this problem mentally. $9.20

Chapter 11: Sets

Worksheet 11.1

1. a. "Dinner" is a subset of "Recipes"
   b. "Dinner", "Crowd Pleasers"
   c. Some recipes in "Dinner" are also part of "Crowd Pleasers."
2. a. Fiction
   b. Nonfiction, Computers, Computer Software
   c. Nonfiction, Gardening, Vegetable Gardening
   d. Nonfiction, Gardening, Flower Gardening
3.
4. *Note*: Which animals go in which category is subjective, since we didn't define exactly what we met by a pet, etc. The important thing is to understand that we can organize animals based on characteristics.
   Dog – Animals, Pets, Farm Animals, Moderate Climate Animals
   Crocodile – Animals, Tropical Animals
   Walrus – Animals, Arctic Animals
   Wolf – Animal, Arctic Animals, Moderate Climate Animals
   Horse – Animals, Pets, Farm Animals, Moderate Climate Animals
5. a. 2
   b. $-4 + -3 = -7$
   c. $1\frac{2}{8} - 7\frac{7}{8} = \frac{10}{8} - \frac{63}{8} = -\frac{53}{8} = -6\frac{5}{8}$
   d. $4 \cdot \frac{-3}{-1} = 4 \cdot 3 = 12$

Worksheet 11.2

1. a. Whole Numbers, Even Integers
   b. Odd Integers, Negative Numbers
   c. Positive Numbers
   d. Negative Numbers

2. Flashcards should have been made.

3. a. –18
   b. 10
   c. –10
   d. –5
   e. 10
   f. $\frac{27}{12} + -\frac{64}{12} = -\frac{37}{12} = -3\frac{1}{12}$

4. 1.035 • $87.45 = $90.51

Worksheet 11.3

1. a. {< $20}
   b. {≠9}

2. a. yes
   b. no
   c. yes

3. Answers will vary based on how the family organizes silverware. Look for a list of sets and a Venn diagram showing how the sets relate.

4. a. 4
   b. –3
   c. 23
   d. –7

5. Students should have reviewed set flashcards.

6. a. 1.075 • $5.99 = $6.44
   b. Students were told to solve this problem mentally. $5.80

7. One example of sets used in programming is to check what's inputted on online forms.

Worksheet 11.4

1. a. 12, 15 *Note*: Each number increased by 3.
   b. –5, –7 *Note*: Each number decreased by 2.
   c. 16, 22 *Note*: Each number increased by 1, more than the previous number increased by.
   d. 27, 29 *Note*: There are 3 numbers that are 2 apart, and then a number 5 greater. This pattern repeats.
   e. 128, 256 *Note*: Each consecutive number is double the preceding number.
   f. –10, –12 *Note*: Each number decreased by 2.

2. a. 1a (common difference was 3), 1b (common difference was –2), and 1f (common difference was –2)
   b. 1e (common ratio was 2)

3. approximately 1.62

4. Students should complete hands-on assignment of counting change.

5. a. –3
   b. –4 + 2 = –2

6. Students should have reviewed flashcards.

Worksheet 11.5

1. a. 5 • $1.48 = $7.40
   b. 1.055 • $7.40 = $7.81
   c. $14.95 + $8.99 = $23.94
      0.85 • $23.94 = $20.35
   d. $0.65
   e. $\frac{5}{\$2.50} = \frac{18}{?}$; $9.00
   f. –$20
   g. 20 mi – $10\frac{1}{3}$ mi = $9\frac{2}{3}$ mi; $|9\frac{2}{3}$ mi$|$ = $9\frac{2}{3}$ miles
   h. 20 mi – $-10\frac{1}{3}$ mi = $30\frac{1}{3}$ mi; $|30\frac{1}{3}$ mi$|$ = $30\frac{1}{3}$ miles

2. a. $\frac{\$5.59}{20}$ = $0.28 or 28 cents per pound
   b. $21.31 – $5.87 = $15.44

3. width = $\frac{4.4 \text{ ft}}{?} = \frac{1 \text{ ft}}{0.5 \text{ in}}$; 2.2 in
   height = $\frac{8.5 \text{ ft}}{?} = \frac{1 \text{ ft}}{0.5 \text{ in}}$; 4.25 in

4. a. 25, 32
   b. 7
   c. Integer, Whole (or Natural/Counting), Odd, Positive, Prime
   d. –48
   e. 9
   f. –3
   g. 1.1
   h. 10° – –5° = 15°; $|15°|$ = 15°; it changed by 15° C
   i. $\frac{\$8}{3 \text{ lb}}$ or 8:3
   j. 2.67
   k. Students were told to solve this problem mentally. $11.40

5. a. No

   b. The base is the number off which we want to find the percent.
   c. A set is a collection or group.

Chapter 12: Statistics and Graphing

Worksheet 12.1

1. a. & b.

Age	a. Frequency	b. Relative Frequency
12	6	60%
13	2	20%
15	1	10%
16	1	10%

Note: We found relative frequency by dividing the frequency by the total number of students (10) and expressing the quotient as a percent.

2. Students were told to solve these problems mentally.
   a. $7.20
   b. $3.00
   c. $3.80

3. a. 10
   b. −12
   c. $\frac{29}{\cancel{7}_1}(\frac{\cancel{7}^1}{12}) = \frac{29}{12} = 2\frac{5}{12}$
   d. $\frac{34}{65} = \frac{?}{234}$; 122 employees
4. First off, we don't know if the unemployment decrease was due to something the current governor did or whether it just happened to occur during his term. It could be that the previous governor passed favorable laws that began taking effect during the current governor's term…or it could be that the economy as a nation is improving due to federal law changes or other events…or due to state legislatures' work…or to a large manufacturing plant opening in the state. Although the current governor might have done a good job in office, we need more information before we would really know. There are other issues to consider besides unemployment.
5. interpretation

Worksheet 12.2

1. a. iii. Everyone living in the neighborhood (i.e., the entire population of which we want to know the opinion of)
   b. v. Everyone who returned a survey (i.e., the people whose opinion we actually obtained)
2. random and large enough
3. a. 49%–53%
   b. 53%–57%
   c. 45%–65%
4. 38%–46%
5. Answers will vary.
   a. We need to know if a random set of testers was chosen. The company could have only selected good spellers to test the curriculum!
   b. The football team does not include a random set of teens with varying interests.
6. 0.75 • $56.78 = $42.59
7. a. −6
   b. −3
   c. $\frac{14}{7} = \frac{?}{2}$; 4
   d. 5

Extra Credit—Students can get extra credit by reading *How to Lie with Statistics* by Darrell Huff.

Worksheet 12.3

1. a. 25%
   b. 0.25 • 1,000 = 250
   c. 22%–28%
2. a. Burger by 7%, as 35%−28% = 7%
   b. Families that eat out together could have different tastes than the population as a whole.
   c. no
3. a. 1,800,000
   b. 600,000
4. a. 13 x 1,000 = 13,000
   b. 1,920 x 1,000 = 1,920,000
5. Groupings must be logical, but can vary. Possible groupings include these:
   Foundation of America
   - The Declaration of Independence Signed
   - Plymouth Plantation Founded
   - Patrick Henry's "Give Me Liberty or Give Me Death"
   Civil War
   - Lincoln's Assassination
   - The Gettysburg Address
   - President Lincoln Elected
   or
   Important Speeches
   - The Gettysburg Address
   - Patrick Henry's "Give Me Liberty or Give Me Death"
   Important Events
   - President Lincoln Elected
   - Lincoln's Assassination
   - The Declaration of Independence Signed
   - Plymouth Plantation Founded
6. a. 30
   b. 9
   c. $\frac{3}{16}$
   d. $\frac{50}{75} + \frac{72}{75} = \frac{122}{75} = 1\frac{47}{75}$
   e. $\frac{1}{4} \cdot 86 = \frac{86}{4} = 21\frac{1}{2}$
7. a. Pie graphs make it easy to see what portion of a whole each number represents.
   b. Bar graphs make it easy to see how data compares with other data.
   c. False. Bar graphs with more than 6 or so bars are harder to take in at a glance.

Worksheet 12.4

1. Check that the graph is drawn to scale, labeled, and both accurately and clearly reflects the data.

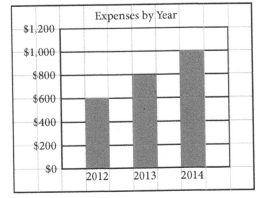

2. Students were told to round to the nearest whole percent.

Score	a. Frequency	b. Relative Frequency
Win	5	31%
Loss	9	56%
Tie	2	13%

3. a. the scale used
   b. The first one; its scale makes it easier to see the difference in costs.
4. a. 5.27

PAGE 413

b. 21
c. $5 + 5(\frac{5}{15}) = 5 + \frac{5}{3} = 5 + 1\frac{2}{3} = 6\frac{2}{3}$
d. $\frac{\$20}{\$80} = \frac{?}{\$240}$; $60

Extra Credit — Students can get extra credit by generating on the computer a pie and bar chart for the data in problem 2. Instructions for generating charts in a computer program can be easily found online.

Worksheet 12.5

1. (4, B)

2. a. (4,4)
   b. (2,2)
   c. (6,−2)
   d. (2,−4)
   e. (−3, 2)
   f. (−5,−4)

3.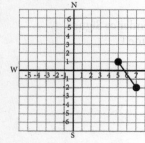

4.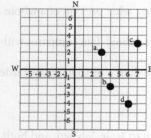

5. a. $5 \div \frac{2}{3} = 5 \cdot \frac{3}{2} = \frac{15}{2} = 7\frac{1}{2}$; you can get $7\frac{1}{2}$ sections.
   b. 45.6 (44.8 + 56.2) = 45.6(101) = 4,605.6
   c. −2
   d. −32, 64

Worksheet 12.6

1. a. the number of people on pension roll in the U.S. by year
   b. Each box represents 10 years.
   c. Each box represents 200,000 people.
   d. Some years were skipped.
   e. The number of people on pension roll remained basically the same.

2. a. −$10 or a $10 loss
   b. $26 or a $26 gain

3. Students were told to round as shown.
   a.

A. Year	B. Average annual expenditure (spending)	C. Entertainment	D.	E.
2003	40,817	2,060	$\frac{2,060}{40,817}$	5.05%
2004	43,395	2,218	$\frac{2,218}{43,395}$	5.11%
2005	46,409	2,388	$\frac{2,388}{46,409}$	5.15%
2006	48,398	2,376	$\frac{2,376}{48,398}$	4.91%
2007	49,638	2,698	$\frac{2,698}{49,638}$	5.44%
2008	50,486	2,835	$\frac{2,835}{50,486}$	5.62%

   b.

A. Year	B. Average annual expenditure (spending)	F. Reading	G.	H.
2003	40,817	127	$\frac{127}{40,817}$	0.31%
2004	43,395	130	$\frac{130}{43,395}$	0.30%
2005	46,409	126	$\frac{126}{46,409}$	0.27%
2006	48,398	117	$\frac{117}{48,398}$	0.24%
2007	49,638	118	$\frac{118}{49,638}$	0.24%
2008	50,486	116	$\frac{116}{50,486}$	0.23%

   c. No
   d. No. We don't know if reading is just cheaper (after all, there are libraries) or if people do not read a lot.

4.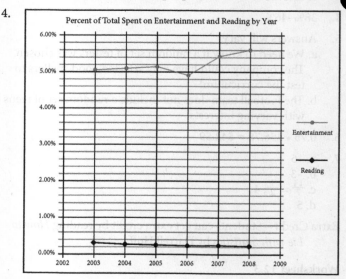

5. a. $16 = 2 \cdot 2 \cdot 2 \cdot 2$
      $24 = 2 \cdot 2 \cdot 2 \cdot 3$
      LCM $= 2 \cdot 2 \cdot 2 \cdot 2 \cdot 3 = 48$
   b. $\frac{15}{48} + \frac{18}{48} = \frac{33}{48} = \frac{11}{16}$
   c. −0.75
   d. 72.75, 72

6. True

Worksheet 12.7

1. a. 41 years old *Note*: We added all ages together and divided by total number of students (10). We rounded the quotient to the nearest whole number.
   b. 504
   c. $240,400
   d.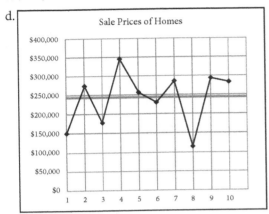
   e. No. It means the average was $116.

2. Students were told to round to the thousandths place.
   a. 0.125
   b. 0.200
   c. 0.231

3. Every child is unique. An average is just a number that represents the middle age—children actually learn to walk both earlier and later.

4.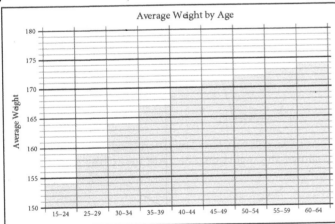

5. Students were told to solve these problems mentally.
   a. 17
   b. 183 *Mental Process*: Think of it as 87 + 100 = 187. Now subtract the 4 we added when we rounded 96. 187 − 4 = 183

6. a. $3\frac{1}{3}$, 4, $4\frac{2}{3}$ *Note*: The pattern was to add $\frac{2}{3}$.
   b. 16, 14, 19 *Note*: The pattern was to add 5 and then subtract 2.

Worksheet 12.8

1. a. 3, 5, 7, 7, 7, 8, 9, 10, 13, 14, 15
      Average: 8.91
      Median: 8
      Mode: 7

   b. $101, $120, $120, $122, $125, $128, $130, $134
      Average: $122.5 ($122,500)
      Median: $\frac{(\$122 + \$125)}{2}$ = $123.5 ($123,500)
      Mode: $120 ($120,000)

   c. 55, 60, 65, 68, 69, 70, 70, 71, 72, 73, 75, 79, 80, 95
      Average: 71.57
      Median: $\frac{(70 + 71)}{2}$ = 70.5
      Mode: 70

   d. 76, 77, 80, 82, 87, 90, 91
      Average: 83.29
      Median: 82
      Mode: There is no mode for this data, as all the numbers occur the same number of times.

   e. 18, 18, 19, 19, 25, 30
      Average: 21.5
      Median: $\frac{(19 + 19)}{2}$ = 19
      Mode: 18 and 19

2. Students were told to round the relative frequency to a whole percent.

Grade	Frequency	Relative Frequency
0–60 (F)	2	14%
61–70 (D)	5	36%
71–80 (C)	6	43%
81–90 (B)	0	0%
91–100 (A)	1	7%

3. a. 0 ft, 5 ft, 5 ft, 5 ft, 5 ft
      Median: 5 ft
   b. Average: (0 ft + 5 ft + 5 ft + 5 ft + 5 ft) ÷ 5 = 4 ft
      It would have used 4 ft.

4. a. 1, 2, 2, 3, 3, 3, 3, 3, 3, 3, 6, 6
      Mode: 3
   b. 3.17

5. This means that if you were to take the word count of all books, line each result up in order, and look for the middle number, you'd get 64,000. If they used the average instead, that would mean that it represented the number we'd get by adding all the word counts in each of the books, and then dividing that by the number of books.

6. (−4, 3)

7. Check that graph communicates same data and is clearly labeled.

   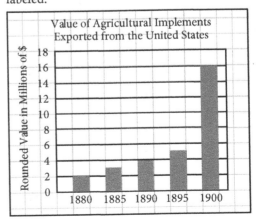

Chapter 13: Naming Shapes: Introducing Geometry

Worksheet 13.1

1. a. −26
   b. 90%
   c. 0.12 • 50 = 6
   d. 0.05 • 25 = 1.25
   e. $\frac{45}{100}$, which reduces to $\frac{9}{20}$, 0.45, 45%
   f. 46.5%
   g. −4
   h. $-14\frac{2}{3}$
   i. 40
   j. $\frac{\$4,500}{\$19,500} = \frac{\$5,400}{?}$; $23,400
   k. $\frac{128}{160} - \frac{108}{160} = \frac{20}{160} = \frac{1}{8}$
   l. $-\frac{1}{5}$

2. $\frac{78 + 90 + 75 + 81 + 82 + 50 + 99 + 64}{8}$; 77.38

3. Students should have added to flashcards and math notebook.

4. a. earth measure
   b. the Greek mathematicians
   c. "breathless length" and a great circle [The "breathless length" definition is from Euclid, "Euclid's Elements: Book 1, Definitions, Definition 2." Found in David E. Joyce, *Euclid's Elements*, Department of Mathematics and Computer Science, Clark University, http://aleph0.clarku.edu/~djoyce/java/elements/bookI/bookI.html (accessed 10/04/14); a great circle is the definition in spherical geometry.]
   d. that each one is a tool

Worksheet 13.2

1. a. parallel lines; not applicable (no angles formed)
   b. perpendicular lines; right angle
   c. neither parallel nor perpendicular; acute angle

2. Students should have marked the right angle in problem 1b as shown.

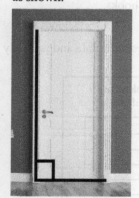

3. We can assume the angle formed is a right angle.

4. Marked angles/lines should be traced in the color listed.
   a. purple
   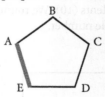
   b. blue
   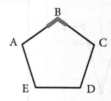
   c. red
   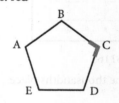
   d. green
   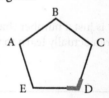
   e. orange
   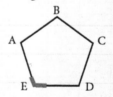

5. Answers will vary based on letters used. If labeled as shown, the angle would be ∠ABC or ∠CBA.

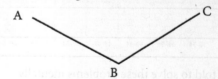

6. Students should have added to and reviewed flashcards and added to math notebook.

7.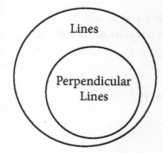

8. (45 + 50 + 62 + 55 + 40 + 59 + 60) ÷ 7 = 53
9. Students were told to solve these problems mentally.
   a. $4.22; 4 dollars, 2 dimes, 2 pennies
   b. $4.25; 4 dollars, 1 quarter
   c. 445
   d. $19.95
   e. 11
   f. 172

Worksheet 13.3

1. a. square
   b. decagon
   c. rectangle
   d. octagon
   e. hexagon *Note*: This is a photo of graphene.
2. a. all (1a–1e)
   b. 1a and 1c
   c. 1a and 1c
   d. 1b
   e. No; it is not a closed figure.
3.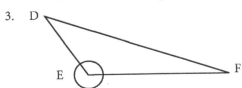
4. a. (0,0)
   b. (6,0)
   c. (0,−4)
   d. (6,−4)
5.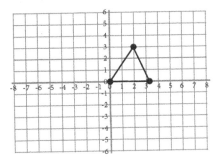
6.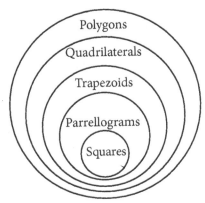
7. Students should have added to and reviewed flashcards.
8. Students were told to round to the thousandths place.
   a. 0.167
   b. 0.212
   c. 0.317
9. a. −4 + −5 = −9
   b. 18
   c. −8

Worksheet 13.4

1. Drawing will vary. Art books from the library can supply further guidance if needed.
2. Possibilities include bowls, door knobs, sun, moon, circular light switches, screw heads, hammer heads, erasers on pencils, CDs, DVDs, lights indicating power on many appliances, etc.
3. a. right scalene triangle
   b. obtuse isosceles triangle
   c. acute equilateral triangle
4. Circles should have been drawn with a compass.
   a.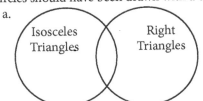
   b.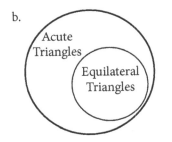
5. a. cylinder
   b. rectangular prism
   c. rectangular prism
   d. cylinder
6. *Note*: Fractions in 6e, 6g, and 6h should not be simplified.
   a. 8 + 10 = 18
   b. 62%
   c. 25.7%
   d. 342.9%
   e. 0.437; $\frac{437}{1,000}$
   f. 0.255 • 60 = 15.3
   g. 0.4715; $\frac{4,715}{10,000}$
   h. 0.075; $\frac{75}{1,000}$
   i. 1.085 • $38.99 = $42.30
   j. 2.6 *Note*: Students were told to round to the nearest tenth.
7. ∠DEF or ∠FED
8. Students should have added to and reviewed flashcards.

Worksheet 13.5

1. a-b. Students should have traced triangles and noticed that they are the exactly the same.

c. Students should have formed a rotation with the triangles (one possible example is shown).

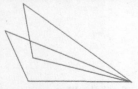

d. Students should have formed a reflection with the triangles (one possible example is shown).

e. Students should have formed a translation with the triangles (one possible example is shown).

f. obtuse, scalene

2. rotation

3.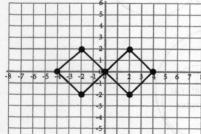

4.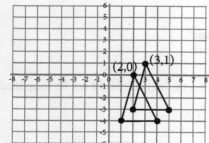

5. $\frac{7 \text{ in} + 5 \text{ in} + 10 \text{ in} + 6.5 \text{ in} + 7.8 \text{ in} + 9 \text{ in}}{6}$ = 7.6 in *Note*: Students were told to round to the nearest tenth.

6. Answers will vary. Students should have picked one of listed assignments.

7. Students should have added to and reviewed flashcards and added to math notebook.

Worksheet 13.6

1. a. the fear of the Lord
   b. God's Word
   c. "...changed the truth of God into a lie, and worshiped and served the creature more than the Creator..." (Romans 1:25).
   d. They worshiped and served their own intellect instead of God.
   e. We're conceived in sin, we don't know everything (God does), and the wisdom of this world is foolishness to God. *Note*: Only 2 reasons need to be listed.

2. Check for a list of 10 household objects you could describe with a rectangle. Possibilities include windows, computer monitors and keyboards, pictures, doors, papers, notebooks, tissue boxes, outlets, appliances, TVs, tables, door latches, and envelopes.

3. a. hexagon
   b. acute triangle
   c. equilateral triangle

4. Letters used to label triangle may vary. The top left angle and the top line should be described using angles. $\overline{AB}$ or $\overline{BA}$; ∠CAB or ∠BAC

5. Students need quizzed on flashcards.

Chapter 14: Measuring Distance

Worksheet 14.1

1. a. 4 ft − 2$\frac{1}{2}$ ft = 1$\frac{1}{2}$ ft
   b. Answer may vary based on doorway, but will probably be yes. Students were asked to see if a 25-inch wide desk would fit through the doorway to their bedroom.

2. a. Should be the width of a window.
   b. Should be the window width plus 4 inches—2 inches for each side).

3. a. <
   b. =
   c. <

4. a. 0.2 • 8 ft = 1.6 ft
   b. 2$\frac{1}{4}$% = 2.25% = 0.0225
   c. 0.0225 • 60 yd = 1.35 yd
   d. 0.35 • 20 mi = 7 mi
   e. 0.65 • 20 mi = 13 mi
      or 20 mi − 7 mi = 13 mi
   f. 0.4 • 500 mi = 200 mi
   g. 0.6 • 500 mi = 300 mi
      or 500 mi − 200 mi = 300 mi

5. a. inches, feet, and meters
   b. feet, yards, and meters
   c. inches and centimeters

6. a. 4
   b. $\frac{4}{5} + \frac{11}{60} = \frac{48}{60} + \frac{11}{60} = \frac{59}{60}$
   c. $\frac{3}{10}$
   d. 11,2
   e. $\frac{8 \text{ ft} + 10 \text{ ft} + 12 \text{ ft} + 11 \text{ ft} + 9 \text{ ft} + 7 \text{ ft}}{6}$ = 10 ft *Note*: Students were told to round to the nearest foot.

7. Students should have added to and reviewed flashcards.
8. a. The meter is currently defined in terms of "the speed of light in a vacuum"—specifically as "the length of the path traveled by light in a vacuum during an interval of $\frac{1}{299,792,458}$ of a second." [Tina Butcher, Linda Crown, Rick Harshman, and Juana Williams, eds. *NIST Handbook 44: 97th National Conference on Weights and Measures 2012*, 2013 ed. (Washington: U. S. Department of Commerce, 2012), B-6, found on http://www.nist.gov/pml/wmd/pubs/h44-13.cfm (accessed 10/6/14).]
   b. We fail.
   c. The length of a cubit varied because it was based on arm length, and arm lengths varies.

Worksheet 14.2

1. a. $\frac{6 \text{ ft}}{? \text{ in}} = \frac{1 \text{ ft}}{12 \text{ in}}$; 72 in *Note*: The numerator changed from 1 ft to 6 ft (i.e., was multiplied by 6), so we had to multiply the denominator by the same quantity.
   b. $\frac{2 \text{ mi}}{? \text{ ft}} = \frac{1 \text{ mi}}{5,280 \text{ ft}}$; 10,560 ft *Note*: We found this by seeing that our numerator of 1 mi was doubled (i.e., multiplied by 2) to form 2 mi, so our denominator of 5,280 ft also needs to be doubled (i.e., multiplied by 2) to form an equivalent ratio.
   c. $\frac{87 \text{ in}}{? \text{ yd}} = \frac{36 \text{ in}}{1 \text{ yd}}$; 2.42 yd *Note*: We found this by dividing 87 in by 36 in to see how much 36 had to be multiplied by to get to 87; we then multiply the denominator of 1 yd by the same quantity to form an equal ratio.
   d. $\frac{1,000 \text{ ft}}{? \text{ mi}} = \frac{5,280 \text{ ft}}{1 \text{ mi}}$; 0.19 mi *Note*: We found this by dividing 5,280 ft by 1,000 ft and finding that the numerator was divided by 5.28. We then divided our denominator of 1 mi by the same amount to form an equivalent ratio.

2. 100 in + 100 in + 40 in + 10 in = 250 in
   $\frac{? \text{ yd}}{250 \text{ in}} = \frac{1 \text{ yd}}{36 \text{ in}}$; 6.94 yd

   Note: In real life, you would probably round the yardage you asked for up to 7, or convert the 0.94 part back to inches, so as to ask for a yardage that can easily be measured with a yardstick.

3. a. 20 • 2 ft = 40 ft
   b. $\frac{?}{40 \text{ ft}} = \frac{1 \text{ yd}}{3 \text{ ft}}$; $13\frac{1}{3}$ yd

4. a. 8% + 8% = 16%
   0.16 • 21 mi = 3.36 mi
   b. 21 mi − 3.36 mi = 17.64 mi

5. Students should have reviewed flashcards.

Worksheet 14.3

1. Check that student solved using method shown.
   a. 7 ft • $\frac{12 \text{ in}}{1 \text{ ft}}$ = 84 in
   b. 5 mi • $\frac{5,280 \text{ ft}}{1 \text{ mi}}$ = 26,400 ft
   c. 105.75 in • $\frac{1 \text{ yd}}{36 \text{ in}}$ = 2.94 yd
   d. 500 ft • $\frac{1 \text{ mi}}{5,280 \text{ ft}}$ = 0.09 mi

2. Students were told to solve these problems mentally.
   a. 15 ÷ 3 = 5; 5 yd
   b. 3 • 12 = 36; 36 in

3. a. The in from the numerator cancels out the in from the denominator, leaving just 5 ft.
   b. The a in the numerator and the a in the denominator cancel out, leaving just the b.

4. a. $\frac{100 \text{ cm}}{1 \text{ m}}$, $\frac{1 \text{ m}}{100 \text{ cm}}$
   b. $\frac{5,280 \text{ ft}}{1 \text{ mi}}$, $\frac{1 \text{ mi}}{5,280 \text{ ft}}$
   c. $\frac{36 \text{ in}}{1 \text{ yd}}$, $\frac{1 \text{ yd}}{36 \text{ in}}$

5. $\frac{65.57 \text{ mi}}{60 \text{ min}} = \frac{? \text{ mi}}{1 \text{ min}}$; 1.09 mi

6. Students should have reviewed flashcards.

Worksheet 14.4

While the answer on problems 1–3 are shown using conversion via ratio shortcut method, using another method is fine so long as the correct answer is obtained.

1. a. $405 • $\frac{0.63 \text{ £}}{\$1}$ = 255.15 £
   b. $33.46 • $\frac{0.63 \text{ £}}{\$1}$ = 21.08 £
   c. 56.90 £ • $\frac{\$1}{0.63 \text{ £}}$ = $90.32
   d. 890.76 £ • $\frac{\$1}{0.63 \text{ £}}$ = $1,413.90

2. a. *starting bahts* = 80 dollars • $\frac{31 \text{ bahts}}{1 \text{ dollar}}$ = 2,480 bahts
   bahts spent = 100 bahts + 50 bahts + 620 bahts = 770 bahts
   remaining bahts = 2,480 bahts − 770 bahts = 1,710 bahts
   b. 250 bahts • $\frac{1 \text{ dollar}}{31 \text{ bahts}}$ = 8.06 dollars
   c. 1.20 • 40 bahts = 48 bahts

3. a. 80 dollars • $\frac{1 \text{ euro}}{1.3 \text{ dollar}}$ = 61.54 euros
   b. 1.07 • 13 euros = 13.91 euros
   1.15 • 13.91 euros = 16 euros
   c. 16 euros • $\frac{\$1.30}{1 \text{ euro}}$ = $20.80

4. $\frac{\$4.50 + \$0.75 + \$0.20 + \$5.75 + \$47.23 + \$2.30 + \$56.21 + \$16.78 + \$3.45 + \$2.60}{10}$ = $13.98

5. Students should have reviewed flashcards.

Worksheet 14.5

If measurements vary slightly, that's okay. It's hard to get the measurement to the exact decimal, although they shouldn't be off by much.

1. a. Check for a line 6 centimeters long.
   b. Check for a line 40 millimeters long.
   c. Check for a triangle with three 2.5 centimeter sides.

2. a. rectangle
   b. 3.7 cm
   c. trapezoid
   d. 5.1 cm
   e. irregular decagon
   f. 1 cm
   g. equilateral acute triangle
   h. 3 cm
   i. 4.7 cm

3. Students were told to solve these problems mentally.
   a. 150 ÷ 100 = 1.5 m
   b. 7 cm • 10 = 70 mm
   c. −7 cm • 10 = −70 mm
   d. 8 • 1,000 = 8,000 m

4. 2 m, 7 m, 7 m, 8 m, 8 m, 8 m, 10 m, 11 m, 18 m
   The mode is 8 m.
5. Answers will vary based on heights measured.
6. a. See text for ideas.
   b. It is easy to convert between units in the system because each unit is 10 times the previous one.

Worksheet 14.6

1. Check that the method shown was used.
   a. $300{,}000 \text{ in} \cdot \frac{1 \text{ ft}}{12 \text{ in}} \cdot \frac{1 \text{ mi}}{5{,}280 \text{ ft}} = 4.73 \text{ mi}$
   b. $120 \text{ yd} \cdot \frac{3 \text{ ft}}{1 \text{ yd}} \cdot \frac{1 \text{ mi}}{5{,}280 \text{ ft}} = 0.07 \text{ mi}$
   c. $1 \text{ mi} \div 0.07 = 14.29$; we would have to run the length 14.29 times.
   d. $2 \text{ league} \cdot \frac{18{,}228.3465 \text{ ft}}{1 \text{ league}} \cdot \frac{1 \text{ yd}}{3 \text{ ft}} = 12{,}152.23 \text{ yd}$
2. $\frac{15 \text{ mi} + 10 \text{ mi} + 20 \text{ mi} + 7 \text{ mi} + 0 \text{ mi} + 9 \text{ mi} + 45 \text{ mi}}{7} = 15 \text{ mi}$
   Note: Students were told to round to the nearest whole number.
3. a. 66 millimeters
   b. 4.2 centimeters
   c. right angle
   d. rectangle
   e. rectangular prism
4. Students were told to solve these problems mentally.
   a. $3.96
   b. 63
   c. 24 cents; 2 dimes, 4 pennies
   d. 25 cents; 1 quarter
5. Students should have reviewed flashcards.
6. length = $300 \text{ cu} \cdot \frac{18 \text{ in}}{1 \text{ cu}} \cdot \frac{1 \text{ ft}}{12 \text{ in}} = 450 \text{ ft}$
   breadth (width) = $50 \text{ cu} \cdot \frac{18 \text{ in}}{1 \text{ cu}} \cdot \frac{1 \text{ ft}}{12 \text{ in}} = 75 \text{ ft}$
   height = $30 \text{ cu} \cdot \frac{18 \text{ in}}{1 \text{ cu}} \cdot \frac{1 \text{ ft}}{12 \text{ in}} = 45 \text{ ft}$

Extra Credit — Students could earn extra credit by writing a paragraph after watching the suggested online video.

Worksheet 14.7

1. Check that the method shown was used.
   a. $20 \text{ ft} \cdot \frac{30.48 \text{ cm}}{1 \text{ ft}} = 609.6 \text{ cm}$
   b. $280 \text{ km} \cdot \frac{1 \text{ mi}}{1.60934 \text{ km}} = 173.98 \text{ mi}$
   c. $80.78 \text{ cm} \cdot \frac{1 \text{ in}}{2.54 \text{ cm}} = 31.8 \text{ in}$
2. a. $108 \text{ km} \cdot \frac{1 \text{ mi}}{1.60934 \text{ km}} = 67.11 \text{ mi}$
   b. $90 \text{ km} \cdot \frac{1 \text{ mi}}{1.60934 \text{ km}} = 55.92 \text{ mi}$
   c. $3 \text{ yd} \cdot \frac{0.9144 \text{ m}}{1 \text{ yd}} = 2.74 \text{ m}; 3 \cdot 2.74 \text{ m} = 8.22 \text{ euros}$
3. a. 57 mm
   b. 30 mm
   c. parallelogram
   d. acute
   e. obtuse
4. Students should have added to and reviewed flashcards.

Worksheet 14.8

1. a. $2 \text{ d} \cdot \frac{24 \text{ hr}}{1 \text{ d}} = 48 \text{ hr}$
   b. $1 \text{ wk} \cdot \frac{7 \text{ d}}{1 \text{ wk}} \cdot \frac{24 \text{ hr}}{1 \text{ d}} = 168 \text{ hr}$
   c. $1 \text{ d} \cdot \frac{24 \text{ hr}}{1 \text{ d}} \cdot \frac{60 \text{ min}}{1 \text{ hr}} = 1{,}440 \text{ min}$
   d. $-30 \text{ ft} \cdot \frac{12 \text{ in}}{1 \text{ ft}} \cdot \frac{2.54 \text{ cm}}{1 \text{ in}} = -914.4 \text{ cm}$
   e. $7{,}850 \text{ ft} \cdot \frac{1 \text{ mi}}{5{,}280 \text{ ft}} \cdot \frac{1.60934 \text{ km}}{1 \text{ mi}} = 2.39 \text{ km}$
   f. $7.62 \text{ cm} \cdot \frac{1 \text{ in}}{2.54 \text{ cm}} \cdot \frac{1 \text{ ft}}{12 \text{ in}} = 0.25 \text{ ft}$
   g. $34 \text{ hours} \cdot \frac{1 \text{ day}}{24 \text{ hours}} = 1.42 \text{ days}$
2. a. $\frac{60 \text{ mi}}{\text{hr}} \cdot 4 \text{ hr} = 240 \text{ mi}$
   b. $\frac{60 \text{ mi}}{60 \text{ min}} \cdot 35 \text{ min} = 35 \text{ mi}$
   Note: We substituted 60 min for 1 hr so as to have the same units.
   c. $\frac{-35 \text{ mi}}{60 \text{ min}} \cdot 35 \text{ min} = -20.42 \text{ mi}$
   d. $\frac{30 \text{ mi}}{60 \text{ min}} = \frac{5 \text{ mi}}{?}$; 10 min
3. a. $\frac{120 \text{ min} + 60 \text{ min} + 0 \text{ min} + 30 \text{ min} + 120 \text{ min} + 0 \text{ min} + 30 \text{ min}}{7} = 51.43 \text{ min}$
   b. $51.43 \text{ min} \cdot \frac{1 \text{ hr}}{60 \text{ min}} = 0.86 \text{ hr}$; which rounds to 1 hour
   c. 120 min + 60 min + 0 min + 30 min + 120 min + 0 min + 30 min = 360 min
4. Students were told to solve these problems mentally.
   a. 9:30 a.m. + 2 hr 45 min = 12:15 p.m.
   b. 8:15 a.m. + 5 hr = 1:15 p.m.
   c. 8:15 a.m. + 5 hr + 30 min + 30 min = 2:15 p.m.
   d. 20 min + 15 min + 45 min = 80 min;
      80 min = 1 hr 20 min
      4 p.m. − 1 hr 20 min is 2:40 p.m.
5. Students were told to solve these problems mentally.
   a. 2 hr 24 min
   b. 7 p.m. + 2 hr = 9 p.m.
6. Students should have added to and reviewed flashcards.

Chapter 15: Perimeter and Area of Polygons

Worksheet 15.1

1. a. $9 \cdot 34 \text{ in} = 306 \text{ in}$
   b. $3 \cdot 2 \text{ ft} = 6 \text{ ft}$
   c. 10 ft + 10 ft + 10 ft + 10 ft + 10 ft + 13 ft + 13 ft = 76 ft
   d. $4 \cdot 3 \text{ in} = 12 \text{ in}$
   e. $2(20 \text{ ft}) + 2(5 \text{ ft}) = 40 \text{ ft} + 10 \text{ ft} = 50 \text{ ft}$
2. a. $2(7 \text{ ft}) + 2(11 \text{ ft}) = 14 \text{ ft} + 22 \text{ ft} = 36 \text{ ft}$
   b. $36 \text{ ft} \cdot \frac{1 \text{ yd}}{3 \text{ ft}} = 12 \text{ yd}$
   Price of fencing before tax: $\frac{\$3.99}{\text{yd}} \cdot 12 \text{ yd} = \$47.88$
   Note: Notice how "yd" crossed out!
   c. $0.80 \cdot \$47.88 = \$38.30$
   d. $1.05 \cdot \$38.30 = \$40.22$
3. a. $\frac{156}{56} - \frac{16}{56} = \frac{140}{56} = 2\frac{28}{56} = 2\frac{1}{2}$
   b. $8 \cdot (-1) = -8$

Worksheet 15.2

1. a. $P = s_1 + s_2 \ldots sn$
   b. $P = n \cdot s$ or $P = n(s)$ or $P = ns$

PAGE 420

c. $P = 2 \cdot l + 2 \cdot w$ or $P = 2(l) + 2(w)$ or $P = 2l + 2w$
   d. Check that flashcards were made.
2. Exact letters used may vary, but these relationships should be shown.
   a. $x = J + \$25$
   b. $P = V \cdot I$
   c. $x = -5 \text{ mi} \cdot h$
3. a. $(2 \cdot 13 \text{ ft}) + (2 \cdot 5 \text{ ft}) = 36 \text{ ft}$
   b. $6 \text{ in} \cdot \frac{1 \text{ ft}}{12 \text{ in}} = \frac{1}{2} \text{ ft}$
   $36 \text{ ft} \div \frac{1}{2} \text{ ft} = 36 \cdot 2 = 72$
   We need 72 tiles.
   c. cost of tile = $72 \cdot \$2.99 = \$215.28$
   cost of tile and grout = $\$215.28 + \$5.99 = \$221.27$
   cost including tax = $1.09 \cdot \$221.27 = \$241.18$
4. perimeter of picture = $4 \cdot 3 \text{ ft} = 12 \text{ ft}$
   cost of the frame = $12 \text{ ft} \cdot \$7.99 = \$95.88$
   cost after discount = $0.60 \cdot \$95.88 = \$57.53$
   cost of materials with tax = $1.045 \cdot \$57.53 = \$60.12$
   total with materials, tax, and labor =
   $\$60.12 + \$10.00 = \$70.12$
5. Students were told to solve these problems mentally.
   a. 72
   b. 23
   c. 3 cents; 3 pennies
   d. 5 cents; 1 nickel

Worksheet 15.3

1. Check to make sure areas are listed in square units (abbreviated here as sq).
   a. $7 \text{ ft}(7 \text{ ft}) = 49 \text{ sq ft}$
   b. $6 \text{ ft} \cdot 4 \text{ ft} = 24 \text{ sq ft}$
   c. $12 \text{ in}(12 \text{ in}) = 144 \text{ sq in}$
   d. $17 \text{ yd} \cdot 24 \text{ yd} = 408 \text{ sq yd}$
2. a. $30 \cdot 30 = 900$ square feet
   $\frac{1 \text{ pound}}{450 \text{ sq ft}} = \frac{? \text{ pounds}}{900 \text{ sq ft}}$; 2 pounds
   b. $6 \cdot 2$ pounds = 12 pounds
   c. $1.05 \cdot \$4.99 \cdot 12 = \$62.87$
3. $120 \text{ km} \cdot \frac{1 \text{ mi}}{1.60934 \text{ km}} = 74.56 \text{ mi}$
4. $8 \text{ ft} \cdot \frac{12 \text{ in}}{1 \text{ ft}} = 96 \text{ in}$
   $96 \text{ in} - 54 \text{ in} = 42 \text{ in}$
5. Students should have added to and reviewed flashcards.

Worksheet 15.4

1. Check to make sure areas are listed in square units (abbreviated here as sq).
   a. $10 \text{ m} \cdot 8 \text{ m} = 80 \text{ sq m}$
   b. $13 \text{ ft} \cdot 5 \text{ ft} = 65 \text{ sq ft}$
2. $13 \text{ ft} + 13 \text{ ft} + 6.5 \text{ ft} + 6.5 \text{ ft} = 39 \text{ ft}$
3. a. $2(6 \text{ ft} \cdot 8 \text{ ft}) = 96 \text{ sq ft}$
   $2(8 \text{ ft} \cdot 8 \text{ ft}) = 128 \text{ sq ft}$
   total area = $96 \text{ sq ft} + 128 \text{ sq ft} = 224 \text{ sq ft}$
   door area = $3 \text{ ft} \cdot 6 \text{ ft} = 18 \text{ sq ft}$
   closet area = $4.5 \text{ ft} \cdot 6 \text{ ft} = 27 \text{ sq ft}$

224 sq ft – 18 sq ft – 27 sq ft = 179 sq ft of area to be painted
   b. $\frac{1 \text{ qt}}{95 \text{ sq ft}} = \frac{? \text{ qt}}{179 \text{ sq ft}}$; 1.88 quarts; I need to buy 2 quarts since I can't buy 0.88 of a quart.
4. a.

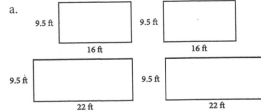

   b. $9.5 \text{ ft} \cdot 16 \text{ ft} = 152 \text{ sq ft}$
   $9.5 \text{ ft} \cdot 16 \text{ ft} = 152 \text{ sq ft}$
   $9.5 \text{ ft} \cdot 22 \text{ ft} = 209 \text{ sq ft}$
   $9.5 \text{ ft} \cdot 22 \text{ ft} = 209 \text{ sq ft}$
   area of all walls = 152 sq ft + 152 sq ft + 209 sq ft + 209 sq ft = 722 sq ft
   c. $\frac{? \text{ sq yd}}{722 \text{ sq ft}} = \frac{1 \text{ sq yd}}{9 \text{ sq ft}}$; 80.22 sq yd
   d. $\frac{4 \text{ sq yd}}{1 \text{ bunch}} = \frac{80.22 \text{ sq yd}}{? \text{ bunches}}$; $80.22 \div 4 = 20.06$; 21 bunches will be needed, as 20 won't quite be enough.
   e. $21 \cdot \$0.30 = \$6.30$
   f. $\frac{100 \text{ pounds}}{5 \text{ sq yd}} = \frac{? \text{ pounds}}{80.22 \text{ sq yd}}$; 1,604.4 pounds
   g. $1,604.4 \div 100 = 16.04$; 17 bags will be needed, as 16 is not quite enough.
   h. $17 \cdot \$0.33 = \$5.61$
   i. $80.22 \cdot \$0.33 = \$26.47$ labor cost
   j. Perimeter of the room = $2(16 \text{ ft}) + 2(22 \text{ ft}) = 76 \text{ ft}$
5. Students should have added to and reviewed flashcards.

Chapter 16: Exponents, Square Roots, and Scientific Notation

Worksheet 16.1

From now on, students should use exponents to express square units. *Example*: 6 sq ft should be written 6 ft^2.

1. a. 8^{12}
   b. 50.6^3
   c. 3^{11}
   d. $(\frac{1}{4})^4$
   e. $(-4)^6$
   f. cm^2
   g. in^2
2. Check both the writing of repeated multiplication and the final answer.
   a. $4 \times 4 \times 4 = 64$
   b. $\frac{1}{2} \cdot \frac{1}{2} = \frac{1}{4}$
   c. $0.2 \cdot 0.2 \cdot 0.2 \cdot 0.2 = 0.0016$
   d. $-10 \cdot -10 \cdot -10 \cdot -10 \cdot -10 = -100,000$
3. a. 87 mi^2
   b. $12 \text{ in} \cdot 16 \text{ in} = 192 \text{ in}^2$
   c. $120 \text{ ft} \cdot \frac{1 \text{ yd}}{3 \text{ ft}} = 40$ yards per side
   Area = $40 \text{ yd} \cdot 40 \text{ yd} = 1,600 \text{ yd}^2$
4. a. $7 + 25 = 32$
   b. $4 \text{ ft}^2 + 81 \text{ ft}^2 = 85 \text{ ft}^2$
5. $2 \cdot 2 \cdot 2 \cdot 2 \cdot 2 = 32$

PAGE 421

6. Students should have added to and reviewed flashcards.

Worksheet 16.2

1. a. ±7
   b. ±14
   c. ±15
   d. ±20

2. a. 12 ft
   b. 9 in
   c. 5 yd

3. a. $2 \cdot 2 = 4$
   b. The number that, times itself, equals 16; it equals 4 or –4, since both $4 \cdot 4$ and $-4 \cdot -4 = 16$.

4. Check for both correct exponent and solution.
   a. $24^6 = 191,102,976$
   b. $-10^3 = -1,000$ *Note:* The answer was negative because we had three negative signs (negative, positive, negative).
   c. $(\frac{1}{5})^3 = \frac{1}{125}$
   d. 60 in^2

5. a. $25 + 96 = 121$
   b. $2 + 81 = 83$

6. a. $30 \text{ in} \cdot \frac{1 \text{ yd}}{36 \text{ in}} = 0.83 \text{ yd}$
   b. $20 \text{ km} \cdot \frac{1,000 \text{ m}}{1 \text{ km}} = 20,000 \text{ m}$
   c. $20 \text{ km} \cdot \frac{1 \text{ mi}}{1.60934 \text{ km}} \cdot \frac{5,280 \text{ ft}}{1 \text{ mi}} \cdot \frac{1 \text{ yd}}{3 \text{ ft}} = 21,872.27 \text{ yd}$

7. a. $4 \cdot 3 - 5 = 12 - 5 = 7$
   b. $5 - 2 + 8 = 3 + 8 = 11$
   c. $4 \cdot 11 = 44$

Worksheet 16.3

1. a. $24 \text{ ft}^2 \cdot \frac{12 \text{ in}}{1 \text{ ft}} \cdot \frac{12 \text{ in}}{1 \text{ ft}} = 3,456 \text{ in}^2$
   b. $84,897 \text{ mi}^2 \cdot \frac{1.60934 \text{ km}}{1 \text{ mi}} \cdot \frac{1.60934 \text{ km}}{1 \text{ mi}} = 219,881.13 \text{ km}^2$
   c. $6,466 \text{ km}^2 \cdot \frac{1 \text{ mi}}{1.60934 \text{ km}} \cdot \frac{1 \text{ mi}}{1.60934 \text{ km}} = 2,488.83 \text{ mi}^2$
   d. $20,770 \text{ km}^2 \cdot \frac{1 \text{ mi}}{1.60934 \text{ km}} \cdot \frac{1 \text{ mi}}{1.60934 \text{ km}} = 8,019.38 \text{ mi}^2$

2. Students were told to round answers to the nearest whole number.
   a. $4 \cdot 230 \text{ m} = 920 \text{ m}$
   b. $920 \text{ m} \cdot \frac{1 \text{ yd}}{0.9144 \text{ m}} = 1,006 \text{ yd}$
   c. $(230 \text{ m})^2 = 52,900 \text{ m}^2$
   d. $52,900 \text{ m}^2 \cdot \frac{1 \text{ yd}}{0.9144 \text{ m}} \cdot \frac{1 \text{ yd}}{0.9144 \text{ m}} = 63,268 \text{ yd}^2$

3.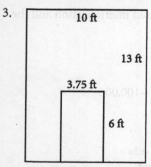

   trim needed = $2(10 \text{ ft}) + 2(13 \text{ ft}) - 3.75 \text{ ft} = 42.25 \text{ ft}$

4. Solve parentheses first: $26 + \sqrt{64}\,(26)$
   Solve exponents and roots, from left to right: $26 + 8(26)$
   Multiply and divide, from left to right: $26 + 208$
   Add and subtract, from left to right: 234

5. Check for both correct exponent and solution.
   a. $(\frac{2}{3})^4 = \frac{16}{81}$
   b. $48^2 = 2,304$
   c. $(-7)^3 = -343$
   d. $\sqrt{100} = \pm 10$
   e. $\sqrt{400} = \pm 20$

6. a. $\sqrt{49 \text{ ft}^2} = 7 \text{ ft}$
   b. $4 \cdot 7 \text{ ft} = 28 \text{ ft}$
   c. $28 \text{ ft} \cdot \frac{1 \text{ yd}}{3 \text{ ft}} = 9.33 \text{ yd}$
   $1.1125 \cdot \$2.50 \cdot 9.33 = \$26$
   Note: 11.25% converts to 0.1125. Students were told to round to the nearest whole dollar.

Worksheet 16.4

1. a. 1
   b. 6

2. a. 10
   b. 100
   c. 1,000
   d. 10,000
   e. 100,000

3. a. 50,000
   b. 5,000,000
   c. 25,000,000

4. a. 4.89×10^{14}
   b. 1.337×10^9
   c. 3.05×10^9

5. a. 1,400,000,000
   b. 17,700,000,000,000
   c. 37,200,000,000,000

6. a. $10 \cdot 45 \text{ in} = 450 \text{ in}$ *Note:* All decagons have 10 sides.
   b. $5 \text{ ft} + 6 \text{ ft} + 8 \text{ ft} + 8.5 \text{ ft} + 8 \text{ ft} + 15 \text{ ft} = 50.5 \text{ ft}$
   c. $(12 \text{ ft})^2 = 144 \text{ ft}^2$
   d. $8 \text{ m} \cdot 6 \text{ m} = 48 \text{ m}^2$

7. a. "Four squared" means $4 \cdot 4$, and it equals 16.
   b. "The square root of four" means the number that, times itself, equals 4; the square root of 4 is ±2.

8. $8 \text{ ft} \cdot 10 \text{ ft} = 80 \text{ ft}^2$; $5 \text{ ft} \cdot 7 \text{ ft} = 35 \text{ ft}^2$; $3 \text{ ft} \cdot 4 \text{ ft} = 12 \text{ ft}^2$;
   $80 \text{ ft}^2 + 35 \text{ ft}^2 + 12 \text{ ft}^2 = 127 \text{ ft}^2$
   $127 \text{ ft}^2 \cdot \frac{1 \text{ yd}}{3 \text{ ft}} \cdot \frac{1 \text{ yd}}{3 \text{ ft}} = 14.11 \text{ yd}^2$
   $14.11 \cdot \$4.50 = \$63.50$

Worksheet 16.5

1. a. <
   b. <
   c. >

2. a. 1,105,000,000
      631,000,000
   b. 8.752×10^{11}

3. Students were told to round to 2 decimal places.
   a. 1.13×10^{17}
   b. 1.3×10^{13}

4. Students were told to round 4b–4d to 3 decimal places.

PAGE 422

a. 1 yr • $\frac{365.25 \text{ d}}{1 \text{ yr}}$ • $\frac{24 \text{ hr}}{1 \text{ d}}$ • $\frac{60 \text{ min}}{1 \text{ hr}}$ • $\frac{60 \text{ sec}}{1 \text{ min}}$ = 31,557,600 seconds
b. $\frac{186,282 \text{ mi}}{1 \text{ sec}} = \frac{? \text{ mi}}{31,557,600 \text{ sec}}$
   31,557,600 • 186,282 = 5.879×10^{12}
   Light travels approximately 5.879×10^{12} miles in a year.
c. 5.879×10^{12} mi • 4.3 = 2.528×10^{13} mi
d. 2.528×10^{13} mi ÷ 3,000 mi = 8.426×10^{9}
e. 5,879,000,000,000
f. 25,280,000,000,000 mi
g. 8,426,000,000

5. a. >
    b. >

6. $2(5 + 2)^2 + 60 ÷ 3$
    Solve inside the parentheses: $2(7)^2 + 60 ÷ 3$
    Exponents and roots, left to right: 2(49) + 60 ÷ 3
    Multiply and divide, left to right: 98 + 20
    Add and subtract, left to right: 118

Worksheet 16.7

1. 21 data points total
    relative frequency = frequency ÷ total data points

Type	Frequency	Relative Frequency
Vanilla Shakes	10	48%
Chocolate Shakes	8	38%
Strawberry Shakes	3	14%

    Note: Students were told to round the relative frequency to the nearest whole percent.

2. Actual may vary; check to make sure data is the same and graph is appropriately labeled.

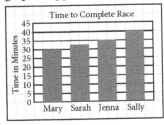

3. The order of these doesn't matter, so long as each parentheses is listed as it appears here. (2,–1), (4,–1), (2,–5), (4,–5)

4. a. $\frac{75 + 77 + 78 + 77 + 52 + 65 + 79 + 75 + 77 + 76}{10} = 73.1$
    b. 52, 65, 75, 75, 76, **77, 77, 77**, 78, 79
       77 appears the most, so 77 is the mode.
    c. 52, 65, 75, 75, **76, 77**, 77, 77, 78, 79
       Since there is not a single middle, we take the average of the two middle numbers.
       $\frac{76 + 77}{2} = 76.5$
       76.5 is the median.

5. a. 2 in • 4 in = 8 in²
    b. 25 in • 14.5 in = 362.5 in²
    c. parallelogram
    d. $(17 \text{ m})^2 = 289 \text{ m}^2$
    e. 9 cm + 8 cm + 7 cm + 12 cm + 7 cm + 8 cm + 9 cm + 40 cm = 100 cm

6. a. $\frac{45 \text{ mi}}{60 \text{ min}}$ • 35 min = 26.25 mi

b. 1.15($2.50 • 26.25) = $75.47

7. a. –5.25 km • $\frac{1,000 \text{ m}}{1 \text{ km}}$ • $\frac{1 \text{ yd}}{0.9144 \text{ m}}$ = –5,741.47 yd
    b. 50.75 dollars • $\frac{0.81 \text{ euros}}{1 \text{ dollar}}$ = 41.11 euros
    c. 70 yd² • $\frac{3 \text{ ft}}{1 \text{ yd}}$ • $\frac{3 \text{ ft}}{1 \text{ yd}}$ = 630 ft²
    d. 2 ft² • $\frac{1 \text{ yd}}{3 \text{ ft}}$ • $\frac{1 \text{ yd}}{3 \text{ ft}}$ = 0.22 yd²

8. a. ±18
    b. 625
    c. 1,269,560,000,000,000,000,000
    d. 4.05897×10^{11}

9. a. 42–46%
    b. none of its sides are equal
    c. dilated

Chapter 17: More Measuring: Triangles, Irregular Polygons, and Circles

Worksheet 17.1

1. a. $\frac{1}{2}$ • 21 ft • 10 ft = 105 ft²
    b. $\frac{1}{2}$ • 16 in • 10 in = 80 in²
    c. $\frac{1}{2}$ • 14 in • 14 in = 98 in²
    d. $\frac{1}{2}$ • 5 ft • 5 ft = 12.5 ft²

2. *area of b in square feet* = 80 in² • $\frac{1 \text{ ft}}{12 \text{ in}}$ • $\frac{1 \text{ ft}}{12 \text{ in}}$ = 0.56 ft²
    area of c in square feet = 98 in² • $\frac{1 \text{ ft}}{12 \text{ in}}$ • $\frac{1 \text{ ft}}{12 \text{ in}}$ = 0.68 ft²
    average area in square feet =
    $\frac{105 \text{ ft}^2 + 0.56 \text{ ft}^2 + 0.68 \text{ ft}^2 + 12.5 \text{ ft}^2}{4}$ = 29.69 ft²

3. a. *area in square feet* = $\frac{1}{2}$ • 8 ft • 9 ft + 8.75 ft(20 ft) = 36 ft² + 175 ft² = 211 ft²
       area in square yards = 211 ft² • $\frac{1 \text{ yd}}{3 \text{ ft}}$ • $\frac{1 \text{ yd}}{3 \text{ ft}}$ = 23.44 yd²
    b. 1.055 • ($12.99 • 23.44) = $321.23
    c. *total cost* = (5 • $50) + $321.23 = $250 + $321.23 = $571.23
    d. *cost* = 1.055($1.29 • 211) = $287.16
       savings = $321.23 – $287.16 = $34.07
       Yes, you'd save $34.07.
       Note: Notice that this pricing was per square foot rather than per square yard, which is why we used 211 (the number of square feet) rather than 23.44 (the number of square yards). We ignored the cost of installation since we were told it stayed the same and all we cared about was the cost difference.

4. *area* = 120 yd • 53.33 yd = 6,399.6 yd²
    area in square feet = 6,399.6 yd² • $\frac{3 \text{ ft}}{1 \text{ yd}}$ • $\frac{3 \text{ ft}}{1 \text{ yd}}$ = 57,596.4 ft²
    bags needed = 57,596.4 ÷ 200 = 287.98
    We would need 288 bags.

5. a. 5,776
    b. –5,776
    c. $\frac{4}{16,807}$
    d. 8 • 4 – 5 = 32 – 5 = 27
    e. 21 ft Note: There's no need to put a plus or minus, as we know we're dealing with area.
    f. $4(13\frac{2}{5}) = 4(\frac{67}{5}) = \frac{268}{5} = 53\frac{3}{5}$
    g. 1 + –6 = –5

6. Students should have added to and reviewed flashcards.

Worksheet 17.2A

1. a. $P = 8 \text{ ft} + 12 \text{ ft} + 6 \text{ ft} + 12 \text{ ft} + 10 \text{ ft} = 48 \text{ ft}$
   b. $P = 2(17 \text{ ft}) + 2(40 \text{ ft}) = 114 \text{ ft}$
   c. $P = 6 \cdot 2 \text{ ft} = 12 \text{ ft}$
2. a. $A = (\frac{1}{2} \cdot 8 \text{ ft} \cdot 6 \text{ ft}) + (12 \text{ ft} \cdot 6 \text{ ft}) = 96 \text{ ft}^2$
   b. $A = 40 \text{ ft} \cdot 15 \text{ ft} = 600 \text{ ft}^2$
   c. $A = 6 (\frac{1}{2} \cdot 2 \text{ ft} \cdot 2.5 \text{ ft}) = 15 \text{ ft}^2$
3. a. $A = (30 \text{ yd} \cdot 40 \text{ yd}) + (\frac{1}{2} \cdot 15 \text{ yd} \cdot 40 \text{ yd}) = 1{,}500 \text{ yd}^2$
   b. $P = 30 \text{ yd} + 40 \text{ yd} + 30 \text{ yd} + 15 \text{ yd} + 50 \text{ yd} = 165 \text{ yd}$

   feet of fencing = $165 \text{ yd} \cdot \frac{3 \text{ ft}}{1 \text{ yd}} = 495 \text{ ft}$

   c. ($\$5 \cdot 495 \text{ ft}) - \$50 = \$2{,}425$
   $\$2{,}425 \cdot 1.065 = \$2{,}582.63$
4. a. $96 \text{ ft}^2 \cdot \frac{30.48 \text{ cm}}{1 \text{ ft}} \cdot \frac{30.48 \text{ cm}}{1 \text{ ft}} = 89{,}186.92 \text{ cm}^2$
   b. $15 \text{ ft}^2 \cdot \frac{12 \text{ in}}{1 \text{ ft}} \cdot \frac{12 \text{ in}}{1 \text{ ft}} = 2{,}160 \text{ in}^2$
   c. $114 \text{ ft} \cdot \frac{1 \text{ yd}}{3 \text{ ft}} \cdot \frac{0.9144 \text{ m}}{1 \text{ yd}} = 34.75 \text{ m}$
   d. $12 \text{ ft} \cdot \frac{12 \text{ in}}{1 \text{ ft}} = 144 \text{ in}$
   e. $3^4 = 81$
   f. ± 16

Worksheet 17.2B

1. a. $P = 14 \text{ in} + 5 \text{ in} + 13 \text{ in} + 13 \text{ in} + 5 \text{ in} + 14 \text{ in} + 24 \text{ in} = 88 \text{ in}$
   b. $P = 2(13 \text{ ft}) + 2(33 \text{ ft}) = 92 \text{ ft}$
   c. $P = 7 \cdot 3 \text{ ft} = 21 \text{ ft}$
2. a. $A = 2 (\frac{1}{2} \cdot 12 \text{ in} \cdot 5 \text{ in}) + (24 \text{ in} \cdot 14 \text{ in}) = 396 \text{ in}^2$
   b. $A = 33 \text{ ft} \cdot 12 \text{ ft} = 396 \text{ ft}^2$
   c. $A = 7 (\frac{1}{2} \cdot 3 \text{ ft} \cdot 3.5 \text{ ft}) = 36.75 \text{ ft}^2$
3. Convert the area of figure a to square feet.
   $396 \text{ in}^2 \cdot \frac{1 \text{ ft}}{12 \text{ in}} \cdot \frac{1 \text{ ft}}{12 \text{ in}} = 2.75 \text{ ft}^2$
   Find the average.
   $\frac{2.75 \text{ ft}^2 + 396 \text{ ft}^2 + 36.75 \text{ ft}^2}{3} = 145.17 \text{ ft}^2$
4. We were told that the shapes were all regular, which means all the sides are the same length.
   a. regular hexagon
   b. $4(3 \text{ in}) = 12 \text{ in}$
   c. $3(4 \text{ in}) = 12 \text{ in}$
   d. $6 \cdot 2 \text{ in} = 12 \text{ in}$
   e. $3 \text{ in} \cdot 3 \text{ in} = 9 \text{ in}^2$
   f. $\frac{1}{2} \cdot 4 \text{ in} \cdot 3.47 \text{ in} = 6.94 \text{ in}^2$
   g. $6(\frac{1}{2} \cdot 2 \text{ in} \cdot 1.73 \text{ in}) = 10.38 \text{ in}^2$
   h. Square: $\frac{9 \text{ in}^2}{12 \text{ in}}$

   Triangle: $\frac{6.94 \text{ in}^2}{12 \text{ in}}$

   Hexagon: $\frac{10.38 \text{ in}^2}{12 \text{ in}}$
   i. Square: 0.75
   Triangle: 0.58
   Hexagon: 0.87
   j. hexagon *Note*: The hexagon is also incredibly sturdy, which is another reason why it is a great shape to hold all that heavy honey! Truly the honeycomb proclaims God's wisdom and care. He created the bee with the knowledge to build its honeycomb in just the right shape.

5. a. $\frac{15}{30} \cdot \frac{15}{30} = \frac{1}{4}$
   b. $-25 \cdot -25 = 625$
   c. $\sqrt{900 \text{ ft}^2} = 30 \text{ ft}$
   d. $12 + 49 \cdot 2 = 12 + 98 = 110$
   e. $4 + (3 - 1) = 4 + 2 = 6$
   f. $24(\frac{9}{75} - \frac{1}{75}) = 24(\frac{8}{75}) = \frac{192}{75} = 2\frac{42}{75}$

Worksheet 17.3

1. a. $C = 3.14 \cdot 100 \text{ yd} = 314 \text{ yd}$
   b. $C = 3.14 \cdot 2(32 \text{ in}) = 200.96 \text{ in}$.
   c. $C = 3.14 \cdot 25 \text{ ft} = 78.5 \text{ ft}$
2. a. $A = 3.14 \cdot (32 \text{ yd})^2 = 3{,}215.36 \text{ yd}^2$
   b. $A = 3.14 \cdot (50 \text{ yd})^2 = 7{,}850 \text{ yd}^2$
   Note: We used 50 yd as we needed to use the radius, and the radius is half the diameter. So since our diameter was 100 yd, our radius was 50 yd.
   c. $A = 3.14 \cdot (12.5 \text{ ft})^2 = 490.63 \text{ ft}^2$
3. a. Since the dog can run in a circle around the pole, he'll have the area of a circle with the chain as its radius.
   $A = 3.14 \cdot (5 \text{ ft})^2 = 3.14 \cdot 25 \text{ ft}^2 = 78.5 \text{ ft}^2$
   b. $C = 3.14 \cdot 2(5 \text{ ft}) = 31.4 \text{ ft}$
   c. $1.05(\$4.50 \cdot 31.4 \text{ ft}) + \$150 = \$298.37$
4. a. $\sqrt{324 \text{ yd}^2} = 18 \text{ yd}$
   b. $12 (\frac{5}{65}) = 12(\frac{1}{3}) = \frac{12}{13}$
   c. $\frac{1}{2} \cdot 6 \text{ m} \cdot 3 \text{ m} = 9 \text{ m}^2$
   d. $7 \cdot 7 \cdot 7 \cdot 7 = 2{,}401$
   e. $18 \text{ yd}^2 \cdot \frac{36 \text{ in}}{1 \text{ yd}} \cdot \frac{36 \text{ in}}{1 \text{ yd}} = 23{,}328 \text{ in}^2$
   f. $3{,}000 \text{ ft} \cdot \frac{1 \text{ mi}}{5{,}280 \text{ ft}} \cdot \frac{1.60934 \text{ km}}{1 \text{ mi}} = 0.91 \text{ km}$
5. Students should have added to and reviewed flashcards.

Worksheet 17.4

1. $5 \cdot 3.14 = 15.7$
2. a. Irrational numbers are numbers that cannot be expressed as a ratio of one integer to another. Irrational numbers 1) never repeat and 2) go on and on for infinity.
   b. No. $1 \div 6$ can be expressed as a ratio of one integer to another ($\frac{1}{6}$) and if it repeats (0.1666666...—the 6 is repeating).
3. a. $C = 3.14 \cdot 2(10 \text{ yd}) = 62.8 \text{ yd}$
   b. $A = 3.14(10 \text{ yd})^2 = 3.14 \cdot 100 \text{ yd}^2 = 314 \text{ yd}^2$
   c. $P = (3.14 \cdot 30 \text{ ft}) \frac{1}{2} + 30 \text{ ft} = 77.1 \text{ ft}$
   $A = 3.14(15 \text{ ft})^2 \cdot \frac{1}{2} = 353.25 \text{ ft}^2$
4. a. A = area of triangle + area of semicircle
   $A = (\frac{1}{2} \cdot 50 \text{ ft} \cdot 125 \text{ ft}) + 3.14(25 \text{ ft})^2 \cdot \frac{1}{2} = 4{,}106.25 \text{ ft}^2$
   b. 90%
   c. $0.90 \cdot 4{,}106.25 \text{ ft}^2 = 3{,}695.63 \text{ ft}^2$
5. Answer should be a circumference and area of a circular object.
6. Students were told to solve these problems mentally.
   a. $10
   b. 46
   c. 89 cents; 3 quarters, 1 dime, 4 pennies
   d. $1; 1 dollar

Chapter 18: Solid Objects and Volume

Worksheet 18.1A

1. a. 2 m • 3 m = 6 m²
   b. 1 m • 3 m = 3 m²
   c. 1 m • 2 m = 2 m²
   d. 2(6 m²) + 2(3 m²) + 2(2 m²) = 12 m² + 6 m² + 4 m² = 22 m²

2. 2(29 ft • 38 ft) = 2,204 ft²
   2(154 ft • 38 ft) = 11,704 ft²
   2(154 ft • 29 ft) = 8,932 ft²
   2,204 ft² + 11,704 ft² + 8,932 ft² = 22,840 ft²

3. 22,840 ft² • $\frac{30.48 \text{ cm}}{1 \text{ ft}}$ • $\frac{30.48 \text{ cm}}{1 \text{ ft}}$ • $\frac{1 \text{ m}}{100 \text{ cm}}$ • $\frac{1 \text{ m}}{100 \text{ cm}}$ = 2,121.91 m²

4. 3(3 in • 12 in) = 108 in²
   2($\frac{1}{2}$ • 12 in • 10 in) = 120 in²
   108 in² + 120 in² = 228 in²

5. Check exponents as well as answer.
   a. $(\frac{15}{60})^2 = (\frac{1}{4})^2 = \frac{1}{4} \cdot \frac{1}{4} = \frac{1}{16}$
   b. $(-2)^4 = 16$
   c. $4.36^3 = 82.88$

Worksheet 18.1B

1. Answers will vary based on the total surface area of the dictionary and tissue box chosen. Check to see that the answer includes the area of all 6 sides.

2. a. 2(10 ft • 10 ft) = 200 ft²
   2(4 ft • 12 ft) = 96 ft²
   2(4 ft • 10 ft) = 80 ft²
   2($\frac{1}{2}$ • 12 ft • 8 ft) = 96 ft²
   Total Surface Area = 200 ft² + 96 ft² + 80 ft² + 96 ft² = 472 ft²
   b. I would need to buy two bottles of waterproofing solution, as one would not quite be enough.

3. 3.14

4. a. 3.14(1.5 in)² = 7.07 in²
   Note: We used 1.5 in as our radius because it is half of our diameter (3).
   7.07 in² • $\frac{2.54 \text{ cm}}{1 \text{ in}}$ • $\frac{2.54 \text{ cm}}{1 \text{ in}}$ • $\frac{10 \text{ mm}}{1 \text{ cm}}$ • $\frac{10 \text{ mm}}{1 \text{ cm}}$ = 4,561.28 mm²
   b. inner hole area = 3.14(1 in)² = 3.14 in²
   circle area = 3.14(4 in)² = 50.24 in²
   area remaining = 50.24 in² – 3.14 in² = 47.1 in²

5. a. 96
   b. 24
   c. 87 cents; 3 quarters, 1 dime, 2 pennies
   d. 90 cents; 3 quarters, 1 dime, 1 nickel

Worksheet 18.2A

1. a. 10 in • 8.5 in • 2.5 in = 212.5 in³
   b. 4 ft • 2 ft • 3 ft = 24 ft³
   c. 8 in • 6 in • 2 in = 96 in³
   d. 3.14(9 in²) • 3 in = 763.02 in³
   e. 3.14 • (1.25 in)² • 3.5 in = 17 in³ Note: Students were told to round to the nearest whole inch.

2. a. 2(4 ft • 2 ft) + 2(4 ft • 3 ft) + 2(2 ft • 3 ft) = 52 ft²
   b. 2(8 in • 6 in) + 2(2 in • 8 in) + 2(2 in • 6 in) = 152 in²

3. Students were told to solve these problems mentally.
   a. 792 Mental Process: 8 times 100 is 800, minus the extra 8 we multiplied by when we multiplied by 100 instead of 99 gets us to 792.
   b. $11
   c. 31 cents; 1 quarter, 1 nickel, 1 penny
   d. 35 cents; 1 quarter, 1 dime

4. Students should have added to and reviewed flashcards.

Worksheet 18.2B

1. Answers will vary based on the filing cabinet/box and tissue box chosen. Check that the answer is the volume.

2. a. 3.14(10 ft)² = 314 ft²
   314 ft² • $\frac{1 \text{ yd}}{3 \text{ ft}}$ • $\frac{1 \text{ yd}}{3 \text{ ft}}$ = 34.89 yd²
   b. 1.055(34.89 • $9.50) = $349.69
   $349.69 + $100 = $449.69
   c. 3.14(4 ft)² • 4 ft = 200.96 ft³
   d. 5 ft • 5 ft • 6 ft = 150 ft³
   7 ft • 5 ft • 4 ft = 140 ft³
   I would get the 5 ft • 5 ft • 6 ft storage unit.
   e. 2 • 2 • 2 = 8

3. a. octagonal prism
   b. area of each triangle = $\frac{1}{2}$(6 in)(6.5 in) = 19.5 in²
   area of entire base = 8 • 19.5 in² = 156 in²
   Note: We thought of the regular octagon as 8 triangles and found the area of each; we then multiplied by the number of triangles (8).
   c. area of each rectangle = 36 in • 6 in = 216 in²
   d. Total Surface Area = 8(216 in²) + 2(156 in²) = 2,040 in²
   e. V = 156 in² • 36 in = 5,616 in³

4. Students should have reviewed flashcards.

Worksheet 18.3

1. a. 18,000 in³ • $\frac{1 \text{ yd}}{36 \text{ in}}$ • $\frac{1 \text{ yd}}{36 \text{ in}}$ • $\frac{1 \text{ yd}}{36 \text{ in}}$ = 0.39 yd³
   b. 5 m³ • $\frac{100 \text{ cm}}{1 \text{ m}}$ • $\frac{100 \text{ cm}}{1 \text{ m}}$ • $\frac{100 \text{ cm}}{1 \text{ m}}$ = 5,000,000 cm³
   c. 270,912 in³ • $\frac{1 \text{ ft}}{12 \text{ in}}$ • $\frac{1 \text{ ft}}{12 \text{ in}}$ • $\frac{1 \text{ ft}}{12 \text{ in}}$ = 156.78 ft³
   d. 500,000 in³ • $\frac{1 \text{ ft}}{12 \text{ in}}$ • $\frac{1 \text{ ft}}{12 \text{ in}}$ • $\frac{1 \text{ ft}}{12 \text{ in}}$ = 289.35 ft³

2. a. Step 1: Convert so the same units.
   30 in • $\frac{1 \text{ ft}}{12 \text{ in}}$ = 2.5 ft
   Step 2: Add the circumference of the circle (the belting covers half the circle on each side, for a total of the entire circumference) to the length of the chain in between.
   C = 3.14(2.5 ft) = 7.85 ft
   chain needed = 7.85 ft + 2(18 ft) = 43.85 ft
   b. $\frac{1}{2}$(6 ft • 8 ft) • 10 ft = 240 ft³
   c. 240 ft³ • $\frac{1 \text{ yd}}{3 \text{ ft}}$ • $\frac{1 \text{ yd}}{3 \text{ ft}}$ • $\frac{1 \text{ yd}}{3 \text{ ft}}$ = 8.89 yd³
   d. The total surface area will equal the sum of the surface area of each side. The top and bottom will be the same triangle, so we will just multiply that area by 2.
   2($\frac{1}{2}$ • 6 ft • 8 ft) = 48 ft²
   6 ft • 10 ft = 60 ft²
   8 ft • 10 ft = 80 ft²
   10 ft • 10 ft = 100 ft²
   48 ft² + 60 ft² + 80 ft² + 100 ft² = 288 ft²

3. a. $-6 \cdot -6 \cdot -6 = -216$
   b. ± 6
4. Students should have reviewed flashcards.

Worksheet 18.4

1. a. $3 \text{ gal} \cdot \frac{4 \text{ qts}}{1 \text{ gal}} = 12$ qts
   b. $\frac{1}{4} c \cdot \frac{16 \text{ Tbsp}}{1 c} = 4$ Tbsp
   c. $8 \text{ gal} \cdot \frac{128 \text{ fl oz}}{1 \text{ gal}} = 1{,}024$ fl oz
   d. 8 fl oz = 1 c
   e. $2.5 \text{ gal} \cdot \frac{4 \text{ qt}}{1 \text{ gal}} \cdot \frac{2 \text{ pts}}{1 \text{ qt}} = 20$ pts

2. a. 1 pt
   b. $1 \text{ pt} \cdot \frac{2 c}{1 \text{ pt}} = 2$ c

3. a. $4 \text{ qt} \cdot \frac{2 \text{ pts}}{1 \text{ qt}} = 8$ pints
   8 pints $\cdot \$0.08 = \$0.64$
   b. $5 \text{ gal} \cdot \frac{4 \text{ qt}}{1 \text{ gal}} \cdot \frac{2 \text{ pts}}{1 \text{ qt}} = 40$ pt; it would take 40 pint bottles.
   c. $\frac{1}{8} c \cdot \frac{16 \text{ Tbsp}}{1 c} = 2$ Tbsp
   d. Since there are 4 pecks in a bushel, a bushel at the peck price would be $\$2.50 \cdot 4 = \$10$. The bushel price of $9.99 is $0.01 cheaper than buying 4 pecks at $2.50 each.

4. a. $\frac{1}{2}(254 \text{ ft})(69 \text{ ft}) = 8{,}763 \text{ ft}^2$
   b. $3.14(9 \text{ ft})^2 = 254.34 \text{ ft}^2$
   c. $4 \text{ m} \cdot 10 \text{ m} = 40 \text{ m}^2$
   d. $8 \text{ in} \cdot 10 \text{ in} \cdot 20 \text{ in} = 1{,}600 \text{ in}^3$
   e. $\frac{1}{2}(12 \text{ ft} \cdot 16 \text{ ft}) \cdot 20 \text{ ft} = 1{,}920 \text{ ft}^3$
   f. $3.14(30 \text{ ft})^2 \cdot 180 \text{ ft} = 508{,}680 \text{ ft}^3$

5. a. $2(10 \text{ in} \cdot 8 \text{ in}) = 160 \text{ in}^2$
   $2(8 \text{ in} \cdot 20 \text{ in}) = 320 \text{ in}^2$
   $2(10 \text{ in} \cdot 20 \text{ in}) = 400 \text{ in}^2$
   $160 \text{ in}^2 + 320 \text{ in}^2 + 400 \text{ in}^2 = 880 \text{ in}^2$
   b. $2(\frac{1}{2} \cdot 12 \text{ ft} \cdot 16 \text{ ft}) = 192 \text{ ft}^2$
   $(12 \text{ ft} \cdot 20 \text{ ft}) = 240 \text{ ft}^2$
   $(16 \text{ ft} \cdot 20 \text{ ft}) = 320 \text{ ft}^2$
   $(19 \text{ ft} \cdot 20 \text{ ft}) = 380 \text{ ft}^2$
   $192 \text{ ft}^2 + 240 \text{ ft}^2 + 320 \text{ ft}^2 + 380 \text{ ft}^2 = 1{,}132 \text{ ft}^2$

6. a. Day 1 = 41 fl oz
   Day 2 = 52 fl oz
   Day 3 = 62 fl oz
   $\frac{41 \text{ fl oz} + 52 \text{ fl oz} + 62 \text{ fl oz}}{3} = 52$ fl oz Note: Answer should be rounded to the nearest ounce.
   b. 60 fl oz − 52 fl oz = 8 fl oz
   c. 41 fl oz + 52 fl oz + 62 fl oz = 155 fl oz
   $155 \text{ fl oz} \cdot \frac{1 \text{ gal}}{128 \text{ fl oz}} = 1.21$ gal

7. Students should have added to and reviewed flashcards.

8. a. wine gallon
   b. no

Worksheet 18.5

1. a. $789 \text{ in}^3 \cdot \frac{1 \text{ gal}}{231 \text{ in}^3} = 3.42$ gal
   b. $5 \text{ qt} \cdot \frac{57.75 \text{ in}^3}{1 \text{ qt}} = 288.75 \text{ in}^3$
   c. $10 \text{ gal} \cdot \frac{231 \text{ in}^3}{1 \text{ gal}} = 2{,}310 \text{ in}^3$
   d. $20 \text{ pt} \cdot \frac{28.875 \text{ in}^3}{1 \text{ pt}} = 577.5 \text{ in}^3$

2. a. $2 \text{ gal} \cdot \frac{231 \text{ in}^3}{1 \text{ gal}} = 462 \text{ in}^3$
   b. $1.5 \text{ ft} \cdot \frac{12 \text{ in}}{1 \text{ ft}} = 18$ in
   $30 \text{ in} \cdot 36 \text{ in} \cdot 18 \text{ in} = 19{,}440 \text{ in}^3$
   $19{,}440 \text{ in}^3 \cdot \frac{1 \text{ qt}}{57.75 \text{ in}^3} = 336.62$ qt
   c. $(13 \text{ in})^3 = 2{,}197 \text{ in}^3$
   1 bushel = $2{,}150.42 \text{ in}^3$
   Yes, it will hold a bushel, as $2{,}197 > 2{,}150.42$.

3. a. volume = $4 \text{ ft} \cdot 3 \text{ ft} \cdot 5 \text{ ft} = 60 \text{ ft}^3$
   $60 \text{ ft}^3 \cdot \frac{12 \text{ in}}{1 \text{ ft}} \cdot \frac{12 \text{ in}}{1 \text{ ft}} \cdot \frac{12 \text{ in}}{1 \text{ ft}} \cdot \frac{1 \text{ gal}}{231 \text{ in}^3} = 448.83$ gal
   b. $18 \text{ in} \cdot \frac{1 \text{ ft}}{12 \text{ in}} = 1.5$ ft
   $\frac{1}{2}(2.76 \text{ ft})(1.5 \text{ ft})6 \cdot 7 \text{ ft} = 86.94 \text{ ft}^3$
   $86.94 \text{ ft}^3 \cdot \frac{12 \text{ in}}{1 \text{ ft}} \cdot \frac{12 \text{ in}}{1 \text{ ft}} \cdot \frac{12 \text{ in}}{1 \text{ ft}} \cdot \frac{1 \text{ gal}}{231 \text{ in}^3} = 650.36$ gal
   c. $20 \text{ ft} \cdot 16 \text{ ft} \cdot 8 \text{ ft} = 2{,}560 \text{ ft}^3$
   $2{,}560 \text{ ft}^3 \cdot \frac{1 \text{ ton}}{80 \text{ ft}^3} = 32$ tons

4. a. $4 \text{ m} \cdot 12 \text{ m} \cdot 8.5 \text{ m} = 408 \text{ m}^3$
   b. $-3 \cdot -3 \cdot -3 \cdot -3 = -81$
   c. ± 11

5. Students should have added to and reviewed flashcards.

Worksheet 18.6

1. a. $586 \text{ mL} \cdot \frac{1 \text{ L}}{1{,}000 \text{ mL}} = 0.586$ L
   b. $4.33 \text{ L} \cdot \frac{1{,}000 \text{ mL}}{1 \text{ L}} = 4{,}330$ mL or 4,333 mL, depending on when rounded
   c. $5 \text{ gal} \cdot \frac{3.78541 \text{ L}}{1 \text{ gal}} = 18.93$ L
   d. $3 \text{ tsp} \cdot \frac{5 \text{ mL}}{1 \text{ tsp}} = 15$ mL

2. $\frac{\$1.19}{1 \text{ L}} = \frac{?}{3.78541 \text{ L}}$; $4.50

3. area = area of semicircle + area of triangle
   $\frac{1}{2} \cdot 3.14(9 \text{ in})^2 + \frac{1}{2}(18 \text{ in} \cdot 25 \text{ in}) = 352.17 \text{ in}^2$
   The diameter of the semicircle (18 in) was also the base of the triangle.

4. a. $V = 40 \text{ ft} \cdot 28 \text{ ft} \cdot 6 \text{ ft} = 6{,}720 \text{ ft}^3$
   Cubic yards to be removed:
   $6{,}720 \text{ ft}^3 \cdot \frac{1 \text{ yd}}{3 \text{ ft}} \cdot \frac{1 \text{ yd}}{3 \text{ ft}} \cdot \frac{1 \text{ yd}}{3 \text{ ft}} = 248.89 \text{ yd}^3$
   b. cost = $248.89 \text{ yd}^3 \cdot \frac{\$5}{\text{yd}^3} = \$1{,}244.44$
   c. volume of pillars = $4 \cdot 3.14(3 \text{ ft})^2 \cdot 7.5 \text{ ft} = 847.8 \text{ ft}^3$
   volume of wall = $36 \text{ ft} \cdot 3 \text{ ft} \cdot 1.5 \text{ ft} = 162 \text{ ft}^3$
   total volume = $847.8 \text{ ft}^3 + 162 \text{ ft}^3 = 1{,}009.8 \text{ ft}^3$
   cost = $1{,}009.8 \text{ ft}^3 \cdot \frac{\$150}{128 \text{ ft}^3} = \$1{,}183.36$

5. a. $\frac{62}{108} \cdot \frac{62}{108} = \frac{31}{54} \cdot \frac{31}{54} = \frac{961}{2{,}916}$
   Note: Notice that we made the problem easier by simplifying the fraction before completing the multiplication.
   b. ± 18

6. Students should have added to and reviewed flashcards.

Worksheet 18.7

1. a. $3{,}579 \text{ lb} \cdot \frac{1 \text{ ton}}{2{,}000 \text{ lb}} = 1.79$ tons
   b. $54 \text{ oz} \cdot \frac{1 \text{ lb}}{16 \text{ oz}} = 3.375$ lb; I would need to purchase 4 packages.

c. 7 lb • $\frac{453.592 \text{ g}}{1 \text{ lb}}$ = 3,175.14 g

  d. 128 lb • $\frac{453.592 \text{ g}}{1 \text{ lb}}$ • $\frac{1 \text{ kg}}{1,000 \text{ g}}$ = 58.06 kg

2. Answers should be listed as whole numbers or fractions, not decimals.

   a. 120 mL • $\frac{1 \text{ tsp}}{5 \text{ mL}}$ • $\frac{1 \text{ Tbsp}}{3 \text{ tsp}}$ • $\frac{1 \text{ c}}{16 \text{ Tbsp}}$ = 0.5 c = $\frac{1}{2}$ c

   b. 256 g • $\frac{1 \text{ c}}{128 \text{ g}}$ = 2 c

   c. 55 g • $\frac{1 \text{ c}}{220 \text{ g}}$ = 0.25 c = $\frac{1}{4}$ c

3. a. 3.14(7 ft)² = 153.86 ft²
   b. 3.14(130 ft)² = 53,066 ft²
   c. 3.14(900 ft)² = 2,543,400 ft²

4. a. *total areas in ft²* = 153.86 ft² + 53,066 ft² + 2,543,400 ft² = 2,596,619.86 ft²

   total acres = 2,596,619.86 ft² • $\frac{1 \text{ acre}}{43,560 \text{ ft}^2}$ = 59.61 acres

   b. 2 pt • $\frac{16 \text{ fl oz}}{1 \text{ pt}}$ = 32 fl oz

   c. 800 ft³ • $\frac{12 \text{ in}}{1 \text{ ft}}$ • $\frac{12 \text{ in}}{1 \text{ ft}}$ • $\frac{12 \text{ in}}{1 \text{ ft}}$ • $\frac{1 \text{ gal}}{231 \text{ in}^3}$ = 5,984.42 gal

   d. 1 gal • $\frac{4 \text{ qt}}{1 \text{ gal}}$ • $\frac{2 \text{ pt}}{1 \text{ qt}}$ • $\frac{2 \text{ c}}{1 \text{ pt}}$ = 16 c

5. a. $A = s^2$
   b. $A = b \cdot h$
   c. $V = B \cdot h$
   d. $A = \frac{1}{2} \cdot b \cdot h$

6. *Note*: Students were told to round to the nearest tenth.

   a. $\frac{6.3 \text{ lb} + 12.8 \text{ lb} + 12.5 \text{ lb} + 24.9 \text{ lb} + 6.4 \text{ lb} + 12.6 \text{ lb} + 12.3 \text{ lb} + 12.5 \text{ lb} + 24.6 \text{ lb} + 8.5 \text{ lb}}{10}$ = 13.3 lb

   b. 6.3, 6.4, 8.5, 12.3, **12.5**, **12.5**, 12.6, 12.8, 24.6, 24.9
   $\frac{12.5 + 12.5}{2}$ = 12.5 *Note*: The median is 12.5, as it's the average of the two middle numbers. We took the average of the two middle numbers in this case since there wasn't a single middle number.

   c. The mode is 12.5, as 12.5 appears the most times.

   d. Service A

7. Answers will vary based on items found in kitchen.

8. Students should have added to and been quizzed on flashcards.

Chapter 19: Angles

Worksheet 19.1

1. a. 90°
   b. 140°
   c. 105°
   d. 65°

2. a. ($\frac{1}{2}$ • 12 ft • 17 ft) • 10 ft = 1,020 ft³
   b. ($\frac{5}{75}$)³ = ($\frac{1}{15}$)³ = $\frac{1}{3,375}$
   c. ±15
   d. (3.14 • 24 in • 24 in) • 36 in = 65,111.04 in³
   e. 65,111.04 in³ • $\frac{1 \text{ gal}}{231 \text{ in}^3}$ = 281.87 gal
   f. 60 fl oz • $\frac{1 \text{ pt}}{16 \text{ fl oz}}$ = 3.75 pt
   g. 10 ℓ • $\frac{1 \text{ gal}}{3.78541 \ell}$ = 2.64 gal

3. a. 2(68 m) + 2(105 m) = 346 m
   3 • 346 m = 1,038 m
   b. 5 • 346 m = 1,730 m
   c. 2 miles • $\frac{5,280 \text{ ft}}{1 \text{ mile}}$ • $\frac{1 \text{ yd}}{3 \text{ ft}}$ • $\frac{0.9144 \text{ m}}{1 \text{ yd}}$ = 3,218.69 m
   3,218.69 m ÷ 346 m = 9.3; you would have to walk around the field 9.3 times

4. a. the opposite direction
   b. Possibilities include gradians or minutes.

Worksheet 19.2

1. a.
   b.
   c.
   d.
   e.
   f.

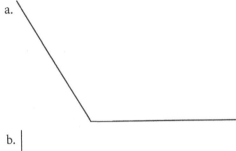

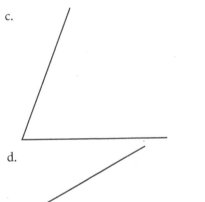
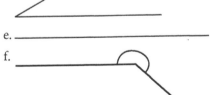

   g. 120° + 90° = 210°
   h. 90° − 40° = 50°
   i. 180° − 40° = 140°

2. a. 70° and 30° *or* (1c and 1d)
   b. 90° *or* (1b)
   c. 120° *or* (1a)
   d. 180° *or* (1e)

3. a.

   b. 55° + 55° = 110°
4. Check to make sure the sides are proportional to 4.5 ft by 3 ft, and that the angles truly are right angles.

   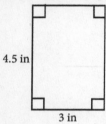

5. a. 4 ft • 5 ft • 7 ft = 140 ft³
   140 ft³ • $\frac{1\ yd}{3\ ft}$ • $\frac{1\ yd}{3\ ft}$ • $\frac{1\ yd}{3\ ft}$ = 5.19 yd³
   b. 2(28 in • 20 in) = 1,120 in²
   2(28 in • 35 in) = 1,960 in²
   2(20 in • 35 in) = 1,400 in²
   1,120 in² + 1,960 in² + 1,400 in² = 4,480 in²
   c. ±21
   d. 7 gal • $\frac{4\ qt}{1\ gal}$ • $\frac{2\ pt}{1\ qt}$ • $\frac{2\ c}{1\ pt}$ = 112 c
   e. 7 gal • $\frac{231\ in^3}{1\ gal}$ = 1,617 in³
   f. 3.14(4 ft)² • 20 ft = 1,004.8 ft³
6. a. *Area of nonagon bases* = 2 • 9($\frac{1}{2}$ • 9 ft • 12.2 ft) = 988.2 ft²
   Area of rectangular sides = 9(28.4 ft • 9 ft) = 2,300.4 ft²
   Total Surface Area = 988.2 ft² + 2,300.4 ft² = 3,288.6 ft²
   b. *Volume = area of base • height*
   ($\frac{1}{2}$ • 9 ft • 12.2 ft • 9)28.4 ft = 14,032.44 ft³

Worksheet 19.3

1 a. 0.4 • 360° = 144°
   b. 0.35 • 360° = 126°
   c. 0.25 • 360° = 90°
   d. Angle of Candidate A Section: 144°
   Angle of Candidate B Section: 126°
   Angle of Undecided Section: 90°
   e. Angle of Candidate A Section: obtuse
   Angle of Candidate B Section: obtuse
   Angle of Undecided Section: right
   f. 360°
2. Check to make sure the angles in the graph are correct and the graph is clearly labeled.
   Rainy = 0.6 • 360° = 216°
   Sunny = 0.3 • 360° = 108°
   Other = 0.1 • 360° = 36°

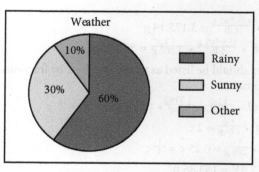

3. a. 3.14 (1.5)² = 7.07 in²
   b. 0.6 • 7.07 in² = 4.24 in²
   c. 3.14 • 3 in = 9.42 in
   d. 0.6 • 9.42 in = 5.65 in
4. a.

Category	Dollar Amount Spent	Ratio of Spending in Category to Total Spending	Percent of Total Spent in Category (Convert Ratio to a Percent)
Household	$160	$\frac{\$160}{\$400}$	40%
Gifts	$80	$\frac{\$80}{\$400}$	20%
Groceries	$100	$\frac{\$100}{\$400}$	25%
Savings	$60	$\frac{\$60}{\$400}$	15%
Total Spending	$400		

b. Check to make sure the angles in the graph are correct and the graph is clearly labeled.
Household = 0.4 • 360° = 144°
Gifts = 0.2 • 360° = 72°
Groceries = 0.25 • 360° = 90°
Savings = 0.15 • 360° = 54°

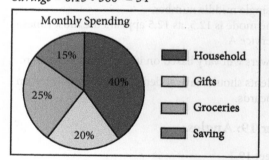

Worksheet 19.4

1. a.

   b.

   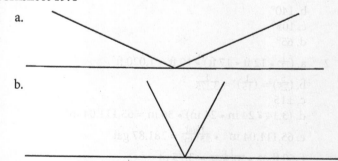

2. Students were instructed to experiment with a flashlight.
3. a. 43°
   b. 53°

4. Check to make sure angles and sides are accurate.
   a. Angles should be 90°; sides can be any length so long as both pairs of opposite ones are the same.

   b. Angles should be 108°; sides can be any length as long as they are all the same length.

   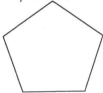

   c. Angles should be 120°; sides can be any length so long as they are all the same length.

   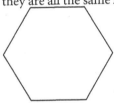

   d. Angle should be 100°.

   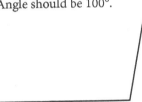

5. a. 5 • 108° = 540°
   b. obtuse
   c. 6 • 120° = 720°
   d. obtuse

6. a. 3.14(20 in)² • 38 in = 47,728 in³
   b. 47,728 in³ • $\frac{1 \text{ gal}}{231 \text{ in}^3}$ = 206.61 gal
   c. 206.61 gal • $\frac{3.78541 \, l}{1 \text{ gal}}$ = 782.1 l
   d. −2.45 • −2.45 • −2.45 • −2.45 • −2.45 = −88.27
   e. $\frac{45 \text{ sec} + 42 \text{ sec} + 44 \text{ sec} + 46 \text{ sec} + 45 \text{ sec}}{5}$ ≈ 44 sec *Note:* Students were told to round to the nearest whole second.

Worksheet 19.5

1. Answers will vary based on real-life angles measured.

2. The size of the triangle isn't important, but check to make sure the angles are all 60°.

   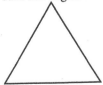

   The sum is 3 • 60° = 180°

3. Students were told to express answers as a whole number, fraction, or mixed number.
   a. 80 mL • $\frac{1 \text{ tsp}}{5 \text{ mL}}$ = 16 tsp
   b. 180 mL • $\frac{1 \text{ tsp}}{5 \text{ mL}}$ • $\frac{1 \text{ Tbsp}}{3 \text{ tsp}}$ • $\frac{1 \text{ c}}{16 \text{ Tbsp}}$ = 0.75 c = $\frac{3}{4}$ c

   c. 840 mL • $\frac{1 \text{ tsp}}{5 \text{ mL}}$ • $\frac{1 \text{ Tbsp}}{3 \text{ tsp}}$ • $\frac{1 \text{ c}}{16 \text{ Tbsp}}$ = 3.5 c = $3\frac{1}{2}$ c

4. a. We can find the trim needed by taking the perimeter minus the openings.
   P = (2 • 5 ft) + (2 • 8 ft) = 26 ft
   openings = (2 • 3.5 ft) + 4 ft = 11 ft
   trim needed = 26 ft − 11 ft = 15 ft of trim
   b. 1.05(15 ft • $5.99) = $94.34
   c. 1.05(15 ft • $0.50) = $7.88
   Add $75 for installation, and you get $82.88.
   Savings: $94.34 − $82.88 = $11.46

5. Students were told to solve these problems mentally.
   a. 54
   b. 13
   c. 48 cents; 1 quarter, 2 dimes, 3 pennies
   d. 50 cents; 2 quarters

Chapter 20: Congruent and Similar

Worksheet 20.1

1. a. 28 in (same as $\overline{AB}$)
   b. 18 in (same as $\overline{FA}$)
   c. 15 in (same as $\overline{EF}$)
   d. irregular hexagon

2. a. ∠BCA = 45°; ∠CAB = 45°
   b.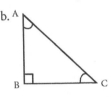
   c. −7 • −7 • −7 = −343

3. a. 40°
   b. 8 ft + 4.6 ft + 4.6 ft = 17.2 ft
   c. $\frac{1}{2}$ • 8 ft • 3 ft = 12 ft²
   d. cost of supplies = 1.085 • 12 • $4.50 = $58.59
   cost of labor = $250 − $15 = $235
   cost of supplies + labor = $58.59 + $235 = $293.59

4. a. $\frac{1}{2}$(3.14)(2.5)² + (5 in • 6 in) = 39.81 in²
   b. 2(6 in) + 5 in + $\frac{1}{2}$(3.14 • 5 in) = 24.85 in

5. Students were told to solve these problems mentally.
   a. 1,200
   b. 12 cents; 1 dime, 2 pennies
   c. 15 cents; 1 dime, 1 nickel

Worksheet 20.2

1. a. and b. Students should have measured all the angles and circled the two 135° ones.

    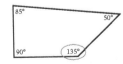

2. 51 ft

3. These problems should be solved using proportions.
   a. $\frac{2 \text{ rows}}{8 \text{ min}} = \frac{20 \text{ rows}}{? \text{ min}}$ = 80 min
   b. $\frac{6 \text{ laps}}{2 \text{ min}} = \frac{24 \text{ laps}}{? \text{ min}}$ = 8 min

PAGE 429

4. a. −8
    b. 4
    c. $(\frac{7}{49})^3 = (\frac{1}{7})^3 = \frac{1}{343}$
    d. ±9
    e. 16 ft
    f. 300 km · $\frac{1 \text{ m}}{1.60934 \text{ km}}$ · $\frac{5,280 \text{ ft}}{1 \text{ mi}}$ = 984,254.41 ft
    g. 29 km² · $\frac{1,000 \text{ m}}{1 \text{ km}}$ · $\frac{1,000 \text{ m}}{1 \text{ km}}$ = 29,000,000 m²

5. a. Congruent
    b. We look at shapes, angles, etc, so that we will be able to figure out information we couldn't otherwise.

Worksheet 20.3

1. a. 4 in
    b. 93
    c. 100 min
    d. 4 hr

2. a. $\frac{30}{1 \text{ min}} = \frac{?}{20 \text{ min}}$; 600 times
    b. $\frac{70 \text{ ft}}{35 \text{ ft}} = \frac{10 \text{ ft}}{? \text{ ft}}$; 5ft
    c. 7 times larger; each dimension of the large rectangle is 7 times the same dimension of the small one.
    d. $\frac{16 \text{ in}}{32 \text{ in}} = \frac{? \text{ in}}{96 \text{ in}}$; 48 in
    e. 3 times larger
    f. 3 · 24 in = 72 in *Note*: We could also have found this via a proportion…only since we already knew how much larger the triangle was, it was easier to simply multiply.
    g. 50°, as angles in similar shapes are congruent.
    h. It doesn't matter what angles have 1, 2, or 3 arc marks, so long as the corresponding angles of both triangles are correctly shown in the drawings given is 2d.

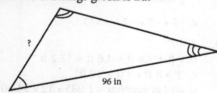

2. Students were told to solve these problems mentally.
    a. 1,500
    b. 1,240
    c. 1,100
    d. −63
    e. 66
    f. 21 cents; 2 dimes, 1 penny
    g. 25 cents; 1 quarter

Worksheet 20.4

1. a. 180°
    b. 180° − 43° − 86° = 51°

2. a. 180° − 55° − 80° = 45°
    b. 180° − 70° − 30° = 80°

3. a. 180°
    b. ∠CAB = 45°, ∠ABC = 90°, as 180° − 45° − 45° = 90°
    c. Yes. It has a right angle.

4. a. A theorem that says that if we know that two sets of their corresponding angles are congruent, two triangles *have* to be similar.

b. The first, third, and forth triangles should be circled.

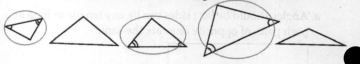

5. $A = \frac{1}{2}$ · 3.14 (2.5 ft)² + (5 · 7) = 44.81 ft²
    1.11(44.81 · $5) = $248.70
    total cost = $248.70 + $100 = $348.70

6. a. 50 · 6.23570 = 311.79 yuans
    b. 56 yuans · $\frac{1 \text{ dollar}}{6.23570 \text{ yuans}}$ = $8.98
    c. 5 m · $\frac{1 \text{ yd}}{0.9144 \text{ m}}$ = 5.47 yd
    d. 50 m² · $\frac{100 \text{ cm}}{1 \text{ m}}$ · $\frac{100 \text{ cm}}{1 \text{ m}}$ = 500,000 cm²
    e. 3.14(4 m)² = 50.24 m²
    f. 50.24 m² · 5 m = 251.2 m³
    g. $\sqrt{121 \text{ in}^2}$ = 11 in

Worksheet 20.5

1. $\frac{2.6 \text{ ft}}{4 \text{ ft}} = \frac{? \text{ ft}}{24 \text{ ft}}$; 15.6 ft

2. We knew that two corresponding angles in the triangles were the same, so then we knew the triangles were similar.

3. a. 18 ft
    b. 120°
    c. 18 ft

4. a. $\frac{40 \text{ miles}}{1 \text{ hr}} = \frac{200 \text{ miles}}{? \text{ hours}}$; 5 hr
    b. $\frac{20 \text{ miles}}{1 \text{ gal}} = \frac{200 \text{ miles}}{? \text{ gal}}$; 10 gal
    10 · $3.30 = $33

5. a. 2 gal · $\frac{3.78541 \text{ l}}{1 \text{ gal}}$ = 7.57 l
    b. 10 gal · $\frac{3.78541 \text{ l}}{1 \text{ gal}}$ = 37.85 l

6. Students were told to solve these problems mentally.
    a. 4 · 4 = 16; 16 qt
    b. 3 · 4 = 12; 12 qt
    c. $\frac{1}{2}$ of 16 = 8; 8 Tbsp
    d. $\frac{1}{4}$ of 16 = 4; 4 Tbsp

Worksheet 20.6

1. Check that the rectangle is 4 in by 2 in, and that the angles are indeed 90°.

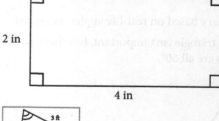

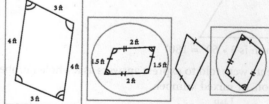

3. a. 3.14(7 in)² · 12 in = 1,846.32 in³

b. 1,846.32 in³ • $\frac{1\,gal}{231\,in^3}$ = 7.99 gal
   c. 2(30 cm • 9 cm) + 2(16 cm • 9 cm) + 2(16 cm • 30 cm) = 1,788 cm²
   d. $\sqrt{64\,ft^2}$ = 8 ft
   e. Check both writing of repeated multiplication and the final answer. 23 • 23= 529
4. a. 180° – 90° – 60° = 30°
   b. yes; AA Similarity Theorem
   c. no; we do not know if the corresponding sides are congruent.

Worksheet 20.7

1. a. 2($\frac{1}{2}$ • 60 in • 80 in) + (32 in • 80 in) + (32 in • 60 in) + (32 in • 100 in) = 12,480 in²
   Note: Each parentheses finds the area of one of the sides, and then we added all those areas together.
   b. ($\frac{1}{2}$ • 60 in • 80 in)32 in = 76,800 in³
   c. 76,800 in³ • $\frac{1\,gal}{231\,in^3}$ = 332.47 gal
   d. 76,800 in³ • $\frac{1\,ft}{12\,in}$ • $\frac{1\,ft}{12\,in}$ • $\frac{1\,ft}{12\,in}$ = 44.44 ft³
2. a. cylinder
   b. 3.14(1.25 in)² = 4.91 in²
   c. 4.91 in² • 4 in = 19.64 in³
   d. $\frac{231\,in^3}{19.64\,in^3}$ = 11.76; we would need 12 containers
3. a. 20 ℓ • $\frac{1\,gal}{3.785417\,\ell}$ = 5.28 gal
   b. $\frac{\$1.60}{1\,\ell}$ = $\frac{\$?}{3.785417\,\ell}$; $6.06
   c. 1 lb = 16 oz
      $2.48 • 16 =$39.68
   d. 4 gal • $\frac{4\,qt}{1\,gal}$ = 16 qt
      $\frac{2\,lemons}{1\,qt}$ = $\frac{?\,lemons}{16\,qt}$; 32 lemons
   e. 4 pt • $\frac{2\,c}{1\,pt}$ = 8 c
      We use 1 tea bag per cup, so we need 8 tea bags.
4. a. 30°
   b. acute
   c. 90° – 30° = 60°
   d. 120°
   e. 180° – 80° – 15° = 85°
5. a. ∠DEF
   b. $\overline{DE}$
   c. *Note*: It doesn't matter what sides have 1, 2, or 3 tick marks, so long as the corresponding sides of both triangles are correctly shown.

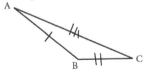 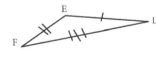

6. a. $\frac{8\,in}{14\,in}$ = $\frac{?\,in}{9.5\,in}$; 5.4 in *Note*: Students were told to round to the nearest tenth.
   b. right
7. a. P = 42% of the circumference of a circle with a 3.5 cm radius + length of left straight line + length of right straight line
      0.42(3.14 • 2 • 3.5 cm) + 4.8 cm + 4.8 cm = 9.23 cm

   b. Distance is currently 3 cm. To enlarge by 3, we'd make it 3 • 3 cm, which equals 9 cm.

Chapter 21: Review

Worksheet 21.1

1. a. three hundred
   b. There are ten digits; each place to the left represents ten times the previous place.
   c. There are two digits; each place to the left represents two times the previous place.
2. a. 3,980 + 3,986 + 4,016 + 4,221 = 16,203
   b. We were told to round $129.99 to nearest dollar ($130).
      price sold = 16,203 • $130 = $2,106,390
      cost = 16,203 • $25 = $405,075
      profit = $2,106,390 – $405,075 = $1,701,315
   c. $8.95 ÷ 20 = $0.45
3. a. Define: *cost per ball of yarn* = 1.05 • $3.99 = $4.19
      ball of yarn per pair of socks = $\frac{1}{3}$
      cost to list per pair of socks = $1.50
      sales price per pair of socks = $9.99
   b. Plan: *amount made = sales price – cost to list per pair of socks – cost of yarn per pair of socks*
      cost of yarn per pair of sock = ball of yarn per pair of socks • cost per ball of yarn
   c. Execute: *cost of yarn per pair of socks* = $\frac{1}{3}$ • $4.19 = $1.40
      Note: The answer would be $1.38 if we converted $\frac{1}{3}$ to a decimal before multiplying.
      $9.99 – $1.50 – $1.40 = $7.09
   d. Check: Yes, $7.09 sounds reasonable given the sales price and cost.
4. a. *cost to list on own website* = $9.50 • 12 + $6 = $120
      cost to list on other website = 200 • $1.50 = $300
      It would be more economical to get the website, as $120 < $300.
   b. $\frac{1}{2}$ • 8$\frac{1}{3}$ ft = $\frac{1}{2}$ • $\frac{25}{3}$ ft = $\frac{25}{6}$ ft = 4$\frac{1}{6}$ ft
5. a. $\frac{1\,congressman}{30{,}000\,people}$ = $\frac{?\,congressmen}{330{,}000\,people}$; 11 congressmen
   b. $\frac{50\,mi}{1\,hr}$ = $\frac{?\,mi}{10\,hr}$; 500 miles
6. a. 3$\frac{1}{2}$ • $\frac{1}{2}$ c = $\frac{7}{2}$ • $\frac{1}{2}$ c = $\frac{7}{4}$ c = 1$\frac{3}{4}$ c. There are 1$\frac{3}{4}$ c in a can.
   b. (1.045 • $1.99) ÷ 1$\frac{3}{4}$ = $1.19
   c. (1.045 • $4.99) ÷ 20 = $0.26
   d. At these prices, it is cheaper to make it.
7. 0.03 • 2 ton = 0.06 ton; 0.065 • 0.06 ton = 0.0039 tons
   0.0039 tons • $\frac{2{,}000\,lb}{1\,ton}$ = 7.8 lb
8. Students were told to solve these problems mentally.
   a. 59
   b. 151
   c. 390
   d. $31.96
   e. 8 cents; 1 nickel, 3 pennies
   f. 10 cents; 1 dime
   g. $9.20
   h. 2 hr

Worksheet 21.2

1. a. trapezoid
   b. rhombus
   c. equilateral triangle; acute triangle
   d. parallelogram
2. a. $3.14(50 \text{ ft})^2 \cdot 4 \text{ ft} = 31,400 \text{ ft}^3$
   $31,400 \text{ ft}^3 \cdot \frac{1 \text{ gal}}{0.13 \text{ ft}^3} \approx 241,538 \text{ gal}$ Note: Students were told to round to the nearest whole number.
   b. $2(3.14 \cdot 50 \text{ ft}) = 314 \text{ ft}$
   c. $3.14(50 \text{ ft})^2 = 7,850 \text{ ft}^2$
   $7,850 \text{ ft}^2 \cdot \frac{1 \text{ yd}}{3 \text{ ft}} \cdot \frac{1 \text{ yd}}{3 \text{ ft}} = 872.22 \text{ yd}^2$
3. a. length = 60 cubits
   $60 \text{ cubits} \cdot \frac{1.5 \text{ ft}}{1 \text{ cubit}} = 90 \text{ ft}$
   breadth = 20 cubits
   $20 \text{ cubits} \cdot \frac{1.5 \text{ ft}}{1 \text{ cubit}} = 30 \text{ ft}$
   height = 30 cubits
   $30 \text{ cubits} \cdot \frac{1.5 \text{ ft}}{1 \text{ cubit}} = 45 \text{ ft}$
   b. $A = 8 \text{ ft} \cdot 20 \text{ ft} = 160 \text{ ft}^2$
   c. $P = 2(10 \text{ ft}) + 2(20 \text{ ft}) = 60 \text{ ft}$
4. a. 3 in (All the side in a square are equal.)
   b. ? in smaller triangle = $\frac{? \text{ in}}{23 \text{ in}} = \frac{37 \text{ in}}{34.5 \text{ in}}$; 24.67 in
   ? in larger triangle = $\frac{? \text{ in}}{34.5 \text{ in}} = \frac{20 \text{ in}}{23 \text{ in}}$; 30 in
   c. $\frac{1}{2} \cdot 23 \text{ in} \cdot 20 \text{ in} = 230 \text{ in}^2$
   d. $230 \text{ in}^2 \cdot 45 \text{ in} = 10,350 \text{ in}^3$
5. a. $20 \text{ km}^2 \cdot \frac{1 \text{ mi}}{1.60934 \text{ km}} \cdot \frac{1 \text{ mi}}{1.60934 \text{ km}} = 7.72 \text{ mi}^2$
   b. $1,330 \text{ in}^3 \cdot \frac{1 \text{ ft}}{12 \text{ in}} \cdot \frac{1 \text{ ft}}{12 \text{ in}} \cdot \frac{1 \text{ ft}}{12 \text{ in}} = 0.77 \text{ ft}^3$
6. $\frac{220 \text{ yd}}{1.75 \text{ in}} = \frac{800 \text{ yd}}{? \text{ in}}$; 6.36 in

Worksheet 21.3

1. a. Students should have listed one lesson from the life of Kepler. Possible answers include God's plans not always being our own, submission to God's plan, perseverance, confidence that God created an orderly universe and math can describe it, and the need to guard against falsehoods in our thinking.
   b. Kepler discovered the laws of planetary motion.
2. Students should have begun a report on how math is used in a career.
3. Students should have begun to study for the final exam.

Worksheet 21.4A

1. Students should have worked on report.
2. a. 0.05
   b. −0.67
   c. 22%
   d. $\frac{17}{25}$
   e. $\frac{25}{3}$
   f. $\frac{25}{100} = \frac{1}{4}$
   g. $(-5.7)^3$
   h. $7.8956 \cdot 10^8$
   i. 125%
   j. 3.5%
3. a. >
   b. =
   c. >
   d. >
   e. <
   f. =
4. a. $4 \cdot 5; 4 \times 5; 4(5)$
   b. $5\overline{)4}; \frac{4}{5}; 4 \div 5$
5. a. The identity property of multiplication is the property that describes that we can multiply by 1 without changing the value.
   b. denominator
   c. rectangle
6. a. 63
   b. 12
   c. $8(7)^2 = 8(49) = 392$
   d. $625 \cdot -3 = -1,875$
   e. $ft \cdot ft \cdot ft$
7. These problems were designed to review fractions and decimals, so make sure the answer was given in the form listed here (fractions as fractions, and decimals as decimals). Fractions should be simplified as much as possible, and decimals should be rounded to the hundredths place.

   a. $\frac{\cancel{4}}{5} \times \frac{\overset{3}{\cancel{6}}}{\underset{2}{\cancel{8}}} = \frac{3}{5}$
   b. $\frac{9}{7} \div \frac{2}{3} = \frac{9}{7} \cdot \frac{3}{2} = \frac{27}{14} = 1\frac{13}{14}$
   c. $\frac{160}{200} + \frac{87}{200} + \frac{14}{200} = \frac{261}{200} = 1\frac{61}{200}$
   d. $\frac{35}{60} - \frac{9}{60} = \frac{26}{60} \div \frac{2}{2} = \frac{13}{30}$
   e. 11.91
   f. 0.8
   g. −32
   h. $0.0425 \cdot \$89.67 = \$3.81$
   i. $50 \text{ in} \cdot \frac{1 \text{ ft}}{12 \text{ in}} \cdot \frac{1 \text{ yd}}{3 \text{ ft}} = 1.39 \text{ yd}$
   j. $8,000 \text{ ft} \cdot \frac{30.48 \text{ cm}}{1 \text{ ft}} \cdot \frac{1 \text{ m}}{100 \text{ cm}} = 2,438.4 \text{ m}$
   k. $5 \text{ gal} \cdot \frac{4 \text{ qt}}{1 \text{ gal}} \cdot \frac{2 \text{ pt}}{1 \text{ qt}} \cdot \frac{2 \text{ c}}{1 \text{ pt}} = 80 \text{ c}$
   l. false
8. The two overarching principles are that 1) math describes real life, and 2) math point us to the Lord.

Worksheet 21.4B

1. a. 0.88
   b. no
   c. Check that solved via the distributive property.
      $4(2) + 4(3) =$
      $8 + 12 = 20$
   d. $44 = 2 \cdot 2 \cdot 11$
      $16 = 2 \cdot 2 \cdot 2 \cdot 2$
      $LCD = 2 \cdot 2 \cdot 2 \cdot 2 \cdot 11 = 176$
   e. $\frac{28}{176} + \frac{88}{176} = \frac{116}{176} = \frac{29}{44}$
   f. $18 = 2 \cdot 3 \cdot 3$
      $48 = 2 \cdot 2 \cdot 2 \cdot 2 \cdot 3$
      $GCF = 2 \cdot 3 = 6$
   g. $\frac{18}{48} \div \frac{6}{6} = \frac{3}{8}$
   h. 12
   i. $\frac{21}{5} \cdot 2 = \frac{42}{5} = 8\frac{2}{5}$

j. Possibilities include negative 9 degrees, negative 9 miles, negative 9 dollars, etc.
k. 6
l. −6

2. a. $\frac{400 \text{ ft}}{?} = \frac{15 \text{ ft}}{1 \text{ in}}$; 26.67 in

   b.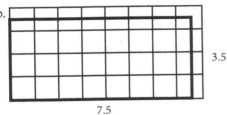

3. Students were told to solve these problems mentally.
   a. 74
   b. 14
   c. $13.80
   d. $23.92

4. a.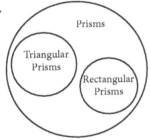

   b. 20, 24
   c. 4

5. a. 62.50%
   b. 0.6250 • 12,300 = $7,687.50

6. 35%–55%

7. (5,2)

8. $\frac{\$567.62 + \$678.25 + \$456.48}{3} = \$567.45$

9. a. 0.60 • $14.99 = $8.99; $8.99 • 1.045 = $9.39
   b. 1.18 • $35 = $41.30

10. a. $1.99 • $\frac{1}{2}$ = $1.00
    b. Bahts per $\frac{1}{4}$ pound: $\frac{1}{4}$ • 500 = 125
       Convert to dollars: 125 bahts • $\frac{0.031 \text{ dollars}}{1 \text{ baht}}$ = $3.88

11. a. 10 d • $\frac{24 \text{ hr}}{1 \text{ d}}$ = 240 hr
    b. 45 Canadian $ • $\frac{1 \text{ US \$}}{1.09 \text{ Canadian \$}}$ = $41.28

12. a. 10^3 = 1,000
    b. ±20

13. a. P = 18 ft + 9 ft + 18 ft + 5 ft + 10 ft = 60 ft
       A = 9 ft • 18 ft + $\frac{1}{2}$(5 ft • 9 ft)
       162 ft² + 22.50 ft² = 184.5 ft²
    b. 60°
    c. 180° − 60° = 120°
    d. Angle should be 120°.

e. 180° − 25° − 43° = 112°

14. 2.25 × 46 = 103.5 gal per minute
    $\frac{103.5 \text{ gal}}{1 \text{ min}}$ • $\frac{60 \text{ min}}{1 \text{ hr}}$ = 6,210 gal per hour

Answers to Quizzes

Quiz 1 (Chapters 1 and 2)

1. a. > or 64 > 62
   b. > or 84 > 66
   c. < or 7 < 9

2. a. In a place-value system, the place, or location of a symbol determines its value.
   b. It means there are 12 digits (including zero) and that each place is worth 12 of the previous place.

3. a. 12:30 p.m.
   b. 10:00 p.m.
   c. 4 p.m.

4. The ending balance is 6,035.

Bonus: Math works because God created and sustains a consistent universe.

Quiz 2 (Chapter 3)

1. a. 2(4) + 8 = 8 + 8 = 16
   b. 2($12) + $7 = $24 + $7 = $31
   c. 4(10 − 9) = 4(1) = 4

2. Students were told to solve these problems mentally.
   a. 38
   b. 105
   c. 73
   d. 750
   e. 800

3. a. $6 + $8 + $10 + $4 = $28
   b. $28 ÷ 7 = $4
   c. $4 + $5 = $9

4. Look for a multiplication word problem and its solution.

5. a. 2012 − 1492 = 520 years
   b. 1492 + 400 = 1892

Bonus: Properties are truths about the ordinances God put in place.

Quiz 3 (Chapter 4)

1. number of books per shelf = 12 x 4 = 48
   number of shelves required for 288 books = 288 ÷ 48 = 6.
   We need 6 shelves.

2. a. $855 ÷ 171 = $5
   b. 135 x $5 = $675

3. a. Define:
   price of first land = $127,980
   price of second land = $50,000
   acres in first land = 160
   acres in second land = 60
   price per acre = ?
   b. Plan: total price ÷ total acres = price per acre
   total price = price of first land + price of second land
   total acres = acres in first land + acres in second land
   c. Execute: total acres = 60 + 160 = 220
   total price = $127,980 + $50,000 = $177,980
   price per acre = $177,980 ÷ 220 = $809
   d. Check: The answer looks reasonable.

   220 x $809 = $177,980

4. a. 8(13) = 104
   b. Check that solved using the distributive property.
   8(4) + 8(3) + 8(6) =
   32 + 24 + 48 = 104

5. a. 179 r 1
   b. 50,232
   c. 0
   d. 85

Quiz 4 (Chapter 5)

1. a. $\frac{26}{8} = \frac{13}{4} = 3\frac{1}{4}$
   b. $\frac{5}{10} = \frac{1}{2}$

2. a. $\frac{55}{8}$
   b. $6\frac{7}{8}$

3. a. 44 = 2 x 2 x 11
   50 = 2 x 5 x 5
   GCF = 2
   b. $\frac{22}{25}$
   c. 16 = 2 x 2 x 2 x 2
   24 = 2 x 2 x 2 x 3
   GCF = 2 x 2 x 2 = 8
   d. $\frac{2}{3}$

4. a. $\frac{10}{84} + \frac{44}{84}$
   b. $\frac{10}{84} + \frac{44}{84} = \frac{54}{84}$
   $\frac{54}{84} ÷ \frac{6}{6} = \frac{9}{14}$

5. a. $\frac{30}{60} - \frac{5}{60}$
   b. $\frac{30}{60} - \frac{5}{60} = \frac{25}{60}$
   $\frac{25}{60} = \frac{5}{12}$

6. a. $\frac{1}{2 \times 2 \times 2 \times 3} + \frac{7}{2 \times 2 \times 5}$
   b. LCM = 2 x 2 x 2 x 3 x 5 = 120
   c. $\frac{5}{120} + \frac{42}{120}$
   d. $\frac{47}{120}$

7. ($30 + $18) x 12 months x 10 years = $5,760

8. $30 − $28 = $2 savings per month
   $2 x 12 x 5 = $120 savings over 5 years
   $2 x 12 x 10 = $240 savings over 10 years

9. $\frac{3}{4} - \frac{2}{3} = \frac{9}{12} - \frac{8}{12} = \frac{1}{12}$ yard

Bonus: See "Fractions in History" box in Lesson 5.1 for possibilities.

Quiz 5 (Chapter 7)

1. a. 2.34
   b. 0.15
   c. 32.72
   d. 1.22

2. a. 0.88 mi
   b. 0.67 c
   c. 1.25 c

3. a. $45.67 − $5.08 = $40.59
   b. (10 • $25) − (5 • $3.25) =

PAGE 434

$250 − $16.25 = $233.75

Note: We multiplied $3.25 by 5 since a gallon lasts 2 lawns.

Quiz 6 (Chapter 8)

1. Students were told to solve these problems mentally.
   a. 3 • $2.50 = $7.50
   b. $20 − $7.50 = $12.50
2. a. $4:12 plates *or* $\frac{\$4}{12 \text{ plates}}$
   b. 4 ÷ 12 = 0.33
3. a. $\frac{4.6 \text{ foot}}{\$35} = \frac{9.2 \text{ foot}}{?}$; $70
   b. $\frac{\$56}{2 \text{ tickets}} = \frac{?}{10 \text{ tickets}}$; $280
   c. $\frac{1 \text{ centimeter}}{5 \text{ inch}} = \frac{2 \text{ centimeter}}{? \text{ in}}$; 10 inch
4. a. $\frac{15{,}000 \text{ seeds}}{\frac{1}{4} \text{ acre}} = \frac{? \text{ seeds}}{3\frac{1}{2} \text{ acres}}$ = 210,000 seeds
   b. We need 7 bags (210,000 ÷ 30,000 = 7), so our cost will be 7 • $30 = $210
   c. The $32 company is cheaper.
   210,000 ÷ 40,000 = 5.25 = 6 bags (5 would not be quite enough.)
   6 • $32 = $192

Bonus: Ideas include exploring sunflowers and how we are created to grow.

Quiz 7 (Chapter 9)

1. Students should have listed all answers to the right of the equal's sign.
   a. $\frac{5 \text{ people}}{10 \text{ people}} = \frac{1 \text{ people}}{2 \text{ people}}$; 0.5; 50%
   b. $\frac{\$250}{\$1{,}000} = \frac{\$1}{\$4}$; $0.25, 25%
2. Students were told to solve this problem mentally. $22.40
3. a. 0.25 • $89.50 = $22.38
   b. 0.15 • $56.15 = $8.42
   c. 100% + 5% = 105%
   1.05 • $14.50 = $15.23
4. a. 40%
   b. 189%
5. 100% − 52% = 48%
   0.48 • $400 = $192
6. 0.20 • $865 = $173

Quiz 8 (Chapter 10)

1. a. −3
   b. −6
2. Answers should be three things that could be represented with negative numbers; possibilities include direction, temperature, and debts.
3.
4. a. −8
   b. −5
   c. −6
   d. 6
   e. −9

   f. 5
   g. 0.0325
   h. −3
5. a. 3
   b. 5.78
   c. 5.78
6. a. 0.70 • $5.50 = $3.85
   Note: We used 0.70 because that's what would be left after 30% was taken: 100% − 30% = 70%.
   b. The first way would give you $1.35 more money.
   Note: profits the second way = $4.25 − $1.75 = $2.50
   comparing the profits = $3.85 − $2.50 = $1.35

Bonus: Each negative sign means *the opposite of*.

Quiz 9 (Chapter 11)

1. Plants: A, B, C, D, E, F
   Trees: C, D
   Flowers: A, B, C, F
2. Yes. Trees and flowers were both subsets of plants.
3. a. Whole Numbers and Odd Integers should be circled.
   b. Positive Numbers should be circled.
4. No. We know because if it were a subset, all of the circle would be drawn within B, as everything in a subset is also a member of the set of which it's a subset.
5. No
6. a. 20, 20, 23
   b. 29, 36, 43
7. a. Yes. Sequence b has a common difference of 7.
   b. A sequence is "an ordered set of quantities." [*The American Heritage Dictionary of the English Language*, 1980 New College Ed., s.v. "sequence."]

Bonus: Answer should be one real-life example of a sequence. Possible answers include the keys on a piano, spirals in a sunflower, and ratios in art.

Quiz 10 (Chapter 12)

Check graph analysis for understanding of statistics and graphs.

Quiz 11 (Chapter 13)

1. a. polygon, quadrilateral, trapezoid, parallelogram, and rectangle (Students should have listed any 3 of these.)
   b. acute triangle
   c. equilateral triangle
2. a. 10
   b. 4
   c. 9
   d. reflection
3.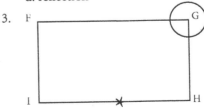

4. a. i
    b. ii

Bonus: Check for an example of a real-life object and shape that describes it; possibilities include toaster (rectangle), Frisbee (circle), honeycomb (hexagon), tables (rectangle or circle), can (cylinder), bowls (circle or cylinder), books (rectangle or rectangular prism), wheels (circle), art (various), etc.

Quiz 12 (Chapter 14)

1. a. 4 mi · $\frac{1,000 \text{ mm}}{1 \text{ mi}}$ = 4,000 mm
    b. 0.5 mi · $\frac{5,280 \text{ ft}}{1 \text{ mi}}$ = 2,640 ft
    c. 24 in · $\frac{2.54 \text{ cm}}{1 \text{ in}}$ = 60.96 cm

2. a. 8,290 km · $\frac{1 \text{ mi}}{1.60934 \text{ km}}$ = 5,151.18 miles
    b. 1.18 · 1,050 rubles = 1,239 rubles
    c. 1,239 rubles · $\frac{\$0.03}{1 \text{ rubles}}$ = $37.17
    d. $\frac{200 \text{ km}}{1 \text{ hr}}$ · $\frac{1 \text{ hr}}{60 \text{ min}}$ = $\frac{200 \text{ km}}{60 \text{ min}}$;
       $\frac{200 \text{ km}}{60 \text{ min}}$ = $\frac{83.33 \text{ km}}{25 \text{ min}}$

3. a. $\frac{1}{4}$ · $3.99 = $\frac{\$3.99}{4}$ = $1.00
    b. 5 in · $\frac{1 \text{ yd}}{36 \text{ in}}$ = 0.14 yd;
       0.14 · $3.99 = $0.56

4. Length = 2.5 cu · $\frac{18 \text{ in}}{1 \text{ cu}}$ · $\frac{1 \text{ ft}}{12 \text{ in}}$ = 3.75 ft
    Breadth (width) = 1.5 cu · $\frac{18 \text{ in}}{1 \text{ cu}}$ · $\frac{1 \text{ ft}}{12 \text{ in}}$ = 2.25 ft
    Height = 1.5 cu · $\frac{18 \text{ in}}{1 \text{ cu}}$ · $\frac{1 \text{ ft}}{12 \text{ in}}$ = 2.25 ft

Quiz 13 (Chapter 15)

1. a. 7 ft · 7 ft = 49 sq ft
    b. 4 · 7 ft = 28 ft
    c. 12 in · 7 in = 84 sq in
    d. 2 (13 in) + 2 (7 in) = 40 in

2. a. 360 · 160 = 57,600 sq ft
       $\frac{57,600 \text{ sq ft}}{1,000 \text{ sq ft}}$ = 57.6, so 58 bags
    b. $70 · 58 = $4,060
    c. $70 − $30 = $40 savings on each bag
       total savings = $40 · 58 = $2,320
    d. 2(360 ft) + 2(160 ft) = 1,040 ft

3. Actual letters may vary; check to make sure that the relationships are accurate.
    a. P = 2l + 2w or P = 2(l) + 2(w) or P = 2 · l + 2 · w
    b. P = S + 2

Bonus: We can use formulas in real life because we live in a consistent universe held together by a consistent God.

Quiz 14 (Chapter 16)

1. Check both writing of repeated multiplication and the final answer.
    a. 7 · 7 · 7 = 343
    b. 88 · 88 = 7,744
    c. 10

2. ±9

3. $\sqrt{144 \text{ ft}^2}$ = 12 ft

4. a. 156,000,000

PAGE 436

b. 5.64 × 10^8
c. 1.78 × 10^7

5. a. 120 yd^2 · $\frac{3 \text{ ft}}{1 \text{ yd}}$ · $\frac{3 \text{ ft}}{1 \text{ yd}}$ = 1,080 ft^2
    b. 132 ft^2 · $\frac{1 \text{ yd}}{3 \text{ ft}}$ · $\frac{1 \text{ yd}}{3 \text{ ft}}$ = 14.67, so 15 yd^2

Bonus: Answer should be an aspect of God's creation scientific notation helps us describe; possibilities include distance to stars or planets, mass of the earth, and bacteria in a pond.

Quiz 15 (Chapter 17)

1. a. 2.5 ft + 2.6 ft + 4.4 ft = 9.5 ft
    b. $\frac{1}{2}$ · 2.6 ft · 2.2 ft = 2.86 ft^2
    c. 2.86 ft^2 · $\frac{12 \text{ in}}{1 \text{ ft}}$ · $\frac{12 \text{ in}}{1 \text{ ft}}$ = 411.84 in^2

2. I would need to know the height of the triangle.

3. ($\frac{1}{2}$ · 150 ft · 60 ft) + ($\frac{1}{2}$ · 50 ft · 70 ft) + (100 ft · 70 ft) = 13,250 ft^2

4. a. C = 3.14 · 2(13 in) = 81.64 in
    b. A = 3.14 · (13 in)^2 = 530.66 in^2
    c. C = 3.14 · 2(8 in) = 50.24 in

5. False; irrational numbers cannot be express as a ratio of one interger to another.

Bonus: Look for a fact about π's history; possibilities include how the Greek mathematician Archimedes explored the ratio between the circumference and diameter of a circle, that π is a Greek letter p, and that the value for π used in describing the molten sea of Solomon's temple was accurate.

Quiz 16 (Chapter 18)

1. a. 2 qt · 8 = 16 qt; 16 qt · $\frac{1 \text{ gal}}{4 \text{ qt}}$ = 4 gal
    b. 10 bu · $\frac{4 \text{ pk}}{1 \text{ bu}}$ · $\frac{8 \text{ qt}}{1 \text{ pk}}$ = 320 qt
    c. 10 oz · $\frac{28.3495 \text{ g}}{1 \text{ oz}}$ = 283.5 g
    d. 10 gal · $\frac{3.78541 \text{ l}}{1 \text{ gal}}$ = 37.85 l

2. a. (1.5 in · 1.5 in · 2) + (3.5 in · 1.5 in · 4) = 4.5 in^2 + 21 in^2 = 25.5 in^2
    b. 1.5 in · 1.5 in · 3.5 in = 7.88 in^3

3. a. (1 in · 2 in)2 + (2 in · 0.8 in) + 2($\frac{1}{2}$ · 0.8 in · 0.9 in)
       = 4 in^2 + 1.6 in^2 + 0.72 in^2 = 6.32 in^2
    b. $\frac{1}{2}$ · 0.8 in · 0.9 in · 2 in = 0.72 in^3

4. a. V = 20 ft · 10 ft · 5 ft = 1,000 ft^3
       1,000 ft^3 · $\frac{12 \text{ in}}{1 \text{ ft}}$ · $\frac{12 \text{ in}}{1 \text{ ft}}$ · $\frac{12 \text{ in}}{1 \text{ ft}}$ · $\frac{1 \text{ gal}}{231 \text{ in}^3}$ = 7,480.52 gal
    b. $\frac{1 \text{ c}}{1,000 \text{ gal}}$ = $\frac{? \text{ c}}{7,480.52 \text{ gal}}$; 7.48 c

Bonus: Surface area helps us see how different birds have the size wings they need.

Quiz 17 (Chapter 19)

1. a. 90°
    b. 130°

2. a. right
    b. obtuse
    c. 90°

3. $90° + 130° = 220°$

4. a. Angle should measure 120°.

b. Angle should measure 45°.

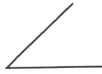

5. a. $0.45 \cdot 360° = 162°$
   b. $0.3 \cdot 360° = 108°$
   c. $0.25 \cdot 360° = 90°$

Bonus: Because math and reasoning were held up as the source of truth, men did not question the Greek proof of an earth-centered universe or see if it matched reality.

Quiz 18 (Chapter 20)

1. a. 3 m
   b. 17°

2. a. yes; AA Similarity Theorem
   b. $\frac{14 \text{ ft}}{? \text{ ft}} = \frac{42 \text{ ft}}{70 \text{ ft}}$; 23.33 ft
   c. 5°
   d. Finding $\overline{AC}$; $\frac{14 \text{ ft}}{? \text{ ft}} = \frac{42 \text{ ft}}{30 \text{ ft}}$; 10 ft
      P = 10 ft + 14 ft + 23.33 ft = 47.33 ft

3. $\frac{2 \text{ ft}}{1.5 \text{ ft}} = \frac{? \text{ ft}}{16.5 \text{ ft}}$; 22 ft

4. $\overline{AC}$ and $\overline{BC}$
   $\overline{AB}$ and $\overline{DE}$ and $\overline{GF}$
   $\overline{DG}$ and $\overline{EF}$

5. a. 180°
   b. 180° − 45° − 45° = 90°

6. a. A and D are similar to each other, and B and C are similar to each other (congruent shapes are also similar).
   b. B and C are congruent.

Bonus: Reasoning and proofs start with assumptions.

Answer to Tests

Test 1 (Chapters 1–6)

1. a. $\frac{3}{2}$ acres + $\frac{1}{3}$ acres = $\frac{9}{6}$ acres + $\frac{2}{6}$ acres = $\frac{11}{6}$ acres = $1\frac{5}{6}$ acres
   b. $\cancel{3} \cdot \frac{5}{\cancel{3}}$ cups = 5 cups
   c. $\frac{1}{2}$ yd + $\frac{2}{3}$ yd = $\frac{3}{6}$ yd + $\frac{4}{6}$ yd = $\frac{7}{6}$ yd = $1\frac{1}{6}$ yard
   d. $12\frac{1}{2}$ in − $5\frac{3}{4}$ in = $\frac{25}{2}$ in − $\frac{23}{4}$ in = $\frac{50}{4}$ in − $\frac{23}{4}$ in = $\frac{27}{4}$ in = $6\frac{3}{4}$ inches

2. *total expenses* = $16 + $60 + $13 + $11 = $100
   number of pounds = 10 x 20 lb = 200 lb
   8 times expenses = 8 x $100 = $800
   price to charge per pound = $800 ÷ 200 lb = $4 per pound

3.

Check Number	Date	Memo	Payment Amount	Deposit Amount	$ Balance
	7/1	Opening Balance			24,587
	7/2	Deposit		1,568	26,155
292	7/2	Farmer Supply Company	120		26,035
293	7/3	Tractor Repair Company	134		**25,901**

4. Students were told to solve these problems mentally.
   a. 27 cents
   b. 53 cents
   c. 97 cents
   d. 60
   e. 40

5. a. $\frac{\cancel{3}^1}{4} \times \frac{7}{\cancel{9}_3} = \frac{7}{12}$
   b. $\frac{\cancel{8}^2}{9} \times \frac{5}{\cancel{4}_1} = \frac{10}{9} = 1\frac{1}{9}$
   c. $\frac{42}{105} + \frac{35}{105} = \frac{77}{105} = \frac{11}{15}$
   d. 4(25 − 12) = 4(13) = 52
   e. 9(8) = 72
   f. 88 = 2 x 2 x 2 x 11
      66 = 2 x 3 x 11
      GCF = 2 x 11 = 22
   g. LCM = 2 x 2 x 2 x 3 x 11 = 264

Bonus: Answer should be a biblical truth that helps shape our view of math; possibilities include that God created and sustains all things, that He created us in His image, and that God never changes and is faithful.

Test 2 (Chapters 7–11)

1. a. $\frac{15 \text{ pictures}}{3 \text{ pages}} = \frac{75 \text{ pictures}}{? \text{ pages}}$; 15 pages
   b. $\frac{\$3.50}{5 \text{ pages}} = \frac{\$?}{15 \text{ pages}}$; $10.50
   c. $\frac{\$0.99}{2 \text{ pages}} = \frac{\$?}{15 \text{ pages}}$; $7.43
      $10.50 − $7.43 = $3.07
   d. *amount to spend on ribbon* = $\frac{1}{2}$ x $75 = $37.50
      spools can buy = $37.50 ÷ $1.99 = 18.84, or 18 spools
      Note: We can't round up, as we don't have enough money to get 19.

2. Students were told to solve these problems mentally.
   a. $7.50
   b. $2.50

3. a. $\frac{1.5 \text{ bags}}{1 \text{ acre}} \cdot \frac{? \text{ bags}}{48 \text{ acres}}$; 72 bags
    b. $72 \cdot \$3 + \$5.99 = \$221.99$
    c. $\$216 \cdot 1.055 = \$227.88$
4. $\$5.67 + \$7.89 + \$10.20 = \$23.76$
    $0.75 \cdot \$23.76 = \$17.82$
5. Granny Smith Apples
6. $-5° - 35° = -40°$; $|-40°| = 40°$
7. a. Negative Numbers
    b. $-\$7.50$
    c. 36,39
    d. 3
    e. no
    f. -5
    g. 8
    h. -2

Test 3 (Chapters 12–16)

1.
Continent	Frequency	Relative Frequency
Europe	750	5%
Asia	900	6%
Africa	750	5%
North America	11,400	76%
Australia	450	3%
South America	750	5%
Antartica	0	0%

2. 31
3. a. $\frac{5+4+4+5+8+3+5+7+4+4}{10} = 4.9$ in
    b. mode = 4 in
    c. 3 4 4 4 4 5 5 5 7 8
       median = $\frac{4+5}{2} = 4.5$ in
4. a. $\$1.40 \cdot 35 = \$49$
    $1.05 \cdot \$49 = \$51.45$
    b. $35 \text{ km} \cdot \frac{1 \text{ mi}}{1.60934 \text{ km}} = 21.75$ mi
    c. $\frac{43.5 \text{ mi}}{1 \text{ hr}} = \frac{21.75 \text{ mi}}{? \text{ hr}}$; 0.5 hr = 30 min
5. a. 200 bahts $\cdot \frac{1}{3} = 66.67$ bahts
    b. 66.67 bahts $\cdot \frac{\$0.031}{1 \text{ baht}} = \$2.07$
    c. $1.08 \cdot 250$ bahts = 270 bahts
       270 bahts $\cdot \frac{\$0.031}{1 \text{ baht}} = \$8.37$
6. a. 300 mi $\cdot \frac{1.60934 \text{ km}}{1 \text{ mi}} = 482.80$ km
    b. 50 ft² $\cdot \frac{1 \text{ yd}}{3 \text{ ft}} \cdot \frac{1 \text{ yd}}{3 \text{ ft}} = 5.56$ yd²
    c. 405 in² $\cdot \frac{1 \text{ ft}}{12 \text{ in}} \cdot \frac{1 \text{ ft}}{12 \text{ in}} = 2.81$ ft²
7. $(-3, 2)$
8. a. 7 ft $\cdot$ 4.5 ft = 31.5 ft²
    b. $2(7 \text{ ft}) + 2(4.5 \text{ ft}) = 23$ ft
    c. At least two of the following should be listed: polygon, quadrilateral, trapezoid, parallelogram.
9. 15 ft $\cdot$ 35 ft = 525 ft²
10. a. ± 15
    b. $8 \cdot 8 \cdot 8 = 512$
    c. 45,690,000,000,000,000

11. a. 28%–38%
    b. 4
    c. rotation

Test 4 (Chapters 17–20)

1. a. $4 \text{ c} \cdot \frac{1 \text{ pt}}{2 \text{ c}} = 2$ pt
    b. 60 fl oz $\cdot \frac{1 \text{ c}}{8 \text{ fl oz}} = 7.5$ c or $7\frac{1}{2}$ c
2. a. $3.14 \cdot 10^2 = 314$ ft²
    b. $\frac{1}{2} \cdot 26$ ft $\cdot 17$ ft = 221 ft²
3. a. 221 ft² $\cdot$ 20 ft = 4,420 ft³
    b. 4,420 ft³ $\cdot \frac{12 \text{ in}}{1 \text{ ft}} \cdot \frac{12 \text{ in}}{1 \text{ ft}} \cdot \frac{12 \text{ in}}{1 \text{ ft}} = 7,637,760$ in³
    c. 7,637,760 in³ $\cdot \frac{1 \text{ gal}}{231 \text{ in}^3} = 33,063.9$ gal
4. a. 314 ft² $\cdot$ 15 ft = 4,710 ft³
    b. 4,710 ft³ $\cdot \frac{1 \text{ yd}}{3 \text{ ft}} \cdot \frac{1 \text{ yd}}{3 \text{ ft}} \cdot \frac{1 \text{ yd}}{3 \text{ ft}} = 174.44$ yd³
5. a. 90°
    b. 140°
    c. 25°
6. a. right
    b. $180° - 65° - 35° = 80°$
7. $\frac{30 \text{ ft}}{? \text{ ft}} = \frac{0.6 \text{ ft}}{3 \text{ ft}}$; 150 ft
8. a. $\frac{3 \text{ in}}{2 \text{ in}} = \frac{9 \text{ in}}{? \text{ in}}$; 6 in
    b. 2 in $\cdot$ 3 in = 6 in²
9. $P = 90$ ft + 49 ft + 30 ft + $\frac{1}{2}(3.14 \cdot 2 \cdot 20$ ft$) + 50$ ft + 40 ft = 321.8 ft
10. a. $\frac{1}{2}(26 \text{ ft})(4 \text{ ft}) = 52$ ft²
    b. 13 ft + 13 ft + 26 ft = 52 ft
    c. $3.14(1.5 \text{ ft})^2 \cdot 15$ ft = 105.98 ft³
    d. $\$24 \cdot 105.98 \cdot 6 = \$15,261.12$
11. Check to make sure angle is 50°.

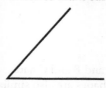

12. 7 ft
13. $2(4 \text{ ft} \cdot 8 \text{ ft}) + 2(2 \text{ ft} \cdot 4 \text{ ft}) + 2(8 \text{ ft} \cdot 2 \text{ ft}) = 64$ ft² + 16 ft² + 32 ft² = 112 ft²

Test 5 (Final Exam)

1. a. 3.75 in $\cdot$ 4.5 in $\cdot$ 9 in = 151.88 in³
    b. $(\frac{1}{2} \cdot 24 \text{ in} \cdot 13 \text{ in}) \cdot 3$ in = 468 in³
    c. $180° - 60° - 60° = 60°$
2. a. total surface area of 1 box =
       $2(12 \text{ in} \cdot 18 \text{ in}) = 432$ in²
       $2(12 \text{ in} \cdot 15 \text{ in}) = 360$ in²
       $2(18 \text{ in} \cdot 15 \text{ in}) = 540$ in²
       432 in² + 360 in² + 540 in² = 1,332 in²
       total surface area of all boxes = 30 $\cdot$ 1,332 in² = 39,960 in²
    b. 39,960 in² $\cdot \frac{1 \text{ ft}}{12 \text{ in}} \cdot \frac{1 \text{ ft}}{12 \text{ in}} = 277.5$ ft²
       277.5 ft² $\div 100 = 2.78$; it will take 3 cans.
3. a. $3.14(4)^2 = 50.24$ in²
    b. $(3.14)(8) = 25.12$ in

4. a. $3.14(5 \text{ ft})^2 \cdot 3 \text{ ft} = 235.5 \text{ ft}^3$
   b. $235.5 \text{ ft}^3 \cdot \frac{12 \text{ in}}{1 \text{ ft}} \cdot \frac{12 \text{ in}}{1 \text{ ft}} \cdot \frac{12 \text{ in}}{1 \text{ ft}} \cdot \frac{1 \text{ gal}}{231 \text{ in}^3} = 1,761.66 \text{ gal}$
5. a. $6,665.16 \text{ in} \div 1,000 = 6.67 \text{ in}$
   b. $1.125 \cdot \$534 = \$600.75$
   c. $0.1 \cdot \$324.23 = \$32.42$
6. $\frac{7 + 8.5 + 4 + 10}{4} = 7.4$ hours *Note:* Students were told to round to the nearest tenth.
7. a. $25 \text{ km} \cdot \frac{1 \text{ mile}}{1.60934 \text{ km}} = 15.53 \text{ miles}$
   b. $15.53 \text{ mi} \cdot \frac{60 \text{ min}}{50 \text{ mi}} = 18.64 \text{ min}$
   c. $6 \text{ L} \cdot \frac{1 \text{ gal}}{3.78541 \text{ L}} = 1.59 \text{ gal}$
   d. $29.5 \text{ rands} \cdot \frac{\$1}{11 \text{ rands}} = \$2.68$
   e. 8,089 rials + 4,567 rials + 1,896 rials + 23,900 rials = 38,452 rials
   $38,452 \text{ rials} \cdot \frac{\$1}{2,684 \text{ rials}} = \$14.33$
   f. $4 \cdot 4,000 \text{ rials} = 16,000 \text{ rials}$
   $16,000 \text{ rials} \cdot \frac{\$1}{2,684 \text{ rials}} = \$5.96$
   g. $100 \text{ g} \cdot \frac{1 \text{ lb}}{453.592 \text{ g}} = 0.22 \text{ lb}$
8. a. $\frac{56}{100} = \frac{14}{25}$
   b. 0.07
   c. 0.88
   d. 0.48 *Note:* Answer should be rounded to the nearest hundredth.
   e. -21
   f. 4
   g. $\frac{90 \text{ mi}}{3 \text{ hr}}$
   h. $44\frac{1}{2}$, 44.5
   i. $\frac{21}{24} - \frac{1}{24} = \frac{20}{24} = \frac{5}{6}$
   j. ± 10
   k. $8 + 6 \cdot 9 = 8 + 54 = 62$
   l. $0.5 \text{ gal} \cdot \frac{4 \text{ qt}}{1 \text{ gal}} \cdot \frac{2 \text{ pt}}{1 \text{ qt}} \cdot \frac{2 \text{ c}}{1 \text{ pt}} = 8 \text{ c}$
9. Point A (3,4)
   Point B (4,−2)
   Point C (−3,−3)

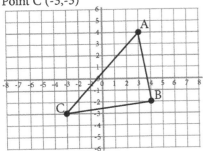

10. Answers should be an additional piece of information to find out about the chart. Possible answers include the following: Who was surveyed and how? Was the survey random? How many people were surveyed?

11.

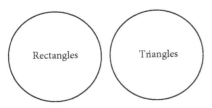

12. a. 5 p.m. EST = 2 p.m. PST; you have 1 hour.
    b. $\frac{2}{3} + \frac{3}{4} = \frac{8}{12} + \frac{9}{12} = \frac{17}{12} = 1\frac{5}{12}$ bushels
    c. $1\frac{1}{2} \cdot \frac{9}{12} = \frac{3}{2} \cdot \frac{9}{12} = \frac{9}{8} = 1\frac{1}{8}$ ft
    d. $\$34,000 \div 12 = \$2,833.33$

Extra Credit
   a. *cost for 50 gal* = $(4 \cdot \$0.25) + (4 \cdot \$0.40) + (0.25 \cdot \$0.75)$
      $= \$1.00 + \$1.60 + \$0.19 = \$2.79$
      cost for 100 gal = $2(\$2.79) = \$5.58$
   b. *cost per tree* = gal • cost per gal
      gal = $2\frac{1}{2}$
      cost per gal = $\$5.58 \div 100 \text{ gal} = \$0.06$
      cost per tree = $2.5 \text{ gal} \cdot \$0.06 = \$0.15$
   c. $P = 2(180 \text{ ft}) + 2(200 \text{ ft}) = 360 \text{ ft} + 400 \text{ ft} = 760 \text{ ft}$
      four lines of wire = $4(760 \text{ ft}) = 3,040 \text{ ft}$
      wire in yards = $3,040 \text{ ft} \cdot \frac{1 \text{ yd}}{3 \text{ ft}} = 1,013.33 \text{ yd}$
      number of pounds = $\frac{1,013.33 \text{ yd}}{? \text{ lb}} = \frac{8 \text{ yd}}{1 \text{ lb}}$; 126.67 lb
      cost for wire = $126.67 \text{ lb} \cdot \$0.08 \text{ per pound} = \$10.13$
   d. *dozen of eggs per year per hen* = $100 \text{ eggs} \div 12 = 8.33$
      dollars made per year per hen = $8.33 \cdot \$0.85 = \$7.08$
      dollars made per month per hen = $\$7.08 \div 12 = \$0.59$

Reference Sheets

Please tear these pages out and place in your math notebook, along with blank paper to use in adding your own notes.

Polygon

(closed, two-dimensional figure with straight lines)

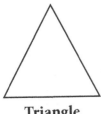
Triangle
(3 sides)

Quadrilateral
(4 sides)

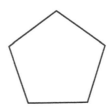
Pentagon
(5 sides)

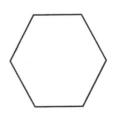
Hexagon
(6 sides)

Heptagon
(7 sides)

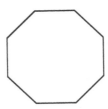
Octagon
(8 sides)

Nonagon
(9 sides)

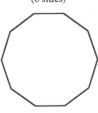
Decagon
(10 sides)

Regular Polygon

(All sides are equal; all edges would touch a circle drawn around the figure, as all angles are the same.)

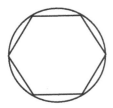
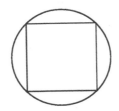

Irregular Polygon

(Polygons that are not regular.)

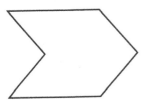

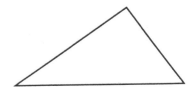

Specific Quadrilaterals

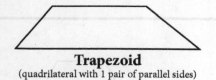
Trapezoid
(quadrilateral with 1 pair of parallel sides)

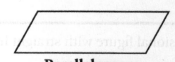
Parallelogram
(quadrilateral with both pairs of opposite sides parallel)

Some books define a trapezoid as a quadrilateral with 1, and only 1, pair of parallel sides, while others as a quadrilateral with 1 (or more) pair of parallel sides. Likewise, some define a rhombus/diamond differently than listed here. Always remember that definitions can — and do — vary!

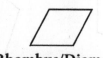
Rhombus/Diamond
(parallelogram with equal-length sides)

Rectangle
(parallelogram with right angles)

Square
(parallelogram with equal-length sides *and* right angles)

Triangles Categorized by Length of Sides

Isosceles
(two equal sides)

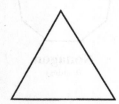
Equilateral
(all equal sides)

Scalene
(no equal sides)

Triangles Categorized by Angles

Right Triangle
(a right angle)

Acute Triangle
(all acute angles)

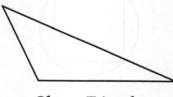
Obtuse Triangle
(an obtuse angle)

Circle

(closed two-dimensional figure; each part of the edge is equally distant from the center)

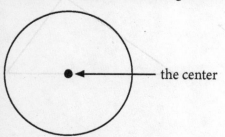
the center

Prism

(A solid with two bases that are parallel polygons, and faces [sides] that are parallelograms; the prism is named after the shape of the bases.)

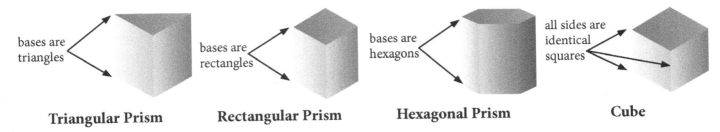

Triangular Prism   **Rectangular Prism**   **Hexagonal Prism**   **Cube**

Cylinder

(A solid with two bases that are equal parallel circles, having an equal diameter in any parallel plane between them.)

Prism and cylinder definitions were based on *Ray's New Higher Arithmetic*, Revised (Cincinnati: Van Antwerp, Bragg & Co., 1880), p. 390.

Formulas

Shape Name	Type of Shape	Perimeter	Area
Polygons		P = *sum of the lengths of each side* or $P = s_1 + s_2 \ldots s_n$	View as multiple triangles or other shapes.
Regular Polygon		P = *(number of sides) • (length of a side)* or $P = n \cdot s$ or $P = n(s)$ or $P = ns$	View as multiple triangles or other shapes.
Rectangle		P = *(2 • length) + (2 • width)* or $P = 2 \cdot l + 2 \cdot w$ or $P = 2(l) + 2(w)$ or $P = 2l + 2w$	A = *length • width* or $A = l \cdot w$ or $A = l(w)$ or $A = lw$
Square		P = *4 • side* or $P = 4 \cdot l$ or $P = 4(s)$ or $P = 4s$	A = *side • side* or $A = s \cdot s$ or $A = s(s)$ or $A = s^2$
Parallelogram		P = *(2 • base) + (2 • side)* or $P = 2 \cdot b + 2 \cdot s$ or $P = 2(b) + 2(s)$ or $P = 2b + 2s$	A = *base • height* or $A = b \cdot h$ or $A = b(h)$ or $A = bh$
Triangle		*Perimeter* = *sum of the lengths of each side* or $P = s_1 + s_2 + s_3$	$A = \frac{1}{2} \cdot base \cdot height$ or $A = \frac{base \cdot height}{2}$ $A = \frac{1}{2} \cdot b \cdot h$ or $A = \frac{b \cdot h}{2}$

A = Area
B = area of the base
b = base
C = circumference
d = diameter

h = height
l = length
n = number of sides
P = Perimeter
π = 3.14

r = radius
s = side
V = Volume
w = width (or height)

PAGE 445

Shape Name	Type of Shape	Perimeter	Area
Circle	(circle with diameter d and radius r)	Circumference = π • diameter or $C = \pi \cdot d$ or $C = \pi(d)$ or $C = \pi d$ Circumference = 2 • π • radius or $C = 2 \cdot \pi \cdot r$ or $C = 2(\pi)(r)$ or $C = 2\pi r$ diameter = 2 • radius or $d = 2 \cdot r$ radius = $\frac{1}{2}$ • diameter or $r = \frac{1}{2} \cdot d$	Area = π • radius • radius or $A = \pi \cdot r^2$ or $A = \pi(r^2)$ or $A = \pi r^2$

Shape Name	Type of Shape	Perimeter	Area
Prism	(triangular prism with base B and height h)	Volume = area of base • height or $V = B \cdot h$	Total surface area = area of all the sides (i.e., surfaces) of a solid object
Cylinder	(cylinder with base B and height h)	Volume = area of base • height or $V = B \cdot h$	did not cover

A = Area
B = area of the base
b = base
C = circumference
d = diameter

h = height
l = length
n = number of sides
P = Perimeter
π = 3.14

r = radius
s = side
V = Volume
w = width (or height)

Units of Measure

Distance

Distance – U.S. Customary
12 inches (in) = 1 foot (ft)
3 feet / 36 inches = 1 yard (yd)
1,760 yard / 5,280 feet = 1 mile (mi)

Distance – Metric/SI
10 millimeters (mm) = 1 centimeter (cm)
10 centimeters = 1 decimeter (dm)
10 decimeters / 100 centimeters / 1,000 millimeters = 1 meter (m)
10 meters = 1 decameters (dam)
10 decameters / 1,000 meters = 1 hectometer (hm)
10 hectometers = 1 kilometer (km)

Conversion Between Systems
1 inches (in) = 2.54 centimeter (cm)
1 foot (ft) = 30.48 centimeter (cm)
1 yard (yd) = 0.9144 meter (m)
1 mile (mi) = 1.60934 kilometer (km)

Time

60 seconds (s) = 1 minute (min)
60 minutes = 1 hour (hr)
24 hours = 1 day (d)
7 days = 1 week (wk)
365 days = 1 year (yr *or* y)

Liquid Capacity

U.S. Customary
3 teaspoons (tsp) = 1 tablespoon (Tbsp)
16 tablespoons = 1 cup (c)
2 cups = 1 pint (pt)
2 pints = 1 quart (qt)
4 quarts = 1 gallon (gal)

2 tablespoons (Tbsp) ≈ 1 fluid ounce (fl oz)
8 fl oz = 1 cup (c)
16 fl oz = 1 pint (pt)
32 fl oz = 1 quart (qt)
128 fl oz = 1 gallon (gal)

Conversion Between Systems
1 teaspoon ≈ 5 milliliters
1 gallon = 3.78541 liters

1 pint = 28.875 in^3
1 quart = 57.75 in^3
1 gallon = 231 in^3

Metric
10 milliliters (ml *or* mL) = 1 centiliter (cl *or* cL)
10 centiliter / 100 milliliter = 1 deciliter (dl *or* dL)
10 deciliter / 100 centiliter / 1,000 milliliter = 1 liter (*l or* L)
10 liters = 1 dekaliter (dal *or* daL)
10 dekaliter = 1 hectoliter (hl *or* hL)
10 hectoliter / 1,000 liters = 1 kiloliter (kl *or* kL)

Dry Capacity

U.S. Customary
2 pints (pt) = 1 quart (qt)
8 quart = 1 peck (pk)
4 peck = 1 bushel (bu) / 32 quarts (qt)

Conversion Between Systems
1 quart = 67.2006 inches^3
1 bushel = 2,150.42 inches^3

Note: The pint and quart here represent a larger capacity than the ones measuring liquid—they should not be used interchangeably. Unless the problem specifically states otherwise, you can assume pint and quart in this course refer to the liquid units.

Mass

U.S. Customary
16 ounces (oz) = 1 pound (lb)
2,000 pounds = 1 ton (called a "short ton")

Conversion Between Systems
1 ounce = 28.3495 grams
1 pound = 453.592 grams
1 U.S. ton (called a short ton) = 0.907185 metric tons

Note: These ounces are different than the fluid ounces listed under liquid capacity.

Metric
10 milligrams (mg) = 1 centigram (cg)
10 centigrams / 100 milligrams = 1 decigram (dg)
10 decigrams / 100 centigrams / 1,000 milligrams = 1 gram (g)
10 grams = 1 dekagram (dag)
10 dekagrams = 1 hectogram (hg)
10 hectograms / 1,000 grams = 1 kilogram (kg)

For more unit details, see the official standards given in Tina Butcher, Linda Crown, Rick Harshman, and Juana Williams, eds. *NIST Handbook 44: 97th National Conference on Weights and Measures 2012*, 2013 ed. (Washington: U. S. Department of Commerce, 2012), Appendix C. Found on http://www.nist.gov/pml/wmd/pubs/h44-13.cfm (accessed 10/6/2014).

Available Fall 2015

Mathematical Curriculum that Adds a Biblical Worldview!

Discover this easy to use text written in a conversational style directly to the student. You will find that all math boils down to a way of describing God's creation and is a useful means we can use to serve and worship Him.

Book 2 offers understanding and focus on:
> Essential principles of algebra
> Coordinate graphing
> Probability and statistics
> Functions, and many more important areas of mathematics!

This curriculum firms up the foundational concepts and prepares students for upper-level math in a logical, step-by-step way. It is aimed at grades 6-8, consists of the *Student Textbook* and *Student Workbook,* which contains all the worksheets, quizzes, and tests, along with an answer key and suggested schedule.

Understanding the core principles of arithmetic and geometry helps students learn even more skills that will allow them to explore and understand the many aspects of God's creation!

Principles of Math Book 2
Package (Student & Workbook)
$64.99 | 978-0-89051-915-8

Student
$34.99 | 978-0-89051-906-6
Paperback 420 pages

Workbook
$29.99 | 978-0-89051-907-3
Paperback 440 pages

Katherine Loop is the owner of Christian Perspective, a family ministry dedicated to encouraging homeschooling families through a variety of methods, including resources, free e-mail newsletters, e-mail/phone support, and speaking engagements. She is a freelance writer, editor, marketer, and video editor with a mission to help families "Seek the LORD, and His Strength" (Psalm 105:4) in every aspect of their lives.

Master Books® Curriculum

masterbooks.com   800-999-3777

HOMESCHOOL

Master Books® Homeschool Curriculum

- Faith-Building Books & Resources
- Parent-Friendly Lesson Plans
- Biblically-Based Worldview
- Affordably Priced

Master Books® is the leading publisher of books and resources based upon a Biblical worldview that points to God as our Creator. Now the books you love, from the authors you trust like Ken Ham, Michael Farris, Tommy Mitchell, and many more are available as a homeschool curriculum.

New! Summer 2015

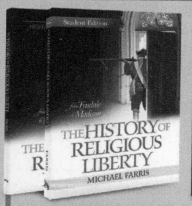

Student: $34.99
978-0-89051-882-3

Teacher: $15.99
978-0-89051-869-4

Student: $17.99
978-0-89051-865-6

Student: $34.99
978-0-89051-875-5

Teacher: $29.99
978-0-89051-876-2

Student: $34.99
978-0-89051-859-5

Teacher: $24.99
978-0-89051-860-1

Visit MasterBooks.com to SAVE and download our 2015 Curriculum Catalog.
800.999.3777

MasterBooks.com
Where Faith Grows!